ELECTRONIC PROPERTIES OF INORGANIC QUASI-ONE-DIMENSIONAL COMPOUNDS

Part I – Theoretical

PHYSICS AND CHEMISTRY OF MATERIALS WITH LOW-DIMENSIONAL STRUCTURES

Series B: Quasi-One-Dimensional Materials

ELECTRONIC PROPERTIES OF INORGANIC QUASI-ONE-DIMENSIONAL COMPOUNDS

Part I – Theoretical

Edited by

PIERRE MONCEAU

Centre de Recherches sur les Très Basses Températures, CNRS, 38000 Grenoble, France

D. REIDEL PUBLISHING COMPANY

A MEMBER OF THE KLUWER ACADEMIC PUBLISHERS GROUP

DORDRECHT / BOSTON / LANCASTER

Library of Congress Cataloging in Publication Data
Main entry under title:

CIP

Electronic properties of inorganic quasi-one-dimensional compounds.

(Physics and chemistry of materials with low-dimensional structures. Series B, Quasi-one-dimensional materials)
Bibliography: p.
Includes indexes.
Contents: pt. 1. Theoretical.
1. One-dimensional conductors. 2. Chemistry, Inorganic. 3. Chemistry, Physical and theoretical. I. Monceau, Pierre, 1942– II. Series.
QC176.8.E4E37 1985 530.4′1 84–27565
ISBN 90–277–1801–6 (set)
ISBN 90–277–1789–3 (pt. 1)

Published by D. Reidel Publishing Company,
P.O. Box 17, 3300 AA Dordrecht, Holland.

Sold and distributed in the U.S.A. and Canada
by Kluwer Academic Publishers,
190 Old Derby Street, Hingham, MA 02043, U.S.A.

In all other countries, sold and distributed
by Kluwer Academic Publishers Group,
P.O. Box 322, 3300 AH Dordrecht, Holland.

Printed in The Netherlands

TABLE OF CONTENTS

TABLE OF CONTENTS TO PART II

PREFACE TO PART I

Intensive experimental and theoretical work has been undertaken these last few years in the understanding of the chemical and physical properties of quasi-one-dimensional compounds. This designation is ascribed to materials which exhibit large anisotropy ratios in some physical properties when measured along one direction with regard to orthogonal directions. The recently-acquired possibility for chemists to synthesize real, inorganic as well as organic, one-dimensional systems with fascinating physical properties has been one of the essential motivations which have attracted many new research workers. The first two volumes in the present series 'Physics and Chemistry of Materials with Quasi-One-Dimensional Structures' are intended to review the experimental and the theoretical progress made in this field in the last years. These volumes deal with electronic properties of inorganic compounds.

The physical properties of the platinum chain compounds, of the superconducting polymeric sulfur nitride and the transition metal tri- and tetrachalcogenides are reviewed in Part II. Some of the latter compounds have aroused much interest because they have been considered as prototypes for the electrical conduction mechanism proposed by Fröhlich in 1954 in his theory of superconductivity based on the motion of a charge density wave. Concepts and theoretical models developed to understand the properties of one-dimensional systems are described in the present volume. Two chapters are devoted to the description of the nature of the ground state resulting from the interplay between lattice instabilities and superconductivity and to the analysis of the superconducting properties of an assembly of weakly-coupled chains. Two different approaches, a classical one and the other involving macroscopic quantum effects, are used to account for the collective transport mechanism induced by a charge density wave motion. One chapter concerns the general properties of the non-linear excitations in the ground state of one-dimensional systems, or solitons, and their contribution to transport properties. This book is intended to be of interest to solid state physicists – both students as well as advanced researchers – specialists in synthetic chemistry as well as scientists concerned with new materials.

I wish to express my sincere gratitude to the authors for their collaboration in this book and I warmly thank Prof. F. Lévy, the managing editor, for his suggestions and his constant encouragement.

Grenoble, January, 1984. PIERRE MONCEAU

P. Monceau (ed.), Electronic Properties of Inorganic Quasi-One-Dimensional Materials, I, xi.

GROUND STATE PROPERTIES OF CONDUCTING TRICHALCOGENIDES

S. BARIŠIĆ

Department of Physics, Faculty of Sciences
POB 162, Zagreb, Croatia, Yugoslavia

1. Introduction

In this review we shall discuss the equilibrium properties of the transition metal trichalcogenides MX_3 (M = Ta, Nb, Hf, Zr, Ti; X = S, Se, Te) paying special attention to the conductors $NbSe_3$ and TaS_3, which exhibit incommensurate structural transitions. $NbSe_3$ not only undergoes structural instabilities [1, 2] but under pressure it also shows [3] BCS-like superconductivity. The question of properties, the coexistence and the competition of structural instabilities and superconductivity in the MX_3 compounds [4] will be the main concern of the present paper.

Transition metal trichalcogenides belong to a wider class of materials in which the existence of quasi-one-dimensional conducting electrons leads to interesting low-temperature phase transitions [5]. Historically, it was the search for high-temperature superconductivity [6] which provided the main motivation to the strong development of the quasi one-dimensional physics which has been so apparent in the two last decades. Even nowadays the superconductivity remains one of the most important goals of one-dimensional (1D) research [7], although many other interesting effects, and in particular the non-linearities related to the structural instability have been found in the meantime [8]. The beginning of the one-dimensional history is thus usually associated with Little's proposal [9] for high-temperature superconductivity. The emphasis of this proposal was on a new mechanism for the effective electron–electron interaction. It was suggested that this can be mediated by the polarization fluctuations on the molecules surrounding the polymer backbone with conducting electrons. The one-dimensionality of the electron motion was merely a consequençe of the geometrical requirements optimizing the interactions. However, it appeared soon on theoretical grounds [10] that the one-dimensionality of the electron motion may lead to novel physical effects. At the same time it was realized that the best known superconductors M_3X (M = Nb, V, . . . ; X = Sn, Si, . . .) have some one dimensional electronic properties [11, 12, 6]. This led to a branching in subsequent developments. One line of research was devoted to constructing models with approximate one-dimensional properties [6]. It turned out that the anisotropy of the electron propagation originates from the directional properties of the tightly-bound wave functions [12], rather than from the anisotropy of the underlying lattice. This was followed by investigation of electron-phonon coupling models in the tight-binding limit [13], including the intra- and interchain Coulomb forces between electrons. The analysis of these rather complicated model hamiltonians was carried out in the simple mean-field (ladder) approximations [6, 12–14], as it was thought that the interchain couplings were large enough [15] to ensure their applicability at least in the M_3X compounds.

P. Monceau (ed.), Electronic Properties of Inorganic Quasi-One-Dimensional Materials, I, 1–40.

The other line of research started by examining the purely one-dimensional models, in spite of the lack of any well-established physical prototypes in these early times. The many-body complications associated with the one-dimensionality of the electron motion were carefully investigated in this approach. At first the weak-coupling, high temperature parquet summation was used [10, 16–18]. To a given order it takes correct account of the nontrivial coupling between the Cooper and density wave channels, characteristic for the one-dimensional electrons. This achievement was followed by an extremely rich development of the 1D theoretical models and methods, including Bethe's ansatz approach [19], renormalization group schemas [17, 20], bosonization [17, 20], mappings to equivalent models [17, 20] etc. The three-dimensional couplings [21, 22] were also progressively included in such considerations, with the general effect of stabilizing some particular types of the strong fluctuations found in the 1D theories [17, 23].

It turned out that the bridging between the tight binding models with their ladder approximations and the one-dimensional model results can be achieved in two ways. One way is to have the retarded (electron–acoustic phonon) interactions dominant close to the transition temperature. As we shall also see here, the strong retardation [14, 24] effects may reduce the parquet theory to the ladder one [18, 25]. The second way is to have the interchain hopping energy large with respect to the transition temperature [17, 23]. The resulting ladder theories retain, however, some of the one-dimensional character, exploited in the early ladder treatments [12, 13] of the tight-binding models [26]. The bridging was completed by noting that the tight-binding hamiltonian with long-range interchain Coulomb forces has similar properties, both in the parquet and ladder approximation, as the short-range hamiltonians, provided that the RPA screening is appropriately taken into account [27].

These theoretical developments went in parallel with the synthesis and characterisation of more and more new quasi-one-dimensional conductors [28, 29, 7]. The actual materials are believed to cover the whole theoretical spectrum, mentioned above. M_3X (Nb_3Sn etc) are certainly sufficiently three dimensional that the mean-field (ladder) approximations [6, 12, 13, 26] apply. TTF–TCNQ and $(TMTSF)_2PF_6$ families are likely to belong to the other extreme. They are nearly one-dimensional [7] and dominated by the unretarded Coulomb- and phonon-mediated interactions. The KCP family is also nearly one-dimensional but there the retardation effects in the dominant, phonon mediated interaction make [25] the simple single ladder approximation (i.e. the Peierls theory) qualitatively valid. As we shall argue here the situation is similar in the MX_3 ($NbSe_3$ etc) compounds at ambient pressure. However, high pressures bring the structural transition to low temperatures, where the retardation effects become unimportant. Nevertheless, the validity of the ladder approximations is maintained by the interchain hopping. In this regime the MX_3 compounds present some similarities with the M_3X materials.

This paper will not present the full spectrum of theories. Rather, we shall select those theoretical results and points of view which we believe to be relevant for the MX_3 compounds. In particular, we shall emphasize the role of the phonon-induced, retarded electron–electron interactions. Such an approach requires a discussion of the pertinent experimental results and represents, in consequence, an attempt at synthesising experiment and theory focussed on a given class of materials. The attempt is an early one, because the crucial experiment [30] showing definitively that $NbSe_3$ is a quasi-one-dimensional conductor rather than the anisotropic three dimensional metal was performed only very recently, confirming some earlier indications in this direction [31, 32], which were not considered as being conclusive. In the same spirit of synthesis a related paper [33] will describe the physics of TTF–TCNQ like compounds as an example where the unretarded Coulomb forces between 1D electrons play an important role.

2. Elementary Excitations

2.1. Electronic Bands

The common feature of all transition metal trichalcogenides [34, 4] is that each transition atom M lies in the centre of the trigonal prism of chalcogen atoms (Figure 1). The

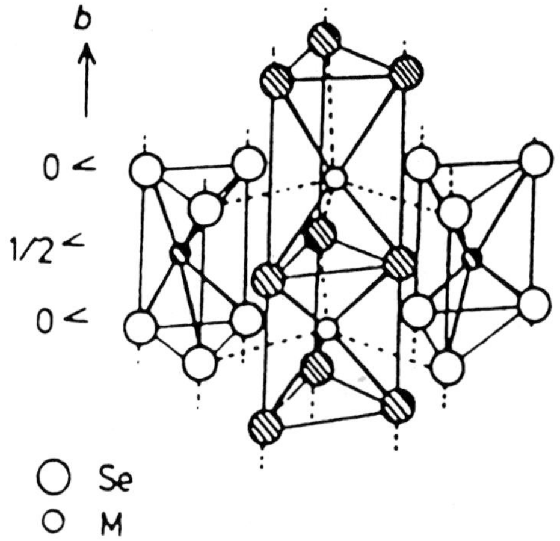

Fig. 1. Stacking of MX_3 prisms along the *b*-axis. Open atoms are at O and shaded atoms at 1/2 *b* (from reference [29]).

prisms form infinite chains along the *b*-axis. However, the form of the prisms and the mutual position of the chains depend on the materials. This difference is accompanied by a large variation of the conducting properties. E.g. NbS_3 is an insulator with an appreciable dimerization along the chain, TaS_3 and $NbSe_3$ are metals exhibiting structural transitions, and $TaSe_3$ is a reasonably good anisotropic metal even at low temperatures.

These differences can be associated with the band structure at the Fermi level. Indeed, recent band calculations [35, 36], based on the tight-binding (TB) or closely related schemes, were able to explain the differences mentioned in the conductivity of all materials considered, namely among $ZrSe_3$, $NbSe_3$ and $TaSe_3$. As the electron band structure is the basis for all later developments let us briefly review some of those results. We shall pay particular attention to $NbSe_3$, since it is the prototype material.

$NbSe_3$ (and one out of two modifications of TaS_3) is monoclinic, with space group $P2_1/m$. The unit cell contains six $NbSe_3$ units, which belong to six chains. From the crystallographic point of view, the chains are usually [4] subdivided in three groups according to the degree of departure of the X–X–X prism basis from the equilateral triangle. The three groups in question are shown in Figure 2. Each equilateral triangle of the prism of the group II can receive six *d*-electrons. The pairing of Se atoms in isosceles triangles, moderate in group I and strong in group III, ejects one antibonding state per triangle above the Fermi level. Therefore only four electrons can go to the Se atoms in such triangles. Out of 6 X 5 *d*-electrons in the unit cell, 4 X 4 thus go to

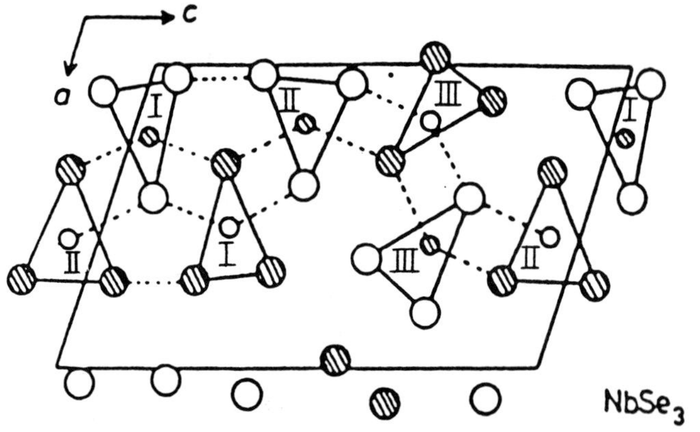

Fig. 2. Projection of the $NbSe_3$ structure in the *ac* plane. Atoms as in Fig. 1 (from reference [29]).

the isosceles triangles and 2 X 6 to the equilateral ones. The remaining two electrons stay in the conducting *d*-bands, presumably in those associated with the chains of the isosceles prisms belonging to the groups I and III. The deformation of the equilateral into the isosceles triangle thus results in a net charge transfer between the chains and the partial (quarter) filling of the bands [4, 35, 36]. The similar reasoning applied to $ZrSe_3$ and $TaSe_3$ explains their respective semiconducting and semimetallic behaviors [4, 35, 36].

The *p, d* separation [4] used here is quite useful for counting the electrons and the orbital degeneracy of the states (bands). However, there seem to be appreciable *p–d* (Se–Nb) hybridisations along the chain, which dominate the longitudinal bandwidth. On the other hand, the question of the transverse (interchain) hopping was controversial for a long time. Especially puzzling in this respect were the small effective number of carriers [4] ($\sim 10^{19}$ e/cm^3) deduced from high-temperature Hall effect measurements [37]. This led to the construction of the band model with small pockets [4] of electrons and holes for temperatures above the structural transitions, at 144 and 59 K. The corresponding supersturcture Bragg reflections occur [31, 34, 39] at (0, 0.243, 0) and (0.5, 0.259, 0.5) positions in the Brillouin zone. The model with pockets was also able to account for such superstructure [4]. However, observation of the one-dimensional sheets by X-ray diffuse scattering [30, 31] rules it out, since it implies that the precursor effects be well-localized around the final superstructure positions (see Section 4.1.).

In the more recent alternative band models [35] the interchain overlaps are much weaker than in the model involving small pockets. The six chains are thus only slightly inequivalent. They contribute six bands of the same nature to the neighbourhood of the Fermi level (Figure 3). As already mentioned, the longitudinal dispersion is due

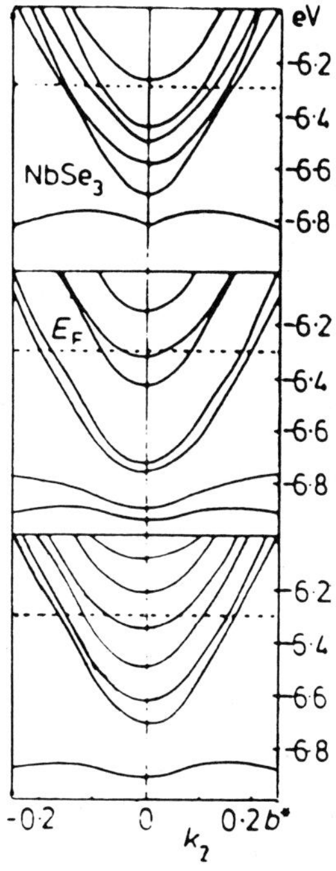

Fig. 3. Dispersion parallel to the chain direction for conducting bands in $NbSe_3$ along $(0, k, 0)$, $(\pi/a, k, 0)$ and $(0, k, \pi/c)$ axes, starting from the top of the Figure (Ref. 35, copyright Institute of Physics, Bristol, England).

to Nb–Se overlaps. The bands group two-by-two, as do the chains themselves. Four bands reach below the Fermi level. Therefore the four wave-vectors k_F fall in two groups, as actually observed through the $2k_F$ reflections. Consistently with small splittings among the bands, these $2k_F$'s are rather close, i.e. all four bands are nearly quarter-filled, in agreement with the preceding electron count.

Figure 3 suggests, further, that the transverse dispersion of 1*d* bands is comparable in both *a* and *c* transverse directions. This is somewhat surprising in view of the existence of the comparatively short contacts along the *c*-axis, but it should be noted that they involve the (group II) insulating chains. The most recent tight-binding band calculations [36] indicate however that the conducting bands have a nonvanishing group II character. This can lead to an effective transverse overlap larger in the *c* than in the *a* direction, and possibly explain the observed anisotropy ratio [38] of conductivities $\sigma_c/\sigma_a \simeq 30$. Above all, the latest Hall effect measurements [37] at room temperature have shifted the estimation of the effective number of carriers from 10^{19} to 10^{21} e/cm^3, reaching qualitative agreement with the picture of quarter filled quasi-1D bands.

In what follows we shall thus use the band structure shown in Figure 3, or rather propose a simplified version of it. The simplification consists in ignoring the inequivalence of the four conducting chains. The longitudinal dispersion can be conveniently parametrized within the simple TB model

$$E(k) = -\frac{2t_{SN}^2}{E_0} - \frac{2t_{SN}^2}{E_0} \cos kb \tag{1}$$

where E_0 measures the difference between the atomic energies of the Nb d-state and the Se p-state. Their overlap t_{SN} is assumed small with respect to E_0. In addition we shall introduce the transverse hopping integral $t_\perp$, without specifying the kind of the transverse dispersion (nesting or not) it leads to. The existing band structure computations are not particularly specific on this account and in what follows we shall discuss different behaviors associated with different transverse dispersions.

Equation (1) also illustrates that the $E_0 \gg t_{SN}$ band model can hardly by distinguished from the direct t_{NN} metal–metal bonding model. Sizeable differences between the two models arise in the corresponding electron – (acoustic, optic) phonon couplings, as will be indicated in Section 3.1. However, due to the lack of relevant information these differences will be mostly ignored there. Further band results (with electron–phonon coupling in mind) are therefore desirable.

2.2. PHONON FREQUENCIES

The vibrational properties of the transition metal trichalcogenides were investigated rather extensively by Raman [40, 41] and infrared techniques [42]. In order to understand the gross features of vibrations this investigation started with a simple ZrS_3-like structure [40, 42], because the mode assignment is easier in this case. The characteristic vibrational frequences ω_{ph} (measured in Kelvins) lie around 300 K. The exception is the mode at 500 K in which the two paired chalcogenide X atoms move in opposition of phase along the prism edge joining them (Fig. 1). The force constant corresponding to $\omega_{ph} \sim 300$ K arises from the M–X bond and amounts to [40]

$$C \simeq 5 \times 10^4 \text{ K/Å}^2. \tag{2}$$

This force constant also determines the dispersion of the acoustic modes. Although the latter has not been measured, to our knowledge, Equation (2) sets the estimated Debye temperature ω_D around 300 K.

In addition to that it was possible to measure directly the transverse dispersion of the Raman and infrared-active modes, i.e. the interchain elastic coupling [40–42]. The effective coupling constant $C_\perp$ turns out [40] to be about three times less than given in (2). Although elastically anisotropic, trichalcogenides are not quasi-one-dimensional from the vibrational point of view.

Due to its lower symmetry, $NbSe_3$ presents the splittings of the modes observed in the ZrS_3 structures [40]. Besides, there is an overall decrease $\Delta\omega_{ph}$ of the phonon frequencies [40]. This decrease is usually imputed to the coupling of phonons to the conducting electrons of $NbSe_3$. Through the usual estimate [43] (see also Section 4.1)

$$\frac{\Delta\omega_{ph}}{\omega_{ph}} \simeq -\lambda_{ph}, \tag{3}$$

where λ_{ph} is the phonon mediated electron–electron coupling constant, this gives us an initial idea of the value of λ_{ph}. We find $\lambda_{ph} \simeq 0.1$, the value we wish to discuss in more detail in Section 3.1, which deals with interactions.

2.3. PLASMA FREQUENCY

In the present context it is also appropriate to mention the reflectivity measurements [44] which determine the plasma edge. The plasma frequency ω_0 is found to be equal to [44]

$$\omega_0 \simeq 1.15 \text{ eV}. \tag{4}$$

From this value it is possible to estimate the band-width W and long-range Coulomb interaction using the usual expression [45, 46, 27] for the plasma frequency (n_F – density of states per M atom, per spin)

$$n_F \frac{e^2}{b} \simeq \frac{1}{4\pi} \left(\frac{\omega_0}{W} \frac{1}{k_F b} \right)^2 \frac{ac}{4b^2}, \tag{5}$$

and knowing the lattice parameters [34] a, b, c. With [44]

$$W = \frac{4t_{SN}^2}{E_0} \simeq 2.5 \text{ eV} \tag{6}$$

we have

$$n_F \frac{e^2}{b} < 0.1. \tag{7}$$

As with Equation (3), this result leads us to the question of interactions discussed in the next Section.

3. Interactions

In the preceding section we have considered the non-critical renormalization of the elementary excitations (plasmons and phonons) by the interactions. However, the interactions also determine the subtle collective behavior of the system. For this reason we wish to proceed here with a more detailed discussion of the interactions. Sections IV and V will deal with the critical behavior induced by those interactions.

3.1. ELECTRON-PHONON COUPLING

It is currently believed that the most important interaction in the conducting trichalcogenides is the coupling of the conducting electrons to the acoustic phonons. The main contribution to this coupling arises from the variation of the overlaps t_{SN} in Equation (1) with acoustic deformation. The magnitude of this variation can be estimated by noting that [14]

$$\frac{dt_{SN}}{db} \simeq -q_{SN}t_{SN} \tag{8}$$

in the TB approximation. Here q_{SN} is the effective Slater coefficient of the Se and Nb wave-functions [14]. The Nb–Se splitting energy E_0 should not be much affected by the acoustic deformations, because it represents an intra-cell property, and the prisms move rigidly in acoustic modes. According to Equation (1) the variation (8) of t_{SN} gives rise to the variations of the effective overlap (the coefficient of cos kb) and the mid-band energy. Equation (1) is thus an example of the band structure in which the electron–acoustic phonon couplings arising from the variations of the effective overlaps and the mid-band energies are of the same order of magnitude. From this point of view the present situation contrasts with the usual metal–metal bonding TB models [14]. The wave-vector dependences of the two electron–phonon couplings are different [47]. However, we shall not be concerned with such details here, i.e. we shall take the couplings for the electrons at k_F through the phonons at $2k_F$ (backward scattering) when both couplings act roughly additively. The corresponding coupling constant is

$$\lambda_{ac} \simeq \frac{4W}{2\pi\, Cq_{SN}^{-2}}\,, \tag{9}$$

with C and W given by Equations (2) and (6), respectively. Due to the large bandwidth W, λ_{ac} is around (below) 0.5. The coupling of electrons via the optical phonons is expected to be somewhat weaker than (9). E.g. it is possible to construct the optical displacements which, to the first order, do not alter t_{SN}^2, in contrast to Equation (8). On the other hand the splitting energy E_0 in Equation (1) does contain some intersite contributions which could be sensitive to optic displacements. It is, however, hard to assess the order of magnitude of these variations. We shall thus proceed with the empirical estimate (3).

3.2. COULOMB INTERACTIONS

The basic picture which we are developing here is that of the TB electrons. The electron–phonon couplings discussed in the preceding subsection are short ranged (cf. Equation (8)). This is consistent with the phonon frequencies ω_{ac}, which vanish in the long-wavelength limit. Despite this, the Coulomb interaction between electrons should be taken as being long-ranged at the outset [14] rather than as described by the Hubbard hamiltonian or its short-range extensions. The long range Coulomb forces lead to qualitatively important effects [48] in Nb_3Sn type of materials, where the structural instability occurs in the long wave-length limit. On the other hand the effects of the long (and short) range Coulomb forces are expected to be mainly quantitative in the present case, as we shall now proceed to argue.

The Coulomb interaction of two parallel strings of atoms carrying the charge density waves (CDW's) of the wave vector q and unit amplitude is [49]

$$\frac{e^2}{b}\sum_g e^{-ig\epsilon b} K_0(R_\perp|q+g|)\, e^{i\phi_\perp}. \tag{10}$$

Here K_0 is the modified zero-order Bessel function, g the longitudinal reciprocal wave vector, $R_\perp$ the distance between the strings, shifted relative to one another by ϵb, a fraction of b, Fig (4). $\phi_\perp$ is the relative phase of the CDW's on the strings. The total Coulomb

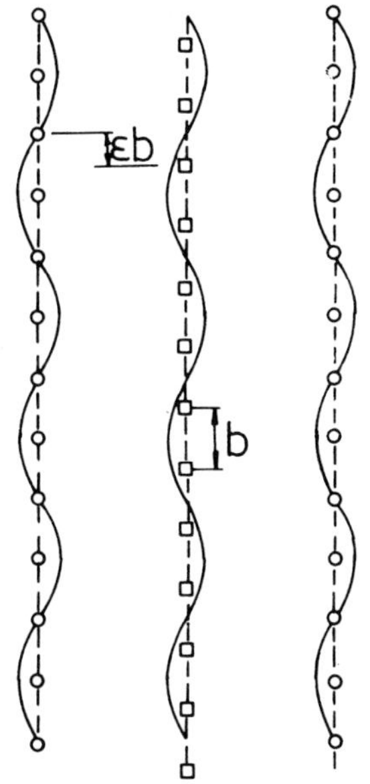

Fig. 4. X(○) and M(□) strings in MX_3, carrying charge density waves.

energy is obtained [49] by multiplying Equation (10) by the appropriate CDW amplitudes and by summing over all strings, after adding the intrastring terms. The latter are given by [49]

$$U_0 - \frac{e^2}{b}\log[2(1-\cos qb)] \tag{11}$$

for the unit CDW amplitude. Here U_0 is the intra-atomic (Hubbard) matrix element for a given string of atoms.

In our case we can decompose each chain in four (1Nb + 3Se) strings, carrying the CDW's. The strong Nb–Se (M–X) hybridization [35, 36] forces all these CDW's to be in phase ($\phi_\perp = 0$ in Equation (10)). The appropriately weighted sum of the contributions (10) and (11) then gives the Coulomb intrachain energy for unit CDW amplitude per chain. The interchain Coulomb energy can be further evaluated as the sum of the interstring interactions between the chains. In this calculation it may be better to keep the relative phase $\phi_\perp$ of the CDW's on two chains as a free parameter. This is particularly indicated when the chains belong to different groups of Fig. 2 (for $NbSe_3$ we shall need only equivalent groups). Then all interstring interchain distances $R_\perp$ satisfy

$2k_F R_\perp > 1$, i.e. all K_0's of Equation (10) are exponentially small. Such weak interchain interaction may lead to the thermally activated ordering of the thus-defined $\phi_\perp$, as is further discussed in Section 4.2. On the other hand, the Coulomb energy between the neighbouring chains which belong to the same group of Fig. 2, although considerably smaller than the intrachain one, is larger than the intergroup energy, because there are some short $2k_F R_\perp < 1$ distances within the group. Together with interchain hybridization and elastic coupling this interaction probably freezes the corresponding transverse phase to a definite value. In any case, we shall continue our analysis with only one phase $\phi_\perp$ per unit cell at a given structural transition and associate with it a small interchain coupling constant $g_1^\perp$. (The generalization to more phases, if necessary, is straightforward).

In contrast to that the intra- and interchain interactions between the long ($q < k_F$) waves is large, according to Equations (10) and (11). In the first instance this requires the introduction of the RPA screening [27]. In Section 4.2 it will be argued that the (logarithmic) theory with the RPA-screened forward ($q < k_F$) Coulomb interaction has much in common with the short-range (Hubbard-like) models, although some important differences persist. It turns out that to the leading order in the small parameter $n_F e^2/b$ (cf. Equation (7)), the effective on-chain forward coupling constant

$$n_F \frac{e^2}{b} \log \frac{1}{n_F \omega_0} < 1 \tag{12}$$

has the same effect as the complicated, screened, long-range Coulomb interaction.

Whereas it is possible to start from the bare long range Coulomb interaction (10, 11), and express the effective coupling (12) in terms of this quantity using the RPA intraband screening, the same procedure does not work for the short range part of the interaction related to U_0 in Equation (11). The preliminary (pre-logarithmic) screening of U_0 is related to the 'interband' processes allowing the electrons to avoid each other on a given 'site' (prism). Although the principles of this screening are quite clear [50], its precise evaluation is lacking for conducting trichalcogenides. It is thus appropriate to take an effective U, which already includes the interband screening, and use it in the subsequent (logarithmic), conducting band theory. Its value can then be evaluated *a posteriori*, comparing the theoretical and the experimental results.

The property which is usually used to get an idea about U is the magnetic susceptibility. The magnetic susceptibility has been investigated by various methods [51, 52], the most recent being the Knight shift measurements [53]. The paramagnetic susceptibility is rather large (of the order of 10^{-4} emu/mole). In the high temperature range it is practically temperature independent. The structural transitions reduce the paramagnetic susceptibility roughly by 20%. Unfortunately, so far, there is no general agreement as to the source of the large low temperature paramagnetism [51–53]. One possibility is the large van Vleck component and the other is the spin susceptibility of those conducting electrons not involved in the transitions. In this situation it seems best to rely upon the high temperature one-dimensional regime. If $n_F U$ were large it would lead to appreciable temperature dependence of the 1D spin paramagnetic susceptibility.

Besides, such $n_F U$ would most likely produce the $4k_F$ phonon softening, which is not observed. All these results are well known from the theories of the strongly-correlated Coulomb gas [17, 20]. As already mentioned in the Introduction, the latter will not be reviewed here but rather in [33], in connection with TTF–TCNQ where this is more appropriate. Here we shall only conclude that

$$n_F U < 1. \tag{13}$$

Comparing Equations (9), (12) and (13) we can reasonably assume that

$$\lambda_{ac} > g_{1,2} > g_1^{\perp}, \tag{14}$$

where $g_{1,2}$ represent the linear combinations of the coupling constants (12) and (13), which appear in backward and forward scattering of the two electrons, respectively;

$$g_1 \simeq n_F U,$$

$$g_2 \simeq n_F U + n_F \frac{e^2}{b} \log \frac{\epsilon_F}{\omega_0}.$$

Simplifying somewhat, we can say that the main reason for the inequality (14) is that λ_{ac} scales as W, in contrast to the $g_{1,2}$'s which go as W^{-1}, and because the trichalcogenides are large bandwidth materials.

4. Peierls Instability

In Sections 2 and 3 we have discussed the high temperature excitations and their interactions. This section is devoted to the description of a phase transition, which results from the interaction of these excitations. We shall start with the structural transition at ambient pressure.

In the preceding section we have discussed the respective strength of the phonon and Coulomb interactions. Beside its strength the interaction is also characterized by its cutoff frequency. The characteristic frequencies of the dominant, phonon mediated interaction are small, according to Section 2.2. This makes retardation effects important [14, 18, 24, 25, 26, 54]. Here we wish to emphasize that this is particularly true for temperatures comparable to or larger than the phonon frequencies. The extreme limit $T \gg \omega_{ph}$ (or ω_D) will turn out to coincide with the Peierls theory [55].

4.1. Harmonic Theory – Kohn Anomaly

The bare electron–electron interaction mediated by a phonon with polarization γ and wave-vector $\mathbf{q}$ is [43]

$$\lambda_\gamma \frac{\omega_\gamma^2(\mathbf{q})}{\omega^2 - \omega_\gamma^2(\mathbf{q})}. \tag{15}$$

The coupling constants λ_γ and the frequencies $\omega_\gamma(\mathbf{q})$ were thoroughly discussed in Sections (3.1) and (2.2), respectively.

The characteristic electron scattering processes involving phonons are shown in Figures 5 to 7. In their evaluation at finite temperatures it is usual [43] to use the Matsubara formulation, where $\omega = i\,2\pi nT$ in Equation (15). If $2\pi T \ll \omega_\gamma$, there is a considerable range of integers n for which the interaction behaves as unretarded, and is given by $-\lambda_\gamma$. This is the familiar BCS limit [56] in which λ_γ is supplemented with the cut-off frequency ω_γ. If however

$$2\pi T > \omega_\gamma(\mathbf{q}) \tag{16}$$

there is only one [18, 25], $n = 0$, value for which the bare interaction (15) is λ_γ and for which it decreases rapidly, as n^{-2}. These small values of the interaction do not appear in the real (phonon momentum and energy conserving) processes such as that of Figure 5 or in the ladder generated by it in Figure 8, provided that $n = 0$ is chosen on the initial phonon line. In contrast to that, the virtual phonon exchanges such as those of Figure 6 and 7 involve summation over n. At high T they can thus be neglected [18, 25] with

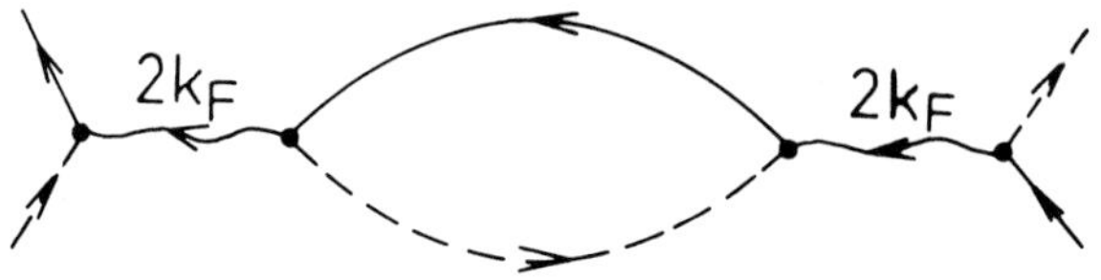

Fig. 5. Electron-hole (bubble) correction to the backward scattering vertex. Full line, electron at k_F; dashed line hole at $-k_F$; dot, electron-phonon coupling; wavy lines, phonons (here at $2k_F$).

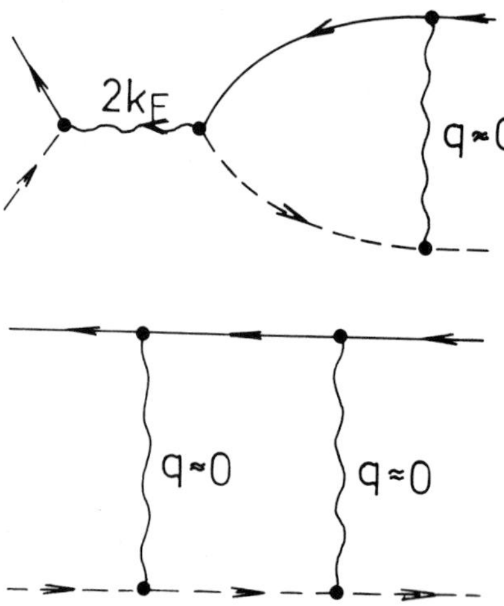

Fig. 6. Electron-hole corrections to (top) backward and (bottom) forward scattering vertices, which involve long wavelength ($q \simeq 0$) interactions (phonons).

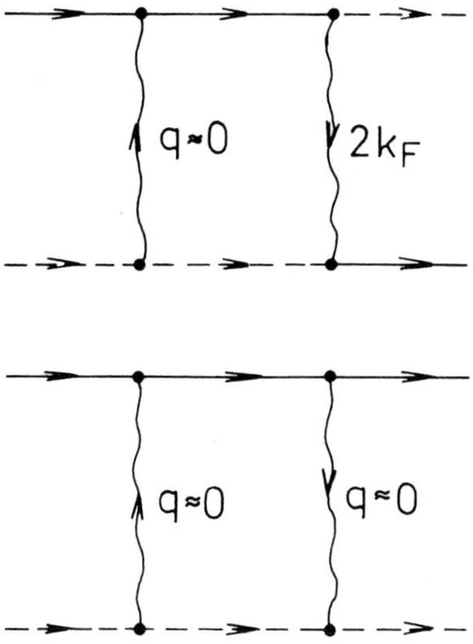

Fig. 7. Electron-electron corrections to (top) backward, and (bottom) forward scattering vertices, which involve long wave-length ($q \simeq 0$) interactions (phonons). Dashed line, electron at $-k_F$.

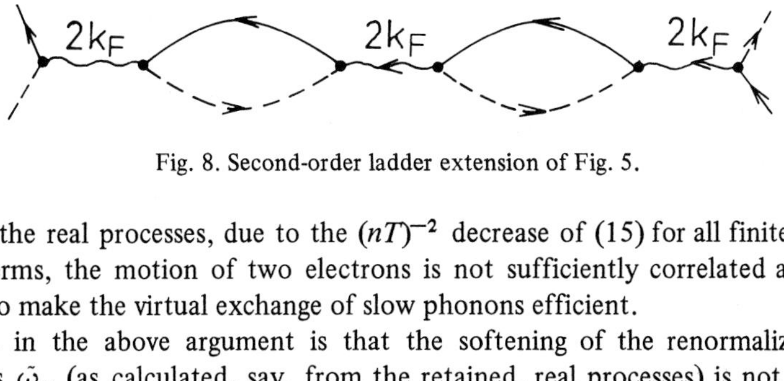

Fig. 8. Second-order ladder extension of Fig. 5.

respect to the real processes, due to the $(nT)^{-2}$ decrease of (15) for all finite n. In more physical terms, the motion of two electrons is not sufficiently correlated at high temperatures to make the virtual exchange of slow phonons efficient.

Implicit in the above argument is that the softening of the renormalized phonon frequencies $\tilde{\omega}_\gamma$ (as calculated, say, from the retained, real processes) is not too strong, so that the renormalized $n = 0$ vertex ($\sim \tilde{\omega}_\gamma{}^{-2}$) is not much larger than λ_γ. This is true for

$$T > T_p^0 \tag{17}$$

where T_p^0 is the temperature at which the renormalization of $\tilde{\omega}_\gamma$ arising from real processes is of the order of ω_γ itself (i.e. $\tilde{\omega}_\gamma \simeq 0$). Whereas the condition (16) eliminates the quantum fluctuations of the phonon field, (17) can be alternatively understood as making the theory harmonic in displacements. This latter interpretation will become clear in Section 4.3.

Here we shall evaluate T_p^0 and discuss the real process contribution in some detail. The irreducible phonon self-energy Π corresponding to the ladder of Figure 8 is given by the simple electron–hole (e–h) bubble of Figure 9 and

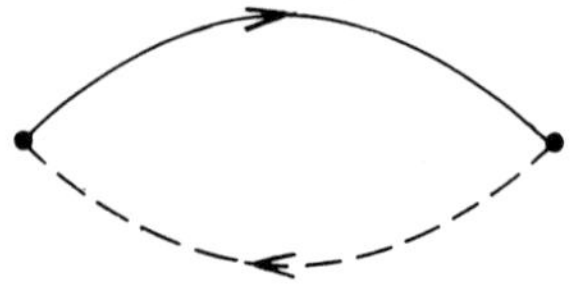

Fig. 9. Phonon self-energy corresponding to ladder extension of Figures 5 and 8.

$$\tilde{\omega}_\gamma^2 = \omega_\gamma^2 - \Pi. \tag{18}$$

The result of the evaluation of the e–h bubble depends on the nesting properties [57–60, 54] of the electron Fermi surface. Good nesting means that there is a wave-vector $\mathbf{q}_0$ such that the Fermi surface translated by $\mathbf{q}_0$ coincides over an appreciable portion with the original Fermi surface. In this case the e–h bubble is logarithmically singular for the phonons in the vicinity of $\mathbf{q}_0$ ($q_{0\mathrm{b}} \simeq 2k_\mathrm{F}$)

$$\omega_\gamma^{-2}\,\Pi(\mathbf{q}_0) \simeq 2\lambda_\gamma \log \frac{\epsilon_\mathrm{F}}{T}. \tag{19}$$

The Kohn anomaly develops around $\mathbf{q}_0$.

T_p^0 of (17) is estimated by requiring that $\tilde{\omega}_\gamma \simeq 0$,

$$T_\mathrm{p}^0 \simeq \epsilon_\mathrm{F} e^{-1/2\,\lambda_{\mathrm{ac}}} \tag{20}$$

assuming that λ_{ac} is the largest among λ_γ's, as was argued in Section 3.1.

Equations (19) and, consequently, (20) hold for [54]

$$T > t_\perp \frac{t_\perp}{\epsilon_\mathrm{F}}. \tag{21}$$

Expanding further $\Pi(\mathbf{q})$ in terms of the longitudinal and transverse component of $\mathbf{q}$–$\mathbf{q}_0$ gives longitudinal [61, 62] and transverse [54] characteristic lengths ξ_0 and $\xi_{0\perp}$ respectively, which characterize the slope of $\omega(\mathbf{q})$ around $\mathbf{q}_0$

$$\begin{aligned} \xi_0 &= \frac{b}{2\pi n_\mathrm{F} T}, \\ \xi_{0\perp} &\simeq \frac{t_\perp}{T} d_\perp, \end{aligned} \tag{22}$$

where $d_\perp$ is of the order of a or c. According to (21) $\xi_{0\perp}$ can be larger or smaller than $d_\perp$, i.e. the softening can be localized around $\mathbf{q}_0$ or extended over the whole transverse Brillouin zone. Although related to it, $\xi_{0\perp}$ should not be confused with the transverse correlation length. The latter involves the critical dependence on temperature in the vicinity of the phase transition, whereas $\xi_{0\perp}$ only fixes the transverse length scale [25]. This will be further discussed in the next Section. The expressions (21) and (22) were derived for cosine transverse dispersion, which probably somewhat overestimates the nesting ability of the actual Fermi surfaces. The other extreme is the complete absence of nesting, when the phonon self-energy behaves as [17]

$$\omega_{\gamma}^{-2}\,\Pi \simeq \lambda_{\gamma} \log \frac{\epsilon_F^2}{T^2 + t_{\perp}^2} \tag{23}$$

irrespective of the transverse wave-vector. Inserting Equation (23) into Equation (18) determines T_p^0 as a fast function of $t_{\perp}$ when $T_p^0 \simeq t_{\perp}$.

For

$$T > t_{\perp} \tag{24}$$

both cases (19) and (23) lead to practically the same behavior. $\omega_{\gamma}^{-2}\,\Pi$ is given by Equation (19), T_p^0 by Equation (20) and the transverse dispersion is either weak $\xi_{0\perp} < d_{\perp}$, as in Equation (22), or completely absent.

The observation [30, 31] of the nearly one-dimensional diffuse sheets at ambient pressure and temperature indicates therefore that the transverse overlaps $t_{\perp}$ are small with respect to the considered temperatures, i.e. Equation (24) holds. From this observation it is however impossible to infer anything about nesting. Admittedly, the diffuse sheets condense [30, 31, 38] into Bragg spots on decreasing the temperature, i.e. some transverse correlations do appear. However, as we shall see below, this is not necessarily due to nesting (22). We shall therefore withold our conclusion concerning this latter until Section 5.1., where it is argued on different grounds that nesting exists in $NbSe_3$.

Let us finally point out that the corrections to the extreme retarded limit (16) do not change the main conclusion reached above, namely that $t_{\perp}$ is small in $NbSe_3$. As will become clear in Sections 5.2. and 5.3, these corrections cannot diminish the transverse correlations without suppressing the $2k_F$ structural instability altogether.

4.2. COULOMB COUPLING OF PEIERLS DEFORMATIONS

Section 3.2. led us to the conclusion (14) that the Coulomb forces in transition metal trichalcogenides are presumably weaker than the phonon-induced couplings. This explains why we have started this section by completely omitting the Coulomb interactions between electrons. At present we wish to justify this approach by showing that nothing qualitatively new happens by introducing the on-chain Coulomb forces (12) and (13) in the limit (14). On the other hand the interchain Coulomb coupling of the $2k_F$ waves $g_1^{\perp}$, although small, may be of qualitative importance.

Let us start by considering the Coulomb forces as perturbations [25]. The two low-order Coulomb corrections to the phonon self-energy Π are shown in Figures 10 and 11.

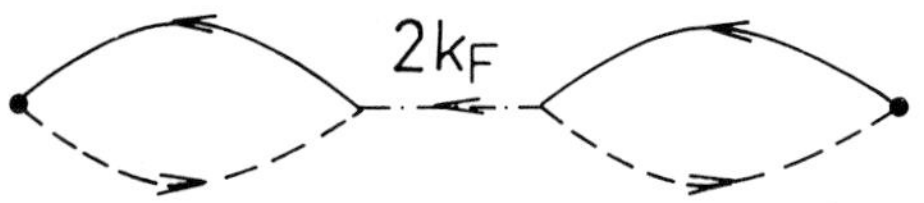

Fig. 10. Coulomb (dot-dashed line) correction to $2k_F$-phonon self-energy involving large momentum transfer.

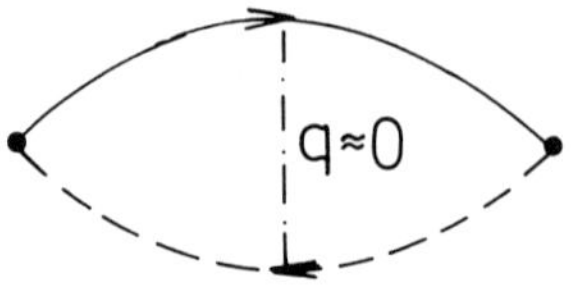

Fig. 11. Coulomb correction to $2k_F$-phonon self-energy involving small momentum transfer.

The two e–h bubbles in Figure 10 may belong to the same chain, and the corresponding Coulomb energy (cf. Section 4.3) is that of the electron redistribution within the chain, induced by the $2k_\mathrm{F}$ deformation. The contribution of Figure 10 to $\omega_{\mathrm{ac}}^{-2}\Pi$ is

$$4\lambda_{\mathrm{ac}} g_1 \log^2 \frac{\epsilon_\mathrm{F}}{T}, \tag{25}$$

where $g_1 \simeq n_\mathrm{F} U$ of Equation (13), and either (21) or (24) is assumed. It should be noted that the contribution (25) is linear in λ_{ac}, just as (19) itself is. We remain thus within the harmonic approximation (cf. Section 4.3).

Similar conclusions hold for the process of Figure 11. Assuming $T > t_\perp$ the internal integration over the transverse momentum in Figure 11 concerns only the interaction, because the transverse hopping is negligible. The averaging of the interaction over the transverse momentum reduces it to the intrachain term. Integration over the longitudinal momentum transfer is dominated by the small (forward scattering) q values. Assuming Equation (11), the on-chain interaction is logarithmically singular in the longitudinal momentum q. The logarithmically singular interaction instead of the usual constant leads to the $\log^3 \epsilon_\mathrm{F}/T$ singularity of the process (11) instead of the usual $\log^2 \epsilon_\mathrm{F}/T$. However, when the three-dimensional RPA screening of the long-range Coulomb interaction is introduced, all interactions beyond the screening distance k_{TF}^{-1} or time ω_0^{-1} ($\omega_0 \simeq v_\mathrm{F} k_{\mathrm{TF}}$, Equation (4)) are screened out [27]. In consequence, the process (11) behaves as

$$-2\lambda_{\mathrm{ac}} g_2 \log^2 \frac{\epsilon_\mathrm{F}}{T} \tag{26}$$

which is similar to Equation (25), with g_1 replaced by [27]

$$g_2 \simeq n_\mathrm{F} U + \frac{n_\mathrm{F} e^2}{b} \log \frac{\epsilon_\mathrm{F}}{\omega_0} \tag{27}$$

of Section 3.2. Although the $\log^3 T$ singularity does not appear in (26), the trace of the logarithmic Coulomb singularity (11) remains present in the second term of Equation (27).

Now comparing Equations (19), (25) and (26) we see that for all $T > T_\mathrm{p}^0$ of Equation (17) the contributions (25) and (26) are smaller than (19), provided that $\lambda_{\mathrm{ac}} > g_{1,2}$ as is found in (14). This shows that, due to the inequality (14), the addition of the on-chain (screened) Coulomb forces does not alter the Peierls theory in an essential way for the conducting trichalcogenides. The conclusion is reached here on evaluting the low-order corrections to the phonon self-energy Π, but it can be justified to any order of the leading logarithmic approximation. It turns out that the whole series for the phonon self-energy Π [18] (parquet or ladder, whichever is appropriate, according to Section 5.1.) can be expressed [27] in powers of $g_{1,2} \log \epsilon_F/T$ and is rapidly convergent for $g_{1,2} \log \epsilon_F/T < 1$ (i.e. at $T > T_p^0$ for $\lambda_{ac} > g_{1,2}$). For such temperatures the well-known (ladder or parquet) singularities [17, 20] which occur in the phonon self-energy [18] for $g \log \epsilon_F/T$ of the order or or much larger than unity (in the ladder and parquet case, respectively), are simply irrelevant [25]. The formation of these singularities is arrested at (by) T_p^0.

Summarizing the conclusions re^ched in this section until now, $2\pi T_p^0 > \omega_{ac}$ and $\lambda_{ac} > g_{1,2}$ appear to be the conditions [25] for the applicability of the Peierls theory.

Let us now turn our attention to the interchain Coulomb coupling of the $2k_F$ CDW's. As was explained in Section 3.2. this coupling is considerably smaller than the intrachain terms. Nevertheless it may be more important than the intrachain terms, provided that it dominates the interchain correlations. Then it plays a crucial role in establishing the long-range order of the $2k_F$ superlattice.

In the perturbative limit under consideration, the leading interchain effect is given [21, 25] by the diagram of Figure 10. The two e–h bubbles in this figure can be attributed to the neighbouring chains rather than to the same chain as in Equation 25. Then this figure represents the interaction of the two CDW's induced by the $2k_F$ deformation on the neighbouring (groups of) chains. Ignoring (for simplicity) $t_\perp$, the corresponding contribution to $\omega_{ac}^{-2}\Pi$ is [25]

$$4\lambda_{ac}\, g_1^\perp \cos\phi_\perp \log^2 \frac{\epsilon_F}{T}. \tag{28}$$

As in Section 3.2, $\phi_\perp$ here represents the relative phase of the two CDW's on the neighbouring (groups of) chains. On expanding (28) around the minimum of the Coulomb energy at $\phi_\perp = \pi$ we find the characteristic transverse distance [25]

$$\xi_{0\perp}^2 \simeq d_\perp^2 g_1^\perp \log^2 \frac{\epsilon_F}{T} \tag{29}$$

analogous to Equation (22), i.e. with λ_{ac} extraced from Equation (28).

For $T = T_p^0$ Equations (20) and (29) give

$$\xi_{0\perp}(T_p^0) \simeq d_\perp \frac{\sqrt{g_1^\perp}}{\lambda_{ac}}. \tag{30a}$$

This value can be compared with Equation (22) at T_p^0

$$\xi_{0\perp}(T_p^0) \simeq d_\perp \frac{t_\perp}{\epsilon_F}\, e^{1/2\lambda_{ac}}, \tag{30b}$$

or with the transverse correlation coming from the elastic coupling

$$\xi_{0\perp} \simeq d_\perp \sqrt{\frac{C_\perp}{\lambda_{ac} C}}, \tag{30c}$$

which involves the elastic constants $C \simeq 3C_\perp$ discussed in Section 2.2. The largest among the values (30) dominates the transverse correlations at T_p^0. In this respect we note that the elastic characteristic distance (30c) is temperature-independent. The existence of the 1D regime at ambient temperature and pressure in $NbSe_3$ thus requires $\xi_{0\perp} < d_\perp$ in Equation (30c). Although $\xi_{0\perp}(T_p^0) < d_\perp$ cannot be ruled out at the higher 144 K transition, the lower 59 K transition seems to be quite 3D. $\xi_{0\perp}(T_p^0) > d_\perp$ at 59K must therefore be sought between Equations (30a) and (30b). It can be noted in this respect that small λ_{ac} enhances the nesting correlations (30b) more than the Coulomb correlations (30a). However, λ_{ac} of Section 3.1 is not too small and the respective role of two interchain couplings is hard to assess. We shall thus postpone this discussion until Section 5.1, where different arguments in favor of the nesting model are given.

Here, we shall only realize that if $\xi_{0\perp}(T_p^0) > d_\perp$ the 1D to 3D crossover occurs in the harmonic $T > T_p^0$ regime. If, however, $\xi_{0\perp}(T_p^0) < d_\perp$ this crossover occurs when anharmonic interactions of fluctuations are important and this is the problem we wish to adress now.

4.3. ANHARMONIC INTERACTIONS OF FLUCTUATIONS

When the temperature is decreased towards T_p^0, i.e. when condition (17) is lifted, the simple bubble approximation of Figure (9) for the phonon self-energy Π ceases to hold. The leading correction [63–65] to the results (19) or (23) is shown in Figure (12).

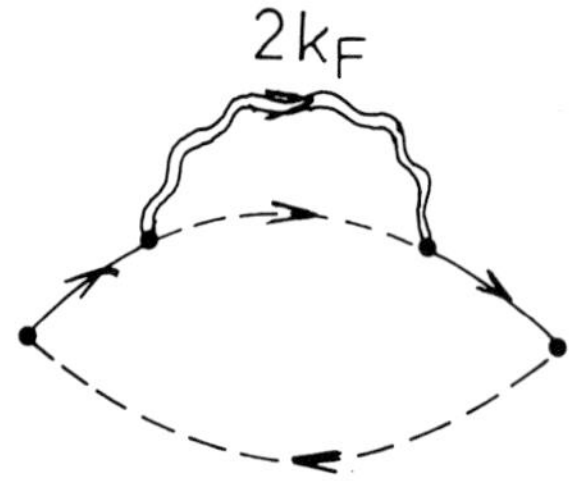

Fig. 12. Renormalization of the electron propagator in the phonon self-energy by $2k_F$ soft phonon, double wavy line, (pseudo-gap effect).

It consists in the renormalization of the electron spectrum by the soft phonon at $2k_F$. The resulting electron pseudogap [64, 65] is the precursor of the real Peierls [55] gap Δ_p.

Rather than going directly through the perturbation calculation of the phonon self energy we shall adopt here a somewhat different approach. Namely, we shall maintain the validity of the condition (16) for $T \simeq T_p^0$ and use $2\pi T_p^0 > \omega_{ac}$ to eliminate the quantum fluctuations of the phonon field. Treating this latter classically as the external field, we shall expand the electron deformation energy in powers of the field.

The lowest order term $E^{(2)}$ is shown in Figure (13). The wavy line with a cross represents the acoustic displacement u_{2k_F+q} and the dot the corresponding deformation potential per unit length (cf. Equation (8)). Comparing Figure (9) and (13) it is clear that, apart from the normalization, $E^{(2)}$ and Π of the preceding Section are the same [14]. This can be illustrated in the one-dimensional regime, when

Fig. 13. Harmonic deformation energy due to coupling of $2k_F$ displacements to electrons. Wavy line with a cross, $2k_F$ displacement.

$$E^{(2)} = \lambda_{ac}\left[\frac{T - T_p^0}{T_p^0} + \xi_0^2 (q - 2k_F)^2\right] C\,|u_{q-2k_F}|^2. \tag{31}$$

For brevity we have included in $E^{(2)}$ the harmonic elastic energy $C|u|^2/2$ of Section 2.2. The only difference with Equations (18), (19) and (20) is that here we have expanded $\log T_p^0/T$ in terms of $(T - T_p^0)/T_p^0$. ξ_0 is of course the same as in Equation (22), with $T = T_p^0$.

The typical next-order term [66, 67] $(1/6)E_N^{(4)}$ is shown in Figure (14) (there are

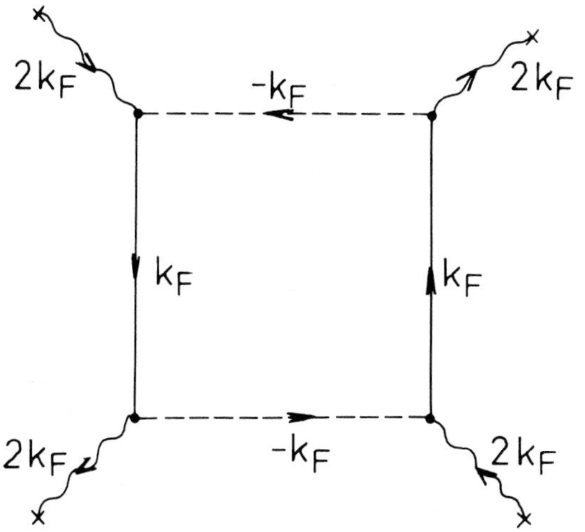

Fig. 14. Quartic deformation energy due to coupling of displacements to electrons.

six equivalent terms of this type). Its connection with the pseudo-gap is best understood by noting that by closing two displacement lines in Figure (14) to form the displacement correlator we get the result which is isomorphous to Figure (12). On the other hand, in the present approach, we shall understand [66, 67]

$$E_N^{(4)} \simeq \frac{0.1\,\lambda_{ac}^2}{2n_F(T_p^0)^2}\,C^2\,|u_{2k_F}|^4 \tag{32}$$

as the anharmonic interaction of the displacive fluctuations, the harmonic energy of which is $E^{(2)}$. As will become clear below, this will allow us to evaluate the correlations of the displacements more accurately than by the perturbation expansion (Figure 12), or by its self-consistent extensions.

Another advantage of the present approach is that the effects of the lattice discreteness can be straightforwardly included in such a calculation. These effects may be important when the wavelength of the deformation π/k_F is a multiple of the lattice constant [68]. This is precisely the case in conducting trichalcogenides, where $\pi/k_F \simeq 4b$. Under such circumstances there is an additional contribution to the quartic interaction. Its electron-phonon part is shown in Figure (15). By using the Umklapp invariance [14] of the (tight-binding) electron propagator and electron-phonon coupling it follows

$$E_U^{(4)} = \frac{\delta}{6} \frac{0.1\lambda_{ac}^2}{n_F(T_p^0)^2} C^2 u_{\pi/2b}^4 \tag{33}$$

with

$$2\delta \simeq \frac{1}{\lambda_{ac}} (n_F T_p^0)^2 \simeq \frac{e^{-2/\lambda_{ac}}}{\lambda_{ac}}.$$

$E_U^{(4)}$ is δ times smaller than $E_N^{(4)}$ due to the appearance of excited states in Figure (15). Such $E_U^{(4)}$ can even be comparable to elastic Umklapp terms.

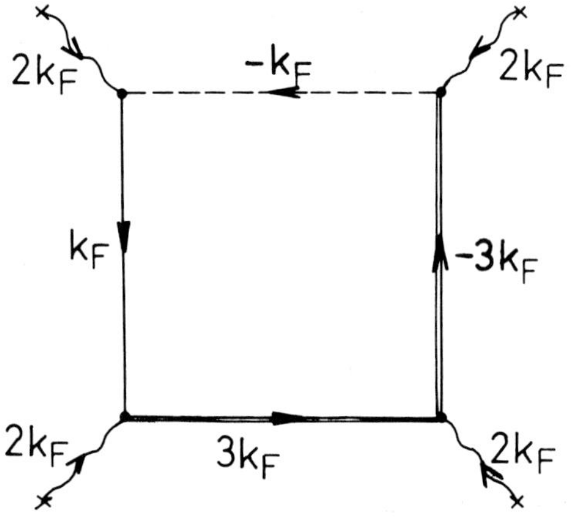

Fig. 15. Umklapp quartic deformation energy of the quarter filled tight-binding band. Double line, excited electron.

It is customary to express [68] the deformation energies (31), (32) and (33) in terms of the slowly-varying complex field $\psi = |\psi| e^{i\phi}$ defined by

$$u(y_j) = \psi \exp\left(i \frac{\pi}{2b} y_j\right) + \text{c.c.}, \tag{34}$$

where $u(y_j)$ is the acoustic displacement of the j-th unit cell on the chain under consideration. The fast component of the displacement is taken out in Equation (34) and ψ

can be thus considered as a slow field. Consequently Equations (31), (32) and (33) can be understood as the Landau expansion [68] in terms of this field.

$E^{(2)}$ is minimal at $2k_F$ rather than at $\pi/2b$, which, according to Equation (34), is the reference point of our expansion in terms of spatial variations of ψ. This mismatch generates [68] a term linear in $\partial\psi/\partial y$, i.e. the Lifshitz invariant. The latter appears in addition to the usual term [61, 62], quadratic in gradient.

$E_N^{(4)}$ is a smooth function of the momentum and it is usual to retain only the part which is local in ψ. $E_U^{(4)}$, when expressed in terms of ψ, defined by Equation (34), is also local. Indeed $[\psi \exp(i(\pi/2b)y)]^4$ is the invariant of the translation group [68].

The resulting Landau energy density is [62, 69]

$$f(y) = A'\left[\frac{T - T_p^0}{T_p^0} |\psi|^2 + i\,\xi_0^2\kappa \left(\psi^* \frac{\partial\psi}{\partial y} - \psi \frac{\partial\psi^*}{\partial y}\right) + \right.$$
$$\left. + \xi_0^2 \left|\frac{\partial\psi}{\partial y}\right|^2\right] + B\left[|\psi|^4 + \frac{\delta}{3}(\psi^4 + \psi^{*4})\right]. \tag{35}$$

A' and B are the shorthand notations for the coefficients of u's in Equations (31) and (32), respectively, and

$$\kappa = \frac{\pi}{2b} - 2k_F \tag{36}$$

measures the distance between $2k_F$ and the commensurability point $\pi/2b$.

For $T < T_p^0$ and sufficiently large κ it is usual to omit [70, 71] the amplitude fluctuations, setting

$$|\psi|^2 \simeq |\psi_0|^2 \simeq \frac{n_F (T_p^0)^2}{0.1\,\lambda_{ac} C}\,\epsilon, \tag{37}$$

where

$$\epsilon = \frac{T_p^0 - T}{T_p^0}.$$

This value of $|\psi_0|^2$ is chosen to minimize the two first terms in the quadratic and quartic bracket of Equation (35), because, with κ large, the others are less important. At $T \simeq 0$ ($\epsilon \simeq 1$) Equation (37) is of the form of the familiar BCS law of corresponding states [36]

$$2\Delta_p \simeq 3.5\, T_p^0, \tag{38}$$

if we set $\Delta_p^2 = \lambda_{ac} n_F^{-1} C |\psi_0|^2$. In fact the actual numerical coefficient in (37) is somewhat different from 3.5 because Equation (37) follows from the expansion. However the superconducting expansion has the same coefficient as (37). The complete analogy [61, 72] between the Peierls and BCS theories can be traced back to the neglect of the elastic contribution to all but the quadratic term (31). Inserting now the value $|\psi_0|$ in Equation (35) reduces this equation to [68, 73].

$$f = 10 n_F T_p^{0\,2} \epsilon \left[\xi_0^2 \left(\frac{\partial \phi}{\partial y} - \kappa \right)^2 - \frac{\epsilon \delta}{3} \cos 4\phi \right]. \tag{39}$$

The functional minimization of this expression gives the well known sine-Gordon equation [68, 73]

$$\xi_0^2 \frac{\partial^2 \phi}{\partial y^2} = \frac{2\epsilon\delta}{3} \sin 4\phi. \tag{40}$$

This equation is usually studied with a more general coefficient of the sine term. In our case such a coefficient would result from the introduction of the elastic anharmonic coupling, but here we shall avoid this inessential complication. The solutions of the sine-Gordon equation are either the constant values $\phi = \pi \cdot n/2$ or the soliton lattices [68, 73] shown in Figure (16). A soliton is the region of the fast phase variation, which joins

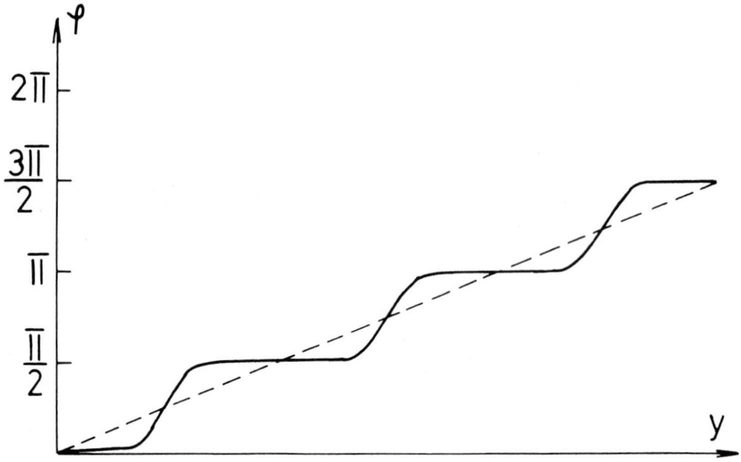

Fig. 16. Soliton lattice in the phase of the order parameter.

two regions where the phase is practically constant and equal respectively to $\pi n/2$ and $\pi(n + 1)/2$. According to Equation (34) the constant values of ϕ correspond to the commensurate displacement, and the soliton corresponds to the discommensuration. The incommensurate phase is lower in energy (39) than the commensurate one for [73–76]

$$\kappa \xi_0 > \sqrt{8\epsilon\delta}/\pi\sqrt{3}. \tag{41}$$

For very large values of κ the soliton lattice reduces [73] to the line $\phi = \kappa y$, i.e. the Umklapp coupling, the right-hand side of Equation (40) is unimportant. The corresponding displacement is the single harmonic. In the opposite limit $\kappa \xi_0 > \sqrt{8\epsilon\delta}/\pi\sqrt{3}$ the solitons become sparse and well-defined. However, in this limit, the coupling between the phase and amplitude fluctuations is relevant [69]. Unfortunately the resulting problem seems to be non-integrable, and has not been entirely elucidated so far. Usually, the periodic sine-Gordon solutions are followed through the commensurate–incommensurate transition, with the addition of some amplitude corrections. Finally the well-stabilized commensurate state is reached [77, 78] for

$$\kappa\xi_0 \ll \sqrt{8\epsilon\delta}/\pi\sqrt{3}.$$

The functional minimization is usually considered sufficient when the fluctuation effects around the ground state are unimportant. This certainly is not the case of the $\xi_{0\perp}(T_p^0) < d_\perp$ limit which requires the inclusion of nonlinear excitations above the state of minimum energy determined above. (The nonlinearities are absent in this case only in the extreme incommensurate limit). Let us turn now to this problem.

The nonlinear excitations can be easily found [79, 78] in the commensurate state. The species which dominates the spatial correlations is the (phase-amplitude) soliton. On the other hand, in the incommensurate case it is hard to go beyond the linearized [74–76] excitations of the solition lattice, and the problem of 1D fluctuations is therefore usually treated by a more powerful but physically less transparent transfer matrix method [70]. Without going into the details of these calculations let us quote the results [80–82] which are best expressed in terms of the structure factor. For $\kappa\xi_0 > \sqrt{8\epsilon\delta}/\pi\sqrt{3}$ and in the 1D regime the main superlattice reflection occurs at [82]

$$q_b^{(0)} \simeq 2k_F - \frac{\delta^2\epsilon^2}{72\xi_0^4\kappa^3}, \tag{42}$$

with the inverse Lorentzian half-width [82]

$$\xi(T) \simeq \frac{20T_p^0}{\pi T}\, b\epsilon \left(1 - \frac{\delta^2\epsilon^2}{24\kappa^4\xi_0^4}\right), \tag{43}$$

This peak could correspond to the one-dimensional sheets observed [30, 31] by X-ray diffuse scattering. These sheets are narrow, i.e. $\xi(T)$ of Equation (43) is large. The incommensurate $\kappa\xi_0 > \sqrt{8\delta\epsilon}/\pi\sqrt{3}$ values of $q_b^{(0)}$ are observed. Independent evidence for large $\xi(T)$ is given by the recent dc current–voltage measurements [84].

In addition to the main reflection, the theory predicts [81, 82] the occurrence of satellites related to the formation of the soliton lattice. The strongest satellite should occur at

$$q_b^{(1)} = q_b^{(0)} \pm \left(4\kappa - \frac{\epsilon^2\delta^2}{18\kappa^3\xi_0^4}\right). \tag{44}$$

$2\pi/(q_b^{(1)} - q_b^{(0)})$ is the period of the soliton lattice in formation. The two satellite peaks differ [82] somewhat in width. The upper sign in Equation (44) goes with the width

$5\xi^{-1}(T)$ and the lower with $3\xi^{-1}(T)$. The intensity of the satellite peaks is $(48\ \xi_0^2\kappa^2/\delta\epsilon)^2$ times lower than that of main reflection. Equations (42–44) describe the 1D anharmonic fluctuative regime. The concept of the soliton lattice remains meaningful even in the presence of 1D fluctuations, provided that the temperature is low enough ($\epsilon \simeq 1$). However the related effects are small due to the smallness of δ (cf. Equation (33)).

The similar analysis is required in the $\xi_{0\perp} > d_\perp$ case, when the soliton lattice competes with the 3D fluctuation effects for $\epsilon \ll 1$. $\epsilon \ll 1$ weakens the already weak ($\delta \ll 1$) commensurability pinning, as is clear from Equation (40) and (41), simultaneously increasing fluctuation effects (to be discussed in Section 5.2, Equation (57)). However, it can be speculated that the notion of the soliton lattice remains meaningful because it corresponds to the effects local in the reciprocal space, whereas the fluctuations produce a comparatively broad diffuse background. This requires further investigations.

The expressions analogous to Equations (42) and (43) can also be derived in the opposite, commensurate limit. They may bear some relevance for the case of orthorhombic TaS_3, as will be discussed in the next subsection. Here we shall focus our attention on another aspect of the incommensurate problem. We know that the electron deformation energies in e.g. Equations (31)–(33) are related to the electron redistribution $\delta\rho$ produced by the deformation field. The electron coordinates are integrated out in these equations, but it is possible to find the explicit expansion of $\delta\rho$ in terms of ψ, analogous to Equation (35). Extending the previously known linear response results [14] to the next (second) order we find [85]

$$\delta\rho^{(2)} \simeq \frac{1}{2\pi i}\,\frac{\lambda_{ac}C}{2n_F(T_p^0)^2}\left(\psi^*\frac{\partial\psi}{\partial y} - \psi\frac{\partial\psi^*}{\partial y}\right) \simeq \frac{\epsilon}{2\pi}\,\frac{\partial\phi}{\partial y}, \qquad (45)$$

consistently with Equations (35) and (39). In contrast to the linear [14] response $\delta\rho^{(1)}$ which contains only the fast ($\simeq 2k_F$) components, the second-order response describes the accumulation of charge in the region of variation of ϕ, e.g. the soliton. Only the slow components of $\delta\rho$ appear in this term.

Equation (45) complements Equation (39). It shows that the change of phase by $\pi/2$ (phase soliton) accumulates the charge

$$e^* \simeq \epsilon\frac{e}{4}\,. \qquad (46)$$

It can be thus attempted to observe the soliton lattice through the charge accumulation [86] by analysing the dc/ac currents [87–92, 67, 8]. For $NbSe_3$ the experiments [92–94, 8] have only been analysed [91] close to 59 K, resulting in a charge proportional to $\sqrt{\epsilon}$. These currents are thus due to the linear response of charge $\delta\rho^{(1)} \sim |\psi| \sim \sqrt{\epsilon}$. It might be that, close to $T_p \simeq T_p^0$, the linear response simply dominates the second order term $\delta\rho^{(2)}$ even if the latter corresponds to the well-formed soliton lattice. However, the soliton lattice may also be ill-defined for several reasons which we now wish to enumerate. The first is suggested by Equation (45) itself. The long-wave components of the charge density interact through the long-range Coulomb interactions. Although the Coulomb energy is (assumed to be) negligible in the main (logarithmic) approximation of Equation (14), the Coulomb energy of the accumulated charge should be compared to the small ($\epsilon\delta$) commensurability term. It is adverse to the formation of the charged soliton lattice, but a detailed analysis is lacking so far. The other possible reasons for the absence of the soliton lattice are ϵ and/or δ in Equation (40). From this point of view the optimal situation for the formation of the well-defined soliton lattice is $\epsilon = 1$. According to the $T = 0$ extrapolation of the result (41), this requires $\sqrt{8\delta}/\pi\sqrt{3} < \kappa\xi_0 < 1$. Unfortunately the condition $\kappa\xi_0 < 1$ does not seem to be satisfied in $NbSe_3$, i.e. the anomaly seems to occur too far from the commensurability point [95]. This manner of reasoning, however, raises the question of the validity of the extrapolation procedure used here ($T = 0$ in ϵ, $T = T_p^0$ in all other coefficients of Equation (39)), the problem to which we now turn.

In fact it is possible to include all powers of $|\psi|$ in the low-temperature analysis of the energy [96, 97] and charge density [98–101]. In the case of energy the results [96] do not differ qualitatively from the straightforward extrapolation of Equation (39) to low temperatures. The reason for this can be well understood in the incommensurate limit. The full solution [96] shows that the phase satisfies essentially the same equation as (39), obtained here from the expansion (35) on setting $T \simeq T_p^0$ in all coefficients except in $\epsilon = 1$. This is important, because the low T regime is dominated by the phase behavior. In other words, the full solution amounts to a more precise determination of $\Delta_p(\epsilon)$ than is given by (37). From the energetic or correlation point of view the difference between Equation (37) and (38) for example, is inessential. Furthermore, there is no difference in the nature of structural fluctuations at $T \simeq 0$ and at T_p^0. At T_p^0 the latter are classical by virtue of condition (16) and at $T \simeq 0$ the temperature in this condition is replaced by Δ_p of Equation (38). Once the quantum fluctuations are eliminated at T_p^0 due to the large mass of ions (small ω_{ac}), they stay out at $T \simeq 0$.

The accurate $T \simeq 0$ solution of the Peierls problem led however to a qualitatively new result. The full [100] summation of the series initiated in (45) shows that the solitons and the antisolitons carry, at $T = 0$, the exact fractional charges $\pm e/4$, respectively, rather than similar but approximate values (46). In fact, results of this kind were first derived either by the numerical analysis of the corresponding discrete model, using a reasonable trial function for solitions [102, 103], or by analytically solving the space-dependent Bogoliubov equations [98].

In conclusion, the discrete Peierls model [14, 102], its appropriate Bogoliubov–like continuation [98, 99], or even the corresponding Landau expansion [62, 68, 92, 69], lead to essentially [104] the same results. However, it appeared recently that the fully discrete formulation is required for solving some special models, one of which is not too far from the model discussed here. In this model the elastic interaction and the electron-phonon coupling are not expanded as in Equation (2) and (8), but rather a particular exponential dependence on displacement of the band-width and the elastic interaction is assumed [105]. This represents a generalization of the Toda chain, and, like the latter, does not show commensurability effects. The latter are introduced [106] through the departures from the original model. This could sometimes be a physically correct starting point of the investigation. Nevertheless we shall proceed here with our model, since the exponential behaviors are idealizations [6] even in the tight-binding limit.

4.4. Ordering by Interchain Coupling

In this subsection we shall consider the explicit effects of the interchain coupling. We shall start with the case $\xi_{0\perp}(T_p^0) < d_\perp$ when the ordering at $T_p < T_p^0$ is preceded by the 1D anharmonic regime of Equation (42–44). This might apply to the observed 1D regime in $NbSe_3$, although the latter could just as well be harmonic (Section 4.1 with $\xi_{0\perp}(T > T_p^0) < d_\perp$, $\xi_{0\perp}(T_p^0) > d_\perp$). Next we shall describe a mean-field ($\xi_{0\perp} > d_\perp$) theory of orthorhombic TaS_3. The small fluctuative corrections to the $\xi_{0\perp} > d_\perp$ results are described in Section 5.2.

In order to determine T_p for $\xi_{0\perp}(T_p^0) < d_\perp$ we shall use the interchain extension of the Landau theory, Section 4.3. We can afford to do that because the latter covers qualitatively correctly the one-dimensional regime, even for $T \ll T_p^0$. The first results for T_p were obtained along these lines in the extreme commensurate [107] or incommensurate [108, 71] limits. It turned out that the most important point is to take correct account of the one-dimensional fluctuations, whereas the interchain coupling could be treated in various approximations. The most popular was the mean-field approximation for the interchain coupling [109–112] when the given chain fluctuates in the average field of the others. However, this mean-field approximation proved inadequate [113] for the ordering of the soliton lattice. It led to different transition temperatures for the main and satellite peaks of the soliton lattice, Equation (42) and (44), respectively.

The solution of the interchain problem requires thereby the use of the transfer matrix method. The latter shows [114] that all the harmonics of the soliton lattice order at the same temperature as the main peak itself, namely at

$$T_p \simeq 40\sqrt{2}n_F(T_p^0)^2 \epsilon_p \frac{\xi_{0\perp}}{d_\perp}\left[1 - \frac{\delta^2 \epsilon_p^2}{3\xi_0^4 \kappa^4}\right], \tag{47}$$

where $\epsilon_p = \epsilon(T_p) \simeq 1$. The first term in Equation (47) represents the classical incommensurate result [71, 109–112], whereas the second arises from the commensurability pinning (33). As with the analogous corrections in Section 4.3, this latter seems to be negligible in $NbSe_3$. If furthermore (22) is assumed for $\xi_{0\perp}$ a particularly simple expression is obtained for T_p:

$$T_p \simeq t_\perp e^{-1/2\,\lambda ac}. \tag{47}$$

The above description of the ordering might be applicable to upper transition in $NbSe_3$ and its monoclinic analog TaS_3. The two transitions [31, 34, 39, 115] occur there at two entirely different wave-vectors $\mathbf{q}_0^{(a)}, \mathbf{q}_0^{(b)}$ and can be therefore treated independently, at least in the approximations corresponding to Equation (47). However, in orthorhombic TaS_3 the situation is somewhat different. Although this material too has to be described in terms of (at least) two types of chains [116], only one unstable wave-vector was observed [117]. In this respect it was recently pointed out that the harmonic (bilinear) interchain coupling can enforce the common wave-vector on both types of chains, if the two types of chains, taken separately, tend to develop the instabilities of similar wave-vector. The argument [116] goes as follows: the first transition occurs at $\mathbf{q}_{0b}^{(a)}$ on one set of chains (transverse components are ignored for simplicity). Next the second type of chains sets in with the instability at $q_{0b}^{(b)} \simeq q_{0b}^{(a)}$. The harmonic interchain coupling of Section (4.1) and (4.2) connects only the same wave-vectors on the two type of chains. Therefore $q_{0b}^{(a)}$ drives the same $-q_{0b}^{(a)}$ component on the second set of chains. $q_{0b}^{(a)}$ is not a natural component for this set of chains. At low temperatures it is thus energetically more favorable to develop at

$$q_{0b} \simeq [q_{0b}^{(a)} + q_{0b}^{(b)}]/2, \tag{48}$$

an instability which is common to both sets of chain. In some respects this is analogous to the case of a Peierls conductor in the magnetic field [98]. Incidentally, q_{0b} of Equation (48) turns out to be (nearly) commensurate, i.e. the 'commensurate' phase occurs here without any help from Umklapp coupling (33). This latter acts in the same direction as the harmonic interchain coupling.

The 'commensurate transition' in TaS_3 occurs at 140 K, well below the incommensurate transition at 215 K, and is accompanied by hysteresis [118]. It has been shown that such hysteresis can arise from nonlinear coupling among the continuous fields [119]. In the present case the fields in question can well be the relative phase of the two kinds of chains and the amplitude of the deformation on that family of chains which sets in around 140 K. In terms of those variables the bilinear interchain coupling is cosine-like in the relative phase and can lead to the soliton lattice in this variable.

The idea that interchain coupling or more precisely, interchain hopping $t_\perp$, can produce soliton lattices was first put forward [60] in the context of CH_x conductors. It has been argued there that in presence of appropriate transverse dispersion it is energetically preferable to associate some electrons with soliton (lattice) states then to put them in the conventional band states. The advantage of such theories is that they do not rely upon small δ (and ϵ) and can thus be relevant even for $NbSe_3$. A careful analysis of the transverse dispersion in conducting trichalcogenides is therefore desirable, for this reason and for reasons which will become clear in the next Section.

5. CDW Versus BCS Instability

On applying pressure of a few kbar to $NbSe_3$ the structural transition is removed [3] as shown in Figure (17). The low-temperature ground state has all the properties of BCS superconductivity [3, 120–122].

Pressure is expected to increase the interchain couplings incorporated in $\xi_{0\perp}$. The straightforward application of Equation (47) would suggest an increase, or at most a weak decrease of T_p, if T_p^0 is given by Equation (20), and the interchain parameters vary slowly with pressure. However, Equation (20) for T_p^0 is valid only if the conditions (21) and (24) are satisfied at T_p^0, i.e. if $T_p^0 > t_\perp^2/\epsilon_F$ or $T_p^0 > t_\perp$ for the nesting or non-nesting cases respectively. When $t_\perp^2/\epsilon_F$ or $t_\perp$ increases towards T_p^0 of Equation (20) the actual T_p^0 decreases steeply to zero. We are thus led to conclude that the decrease of T_p with pressure is due to vanishing of the mean field tendency T_p^0 to the structural instability.

This simple argument, although appealing, cannot be used straightforwardly in the low temperature range, because it is based on the condition (16)

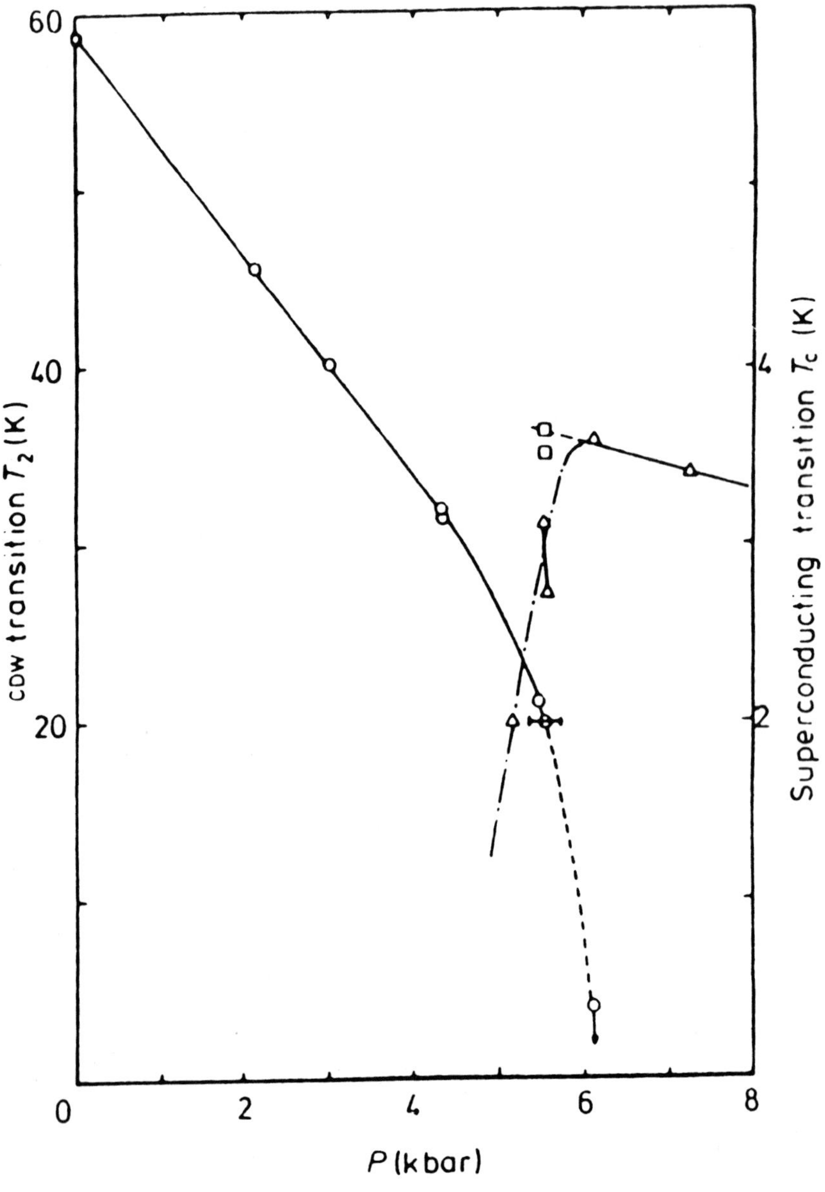

Fig. 17. Variation of T_{CDW}(○) and T_{BCS}(□) of $NbSe_3$ as a function of pressure (Ref. (3b), copyright Institute of Physics, Bristol, England).

$$2\pi T_p^0 > \omega_{ac}, \tag{16}$$

which ceases to hold. This means that the quantum fluctuations of the phonon field are becoming important. The structural transition continues to exist but it is driven by the electron system interacting in absence of the adiabatic parameter (16). Such transition is usually named the charge density wave (CDW) transition to distinguish it from the Peierls transition, which is driven adiabatically by the weakly interacting electron gas. Moreover, the virtual exchange of phonons by electrons can also produce BCS superconductivity beside the CDW transition. The competition of the CDW and BCS phases is discussed in the present chapter.

5.1. Mean-Field Theory

In the temperature range

$$2\pi T < \omega_\gamma(\mathbf{q}) \tag{49}$$

the virtual exchange of phonons introduces the logarithmic singularity [14, 24, 18, 25] of the electron–electron (e–e) Cooper diagram, Figure (7), irrespective of the value of $t_\perp$. In the lowest order this singularity competes [10] with the logarithmic singularity of the e–h channel, which exists provided that the above-mentioned conditions [22]

$$T > \frac{t_\perp^2}{\epsilon_F} \tag{21}$$

or

$$T > t_\perp \tag{24}$$

are fulfilled.

Consider next the coupling between the e–e and the e–h channels [23], e.g. the process shown in Figure (18). For simplicity let us neglect $t_\perp$ on the electron e–e lines. Then the internal transverse integration means simply the averaging of the e–h bubble over the transverse momenta. Assume now that there is nesting and that $T < t_\perp$. According to Equation (22) the log T singularity then only affects a small part ($\sim \xi_{0\perp}^{-2}$) of the transverse Brillouin zone. Therefore the averaged e–h bubble [123] is proportional to log $t_\perp$ rather than to log T. We conclude that this, the so-called parquet coupling between the e–h and e–e channels, is effective only if [23, 123]

$$T > t_\perp. \tag{24}$$

The same conclusion holds trivially in the non-nesting case.

Putting all this together, the parquet coupling between the channels has to be taken into account if the conditions

$$2\pi T < \omega_D \tag{49}$$
$$T > t_\perp \tag{50}$$

and

$$2\pi T_p^0 < \omega_D \tag{51}$$

are fulfilled simultaneously. The last of these conditions ensures the absence of the (pseudo)gap in the electron spectrum below ω_D, i.e. justifies the use of the bare electron lines in Figures 5, 6, 7, 18. Otherwise (i.e. if $2\pi T_p^0 > \omega_D$) the adiabatic Peierls theory is valid down to the lowest temperatures, as explained in the preceding sections.

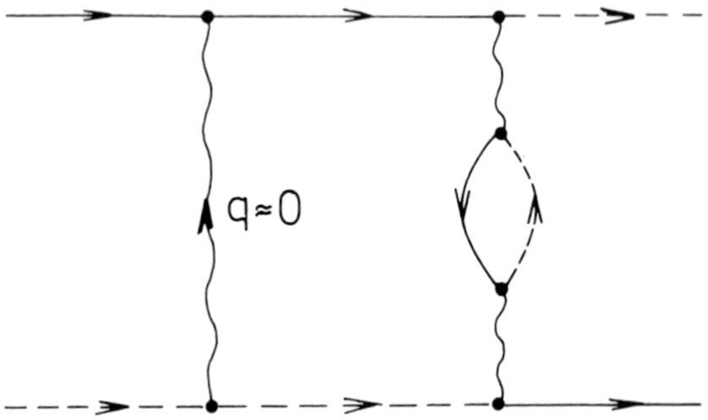

Fig. 18. Parquet coupling between electron-electron (Cooper) channel and electron-hole (density) channel. Low order vertex correction.

Next we wish to argue that these three conditions cannot be made consistent with assumption that $2\pi T_p^0 > \omega_D$ used in the preceding sections at ambient pressure. This is illustrated in Figure (19). In this figure it is shown schematically that when T_p^0 starts to decrease upon meeting (either the $t_\perp$ or) the $t_\perp^2/\epsilon_F$ line it cannot reach the parquet regime. The transition from the regime in which T_p^0 is almost independent of $t_\perp$ to the regime in which T_p^0 falls steeply [22] to zero is too fast for that.

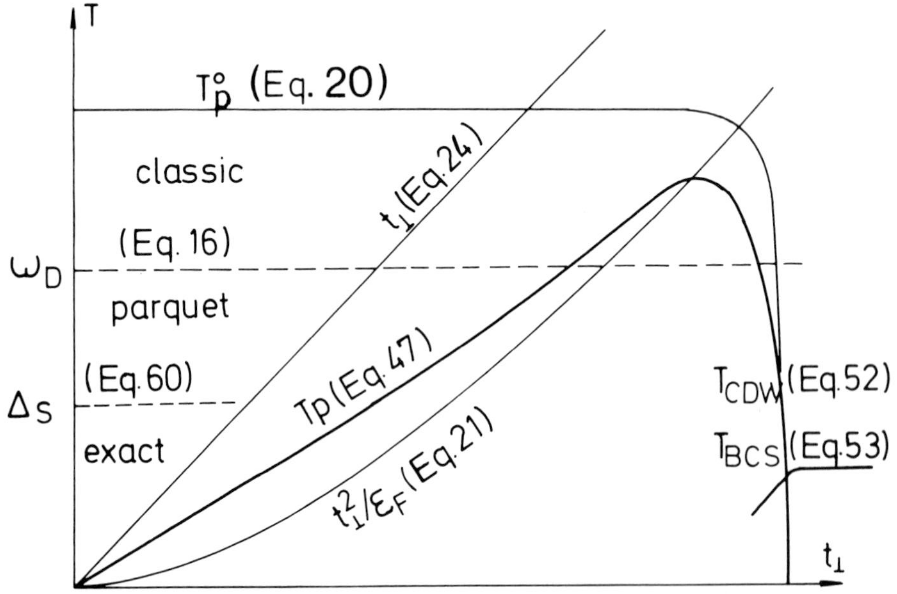

Fig. 19. Hierarchy (schematic) of various temperature and energy scales introduced in the equations of the text, variation with transverse overlap $t_\perp$ (i.e. with pressure).

The parquet corrections are therefore absent when T_p^0 falls below ω_D. However, other electron-hole ladder diagrams such as those shown in Figure (6), are activated in addition to the e–h bubble of Figure (5). These processes involve the forward scattering mediated by phonons, which we shall denote by λ_2 in analogy to the Coulomb g_2 of subsection 3.2. The corresponding CDW transition temperature T_{CDW} is usually [26] written in the form

$$T_{\text{CDW}} \simeq \omega_D \exp -(2\lambda_{ac} + \lambda_2)^{-1} > t_\perp, t_\perp^2/\epsilon_F, \tag{52}$$

where the inequality in (52) symbolizes the fast decrease of T_{CDW} with increasing $t_\perp$. Under pressure T_p^0 continuously transforms in T_{CDW}. Note in this respect that the prefactor ϵ_F of Equation (20) is replaced in (52) by ω_D in order to emphasize that λ_2 contributes only at $T_{\text{CDW}} < \omega_D$.

As already mentioned, the virtual exchange of phonons can produce BCS superconductivity beside the CDW instability. Indeed, the e–e Cooper diagrams of Figure 7 are logarithmically singular for $2\pi T < \omega_D$, irrespective of the value of $t_\perp$. As the parquet corrections are absent, the transition temperature is given by the unmodified BCS expression [26]

$$T_{\text{BCS}} \simeq \omega_D \exp -(\lambda_1 + \lambda_2)^{-1}. \tag{53}$$

Here λ_1 incudes the optical phonon contributions to the e–e coupling additional to λ_{ac}.

In subsection 2.2 we have argued that the optic phonon coupling constants, which determined $\lambda_1 - \lambda_{ac}$ and λ_2 almost entirely (the forward interactions through long-wavelength acoustic phonons are eliminated by retardation [10]) and are not expected to be particularly large. Indeed, Equation (52) is capable of explaining the corresponding observations in $NbSe_3$, without the optic phonon contributions. When T_{CDW} is suppressed by increasing $t_\perp$ it meets a finite T_{BCS}, which depends only weakly on $t_\perp$.

At this stage it is appropriate to discuss the nesting versus the non-nesting (with Coulomb/elastic interchain coupling) picture. Both non-nesting and perfect nesting pictures lead, on the mean-field level, to the complete separation of the BCS and CDW phases. On the other hand, the experiments [3] show (Figure 17) that the sample which has undergone the CDW transition becomes superconducting on further lowering the temperature. It is possible to explain this coexistence by allowing for an imperfect nesting between the Fermi surfaces [124, 125]. The imperfect nesting involves only a part of the Fermi surface. The corresponding dielectric gap opens over that part of the surface, leaving the rest available for the BCS instability. Assuming, for simplicity, that the corresponding density of states n_F^{BCS} is small with respect to n_F, we recover Equation (52) for T_{CDW}, whereas λ's in Equation (53) for T_{BCS} have to be scaled with n_F^{BCS}/n_F. It should be further noted that the CDW and the BCS instability do not play symmetric roles in this argument. Indeed, if the BCS instability occurs at higher temperature, T_{BCS} of Equation (53) larger than T_{CDW} of Equation (52), it involves the whole Fermi surface, being independent of its nesting properties [124, 125]. The BCS transition inhibits therefore the structural instability, in full agreement with experiments [3]. The type of observed coexistence between the structural and the BCS instabilities represents thus a good evidence in favor of the nesting model. In particular this led us to draw in Figure (19) the transition temperature T_p, T_{CDW} for this case. Figure (19) should not however be taken too seriously, because it is based on the cosine transverse dispersion, which has very good nesting properties (i.e. very small n_F^{BCS}). The actual nesting is probably less perfect, i.e. the region between $t_\perp$ and the critical nesting line, $t_\perp^2/\epsilon_F$ in Figure (19), is much narrower.

5.2. Correlation Lengths and Landau Theories

The ladder approximations, being the leading approximations in the low temperature range, provide for the characteristic lengths ξ_0, $\xi_{0\perp}$. In the BCS case, these are given by formally the same expressions as the Equation (22). The CDW lengths might be somewhat different due to the vicinity of the saturation limit (21) or (24), imposed by $t_\perp$, but in the absence of a better estimate we shall continue to use Equations (22). Thus we have

$$\xi_0 \simeq b \frac{\epsilon_F}{T_{\text{CDW, BCS}}} \gg b$$
$$\xi_{0\perp} \simeq d_\perp \frac{t_\perp}{T_{\text{CDW, BCS}}} \gg d_\perp \tag{54}$$

with T_{CDW} or T_{BCS} given by Equations (52) or (53), respectively. In this way the main difference with the Peierls case comes through the values of the transition temperatures at which we define ξ_0, $\xi_{0\perp}$. The addition of the electron–electron Coulomb or elastic interchain coupling omitted here, can only further increase $\xi_{0\perp}^{\text{CDW}}$.

The low temperature X-ray or neutron data under pressure, which in principle could give information about ξ_0 and $\xi_{0\perp}$ of the structural transition are missing so far. The measurement of the anisotropy of the second critical field for superconductivity H_{c_2}, also gives a direct access to these quantities for the superconducting case. The ladder approximation, i.e. Equations (53) and (54), defines the harmonic part of the Landau functional in terms of the usual BCS order parameter. The harmonic part alone can be used to determine H_{c_2}, provided that $\xi_{0\perp} > d_\perp$, as can be expected according to (54). It follows that [126, 127]

$$\frac{H_{c_2}^{\perp}}{H_{c_2}^{\parallel}} = \frac{\xi_{0\perp}}{\xi_0}. \tag{55}$$

The anisotropy of the critical field H_{c_2} was measured [120] on single crystals under pressure up to 7.7 kbar. The ratio (55) turned out to be of the order of 5.5. This can hardly be reconciled with Equations (54) and $\epsilon_F \gg t_\perp$ deduced previously from the high temperature behavior at ambient pressure. The comparatively small anisotropy ratio could be due to the effect of impurities [128]. The latter are expected to shorten

[43] the correlation lengths (54) and wash out [128, 129] some of the anisotropy. It can be also speculated following [130–133] that those few electrons which belong to the curved parts of the Fermi surface contribute more to $\xi_{0\perp}$ than the nearly 1D electrons. However, both possibilities require a more careful theoretical and experimental analysis.

It is interesting in this respect to mention the effect of doping [121, 122] on the superconducting properties. E.g. $NbSe_3$ was doped [121] with Ta and Ti. Ta is isoelectronic with Nb. 5% of Ta in place of Nb suppresses the CDW transition. The BCS transition occurs instead at ambient pressure. The anisotropy of the characteristic lengths measured via the anisotropy of the critical fields is $\xi_0 \simeq 4\xi_{0\perp}$. Thus the (additional) impurities do not alter substantially the anisotropy ratio (55). The fact that in contrast to BCS superconductivity the impurities suppres the CDW instability agrees well with the simple mean-field arguments [134], which consider the effect of impurities on the processes shown in Figures 5, 6 and 7. However at present this fact does not help to uniquely distinguish the source of the low anisotropy ratio (55).

The experimentally determined value [120] of ξ_0 is 350 Å, i.e. $\xi_{0\perp} \simeq 70$ Å $> d_\perp$. This is consistent with the observed mean-field behavior [43] $H_{c_2} \simeq (T_{BCS} - T)$ although the slight upward curvature [120] in $H_{c_2}(T)$ may be taken as an indication of the weak interchain coupling [120] (in this respect only the extreme $\xi_{0\perp} < d_\perp$ case has been considered [135, 7] so far).

Rather than going into the corrections to the linear dependence of H_{c_2} on temperature implied by $\xi_0 > \xi_{0\perp} > d_\perp$ we shall estimate now the corresponding effect on the width of the critical region. With $\xi_{0\perp} > d_\perp$ the interaction of fluctuations is less [25] important than in the case $\xi_{0\perp} < d_\perp$ studied in Section 4.3. The anisotropy (55), however, enhances [15, 25] the corresponding effects with respect to those encountered in the 3D case. In order to estimate this enhancement we remember from Section 4.3. the formal analogy between the Peierls and the BCS Landau expansions. Using this analogy we have the BCS Landau functional

$$f = A' \left[\frac{T - T_{BCS}}{T_{BCS}} |\psi|^2 + \xi_0^2 \left| \frac{\partial \psi}{\partial y} \right|^2 + \xi_{0\perp}^2 \left| \frac{\partial \psi}{\partial z} \right|^2 + \xi_{0\perp}^2 \left| \frac{\partial \psi}{\partial x} \right|^2 \right] + B|\psi|^4, \quad (56)$$

with A' and B from Section 4.3, except that T_p^0 is replaced by T_{BCS}, just as in going from Equation (22) to Equation (54). When $\xi_{0\perp} > d_\perp$ in Equation (56) it is meaningful to symmetrize this equation by rescaling the transverse variables through the transformation [15, 25] $(x', z')/\xi_0 = (x, z)/\xi_{0\perp}$. The ensuing critical behaviors are 3D but enhanced by the scaling factors $\xi_0/\xi_{0\perp}$. E.g. the width ΔT of the critical region is enhanced by [15]

$$\frac{\Delta T}{T_{BCS}} \sim \left(\frac{\xi_0}{\xi_{0\perp}} \right)^4 \quad (57)$$

over its isotropic value. Similar expressions hold for the specific heat, paraconductivity, depression of the transition temperature etc. Usually they are given [15, 25] for the general (rather then BCS) values of coefficients in Equation (56).

The similar type of reasoning applies of course to the CDW transition. The form (35) of the Landau expansion continues to hold for T below ω_D. The harmonic coefficients are the same as in Equation (35) except for the replacement of T_p^0 by T_{CDW} (λ_{ac} by $\lambda_{ac} + \lambda_2/2$), as was already mentioned for ξ's of Equation (54). The effect of the virtual phonon excitations on the anharmonic coefficients has not been discussed in detail so far. However this question is of little importance in conducting trichalcogenides, where the optic phonons seem to be only relatively weakly coupled to the conducting electrons. We can thus safely think that the fluctuation effects are not more important than in the BCS estimation. Needless to say, the estimation (57) holds for the Peierls transition at T_p with $\xi_{0\perp} > d_\perp$.

5.3. ADDITIONAL REMARKS CONCERNING MANY BODY CORRECTIONS

The condition (16)

$$2\pi T_p^0 > \omega_D$$

have led us to the situation in which the one-dimensional many-body (parquet) effects are absent (remember $\lambda_{ac} > g_{1,2}$ too). This condition is certainly fulfilled for the higher transition in $NbSe_3$ at ambient pressure. The lower transition occurs at the temperature $T_p \simeq \omega_D/2\pi$ and $2\pi T_p^0 > \omega_D$ should be satisfied in this case too because $T_p^0 > T_p$. Nevertheless, from this point of view it is of some interest to present briefly a few results of the one-dimensional many-body approach.

As already mentioned in Section 5.1 the one-dimensional many-body effects are important if it turns out formally that

$$2\pi T_p^0 < \omega_D. \tag{51}$$

In this case the parquet coupling between the e–h and e–e channels becomes important provided that [23, 122, 123]

$$\omega_D > T > t_\perp, \tag{49}$$

as discussed in this subsection. The parquet coupling results in the logarithmic renormalization [10, 16, 17, 18, 20] of the coupling constants λ_1 and λ_2

$$\tilde{\lambda}_1(T) = \frac{\lambda_1}{1 - \lambda_1 \log \dfrac{\omega_D}{T}}, \tag{58}$$

$$\tilde{\lambda}_2(T) = \lambda_2 - \frac{\lambda_1}{2} + \frac{\tilde{\lambda}_1(T)}{2}. \tag{59}$$

It can be used for temperatures above

$$\Delta_S \simeq \omega_D \exp -(\lambda_1)^{-1} \tag{60}$$

where the singularity in Equations (58) and (59) occurs. Thus the parquet renormalization can be used at all $T > t_\perp$ if

$$t_\perp > \Delta_S. \tag{61}$$

We can note that this requirement is not inconsistent with (49) and (50). For temperatures below $t_\perp$ the parquet coupling between the channels is absent, i.e. the ladder approximations are valid, as we already know from Figure (19). The effective couplings in these ladders are [23, 123] the values $\tilde{\lambda}_{1,2}(T \simeq t_\perp) = \overline{\lambda}_{1,2}$, which take into account the parquet renormalization in the range $\omega_D > T > t_\perp$. These values appear therefore in Equations (52) and (53) instead of the bare λ's. Also the prefactor ω_D is replaced [23, 123] by $t_\perp < \omega_D$. As is easily seen the parquet renormalization $\lambda \to \overline{\lambda}$ favors the CDW instability with respect to the BCS superconductivity. Furthermore the validity of ladder approximations implies the validity of the Landau expansions. This means that in the present case the fluctuations can be treated in a way analogous to that described in Section 5.2.

The situation is quite different if

$$\Delta_S > t_\perp. \tag{62}$$

In this case the parquet results cannot be used below Δ_S, where the transitions are expected to occur. Usually then the one dimensional problem is solved 'exactly' and $t_\perp$ introduced as a perturbation [109, 112, 7]. In this sense the approach is analogous to that [135] described in Section 4.4 except that it includes [109, 112, 7] quantum fluctuations [7]. (In Section 4.4 we had $T_p^0 > t_\perp$ instead of (62).)

For $T < \Delta_S$ the behavior in the one dimensional regime is determined by the properties of the (long wave) plasmons because Δ_S is in fact the gap in the SDW spectrum [20, 17] so that these excitations are frozen out. The spectrum of plasmons is 1D acoustic [27] for $v_F|\mathbf{q}| > \omega_0$, where ω_0 is the plasma frequency (Coulomb gap) of Section 3.2. In the weak-coupling limit (7), (12) the Coulomb gap is thus important [27] only in the small $k_{TF} = \omega_0/v_F$ region of the Brillouin zone around the Γ point. In the remaining part the 1D acoustic plasmon spectrum is renormalized, first of all by phonons, due to the assumption (15). The zero point motion (quantum fluctuations) of those plasmons enters the behavior of the correlation functions. In particular the BCS and the CDW correlations compete again [112]. The long range order is introduced by the weak interchain coupling and particularly by $t_\perp < \Delta_S$. This latter coupling can be treated analogously to the Josephson coupling in the BCS case, i.e. via the $t_\perp/\Delta_S$ expression [112].

6. Summary

Our main purpose here was to situate the conducting trichalcogenides with respect to the existing theories and to describe briefly the relevant theoretical results. Using the experimental evidences we have thus established a required system of inequalities. The principal role is held by two of them

$$\lambda_{ac} > g \tag{14}$$

and

$$2\pi T_p^0 > \omega_D, \tag{16}$$

at ambient pressure. The first ensures that the (acoustic) phonon mediated electron-electron interaction dominates over the Coulomb interactions. The second says that the dominant interaction is strongly retarded at the characteristic temperature T_p^0. This simplifies considerably the theory, making the parquet coupling of e–h and e–e channels unimportant. The resulting mean-field theory is the familiar Peierls theory in which the weakly interacting electrons are coupled to the classic deformation field. The latter is the natural order parameter and its behavior depends crucially on the interchain coupling, which can be of elastic, nesting and/or Coulomb origin. The mean-field Peierls theory is reasonable provided that the interchain coupling is strong,

$$\xi_{0\perp}(T_p^0) > d_\perp,$$

but important fluctuative corrections occur in the opposite limit. Interesting commensurability effects appear provided that the $2k_F$ instability occurs close enough to the $2k_F$ commensurability point

$$\kappa\xi_0 < 1$$

as seems to be the case in the orthorhombic TaS_3.

When T_p^0 is decreased by applying pressure to $NbSe_3$ the condition (16) ceases to hold. The interchain coupling becomes then important on the mean-field (parquet) level, and not only beyond the mean-field approximation, as was the case in the Peierls limit. At $2\pi T_p^0 \simeq \omega_D$ the continuity (Figure (19)) implies

$$T_p^0 < t_\perp$$

and therefore the parquet coupling between the e–e and e–h channels remains unactivated. Although somewhat more involved than the Peierls limit, this situation can still be described in terms of (the Landau expansions with) classic (CDW, BCS) order parameters. This contrasts with the parquet (or beyond it) situations.

The central inequality (16) and all what follows from it by continuity, has recently received a direct experimental confirmation [136]. The far infrared reflectivity measurements on $NbSe_3$ have demonstrated that the optic phonons fall in the low temperature dielectric gap $\Delta_p \simeq T_p^0 > \omega_D$. Even if the future experiments show that some inequalities used here are weakly or not at all satisfied, the present analysis may still represent a good starting point for the necessary corrections and modifications.

Acknowledgements

Useful correspondence with P. Monceau and discussions with A. Bjeliš and I. Batistić are gratefully acknowledged. Thanks are due to L. P. Gor'kov for several important remarks. This work was partially supported by the Yu–US DOE contract No. 438.

REFERENCES

1. P. Haen, P. Monceau, B. Tissier, G. Waysand, A. Meerschaut, P. Molinié, J. Rouxel, *Proceedings of the 14th Int. Conference on Low Temperature Physics*, Otaniemi, Finland 1975, edited by M. Krusius and M. Vuorio (North Holland, Amsterdam 1975) Vol. 5, p. 445.
2. T. Sambongi, K. Tsutsumi, Y. Shiozaki, M. Yamamoto, K. Yamaya, Y. Abe, *Sol. St. Commun.* **22**, 729 (1977).
3. P. Monceau, J. Peyrard, J. Richard, P. Molinié, *Phys. Rev. Lett.*, **39**, 160 (1977); A. Briggs, P. Monceau, M. Nunez-Regueiro, J. Peyrard, M. Ribault, J. Richard, *J. Phys. C* **13**, 2117 (1980).
4. For an excellent early review and approach see J. A. Wilson, *Phys. Rev.* **19**, 6456 (1979), followed by *J. Phys. F.* **12**, 2469 (1982).
5. For the latest results in this field see *Proceedings of the International Conference on the Physics and Chemistry of Conducting Polymers* and *International CNRS Colloqium on the Physics and Chemistry of Synthetic and Organic Metals*, Les Arcs, France 1982, ed. R. Comés and P. Bernier, *J. Physique C* **44**, C3 (1983).
6. M. Weger, J. B. Goldberg, review in *Sol. St. Phys.* **28**, 1 (1973).
7. D. Jérome, H. J. Schulz, *Adv. in Physics* **31**, 299 (1982) and references therein.
8. P. Monceau, J. Richard, M. Renard, *Phys. Rev. B* **25**, 931 and 948, 1982 and references therein.
9. W. A. Little, *Phys. Rev.* **A 134**, 1416 (1964).
10. Yu. A. Bychkov, L. P. Gor'kov, J. E. Dzyaloshinskii, *Zh. Eksp. Teor. Fiz.* **50**, 738 (1966) [*Sov. Phys. JETP* **23**, 489 (1966)].
11. M. Weger, *Rev. Mod. Phys.* **36**, 175 (1964).
12. J. Labbé, S. Barišić, J. Friedel, *Phys. Rev. Lett.* **19**, 1038 (1968).
13. J. Labbé, J. Friedel, *J. Physique* **27**, 153 and 303 (1966).
14. S. Barišić, J. Labbé, J. Friedel, *Phys. Rev. Lett.* **25**, 919 (1970); S. Barišić, *Phys. Rev. B5*, 932 and 941 (1972).
15. S. Barišić, S. Marčelja, *Sol. St. Commun.* 7, 1395 (1969).
16. I. E. Dzyaloshinskii, A. Y. Larkin, *Zh. Eksp. Teor. Fiz.* **61**, 791 (1971) [*Soviet Physics JETP* **34**, 422 (1972).
17. J. Solyom, *Adv. in Physics* **28**, 201 (1979), and references therein.
18. L. P. Gor'kov, I. E. Dzyaloshinskii, *Zh. Eksp. Teor. Fiz.* **67**, 397 (1974) [*Sov. Phys. JETP* **40**, 1980 (1975)].
19. F. Woynarovich, *J. Phys. C.* **15**, 85 and 97 (1982) and references therein, H. B. Thacker, *Rev. Mod. Phys.* **53**, 253 (1981) and references therein.
20. V. J. Emery, Review, in *Highly Conducting One-Dimensional Solids*, edited by J. T. Devreese, R. P. Evrard and V. E. van Doren (Plenum, N.Y.) 1979.
21. S. Barišić, K. Šaub, *J. Phys. C* **6**, L367 (1973).
22. B. Horovitz, H. Gutfreund, M. Weger, *Phys. Rev. B* **8**, 3174 (1975).
23. V. N. Prigodin, Yu. A. Firsov, *Zh. Eksp. Teor. Fiz.* **76**, 736 and 1602 (1979) [*Sov. Phys. JETP* **49**, 369 and 813 (1979)].
24. G. Bergmann, D. Rainer, *Z. Physik* **263**, 59 (1973).
25. S. Barišić, *Lecture Notes in Physics* (Springer) **65**, 85 (1977).
26. For an updated presentation of the mean-field results see B. Horovitz, H. Gutfreund, M. Weger, *Sol. St. Commun.* **39**, 541 (1981).
27. S. Barišić, *J. Physique* **44**, 185 (1983).
28. G. A. Toombs, *Phys. Reports C* **40**, 183 (1978) and references therein.
29. A. Meerschaut, J. Rouxel, *J. Less Common Met.* **39**, 197 (1975).
30. J. P. Pouget, R. Moret, A. Meerschaut, L. Guemas, J. Rouxel, in Ref. [5], p. 1729.
31. K. Tsutsumi, T. Takagaki, M. Yamamoto, Y. Shiozaki, M. Ido, T. Sambongi, K. Yamada, Y. Abe, *Phys. Rev. Lett.* **39**, 1675, (1977).
32. S. Tomić, K. Biljaković, D. Djurek, J. R. Cooper, *Sol. St. Commun.* **38**, 109 (1981); S. Tomić, *Sol. St. Commun.* **40**, 321 (1981).
33. S. Barišić, A. Bjeliš, in *Theoretical Aspects of Bond Structures and Electronic Properties of Pseudo-One-Dimensional Solids*, edited by H. Komimura, D. Riedel, Dordrecht (in press).

34. J. L. Hodeau, M. Marezio, C. Roucau, R. Ayroles, A. Meerchaut, J. Rouxel, P. Monceau, *J. Phys. C* **11**, 4117 (1978).
35. D. W. Bullet, *J. Phys. C* **12**, 277 (1979); *Sol. St. Commum.* **26**, 563 (1978).
36. R. Hoffmann, S. Shaik, J. C. Scott, M-H Whangbo, M. J. Foshee, *J. Sol. St. Chem.* **34**, 263 (1980); M.-H. Whangbo, P. Gressier, *Inorgan.Chem.* **23**, 1305 (1984).
37. N. P. Ong, P. Monceau, *Sol. St. Commun.* **26**, 487 (1978), revisited by G. X. Tassema, N. P. Ong, *Phys. Rev. B* **23**, 5607 (1981).
38. N. P. Ong, Y. W. Brill, *Phys. Rev. B* **18**, 5265 (1978); N. P. Ong, *Phys. Rev. B* **18**, 5272 (1978).
39. R. M. Fleming, D. E. Moncton, D. B. McWhan, *Phys. Rev. B* **18**, 5560 (1978).
40. T. J. Wieting, A. Grisel, F. Levy, Ph. Schmid, *Lecture Notes in Physics*, Vol. 95, Springer, New York, 1979, p. 354.
41. A. Zwick, M. A. Renucci, R. Carles, N. Saint-Cricq, J. B. Renucci, *Physics B&C* **105**, 361 (1981).
42. S. Jandl, M. Banville, J. -Y. Harbec, *Phys. Rev. B* **22**, 5697 (1980); S. Jandl, J. Deslandes, *Can. J. Phys.* **59**, 936 (1981).
43. E. g. A. A. Abrikosov, L. P. Gor'kov, I. E. Dzyaloshinskii, *Quantum Field Methods in Statistical Physics*, Prentice-Hall, Englewood Cliffs, 1963.
44. H. P. Geserich, G. Scheiber, F. Lévy, P. Monceau, *Sol. St. Commun.* **49**, 335 (1984).
45. I. E. Dzyaloshinskii, E. I. Kats, *Zh. Eksp. Teor. Fiz.* **55**, 338 (1968) [*Sov. Phys. JETP* **28**, 178 (1969)].
46. P. F. Williams, A. N. Bloch, *Phys. Rev. B* **10**, 1097 (1974).
47. T. K. Mitra, *J. Phys. C* **2**, 52 (1969).
48. S. Barišić, *Sol. St. Commun.* **9**, 1507 (1971).
49. K. Šaub, S. Barišić, J. Friedel, *Phys. Lett.* **56A**, 302 (1976).
50. J. Friedel, C. Noguera, *Int. J. Quantum Chem.* **23**, 1209 (1983).
51. F. Y. di Salvo, J. V. Waszczak, K. Yamaya *J. Phys. Chem. Solid* **41**, 1311 (1980).
52. J. D. Kulick, J. C. Scott, *Sol. St. Commun.* **32**, 271 (1980).
53. F. Devreux, *J. Physique* **43**, 489 (1982).
54. D. Horovitz, *Phys. Rev. B* **16**, 3943 (1972).
55. R. E. Peierls, *Quantum Theory of Solids*, Oxford Univ. Press, London, 1955, p. 108.
56. J. Bardeen, L. N. Cooper, J. R. Schrieffer, *Phys. Rev.* **108**, 1175 (1957).
57. H. Fröhlich, *Proc. Roy. Soc.* (*London*) **223A**, 296 (1954).
58. W. Kohn, *Phys. Rev.* **115**, 809 (1959); A. M. Afanas'ev, Yu. Kagan, *Zh. Ekspe. Teor. Fiz.* **43**, 1456 (1962) [*Soviet Physics JETP* **16** 1030 (1963)].
59. G. Beni, *Sol. St. Commun.* **15**, 269 (1974).
60. S. A. Brazovskii, L. P. Gor'kov, J. R. Schrieffer, *Phys. Scripta* **25**, 423 (1982).
61. B. R. Patton, L. J. Sham, *Phys. Rev. Lett.* **31**, 631 (1973).
62. A. Bjeliš, K. Šaub, S. Barišić, *N. Cimento,* **23B**, 102 (1974).
63. P. A. Lee, T. M. Rice, P. W. Anderson, *Phys. Rev. Lett.* **31**, 462 (1973).
64. M. J. Rice, S. Sträsler, *Sol. St. Commun.* **13**, 1389 (1973).
65. A. Bjeliš, S. Barišić, *J. Physique Lettres* **36**, 169 (1975).
66. E. G. Brovman, Yu. Kagan, *Zh. Eksp. Teor. Fiz.* **52**, 557 (1967) [*Soviet Physics JETP* **25**, 365 (1967)]; *Zh. Eksp. Teor. Fiz.* **57**, 1329 (1969) [*Soviet Physics JETP* **30** 721 (1970)].
67. D. Allender, J. W. Bray, J. Bardeen, *Phys. Rev. B* **9**, 119 (1974).
68. I. E. Dzyaloshinskii, *Nobel Symposium* **24**, editors B. and S. Lundquist, Nobel Foundation, Stockholm, Academic Press, N.Y. and London, 1973, p. 143; S. A. Brazovskii, I. E. Dzyaloshinskii, *Low Temperature Physics LT* **14**, Vol. 4, (1975) 337.
69. I. Batistić, S. Barišić, *Sol. St. Commun.* **33**, 603 (1980).
70. D. J. Scalapino, M. Sears, R. A. Ferrell, *Phys. Rev. B* **6**, 3409 (1972).
71. K. Uzelac, S. Barišić. *J. Physique* **36**, 1267 (1975); and *J. Physique – Lettres* **38**, 47 (1977).
72. M. J. Rice, S. Sträsler, *Sol. St. Commun.* **13**, 125 (1973).
73. W. L. Mc Millan, *Phys. Rev. B* **14**, 1496 (1976).
74. L. N. Bulaevskii, D. J. Khomski, *Zh. Eksp. Teor. Fiz.* **74**, 1863 (1978) [*Soviet Physics JETP* **47**, 971 (1978)].

75. V. L. Pokrovskii, A. L. Talapov, *Zh. Eksp. Teor. Fiz.* **75**, 1151 (1978) [*Soviet Physics JETP* **48**, 579 (1978)].
76. G. Theodorou, T. M. Rice, *Phys. Rev. B* **18**, 2840 (1978).
77. P. Lederer, P. Pitanga, *J. Physique* **39**, 993 (1978).
78. K. R. Subbaswamy, S. E. Trullinger, *Phys. Rev. A* **19**, 1340 (1979).
79. J. A. Krumhansl, J. R. Schrieffer, *Phys. Rev. B* **11**, 3535 (1975).
80. N. Gupta, B. Sutherland, *Phys. Rev. A* **14**, 1790 (1976).
81. Y. Okwamoto, H. Takayama, H. Shiba, *J. Phys. Soc. Japan* **46**, 1[illegible]0 (1979).
82. J. Batistić, S. Barišić, *J. Phys. Lettres* **40**, 613 (1979).
83. A. N. Filonov, G. M. Zaslavsky, *Phys. Lett.* **86A**, 237 (1981).
84. G. Mihaly, Gy. Hutiray, L. Mihaly, *Sol. St. Commun.* **48**, 203 (1983).
85. S. Barišić, I. Batistić, to be published.
86. P. Bak, *Phys. Rev. Lett.* **48**, 692 (1982).
87. J. Bardeen, *Sol. St. Commun.* **13**, 357 (1973).
88. P. A. Lee, T. M. Rice, P. W. Anderson, *Sol. St. Commun.* **14**, 703 (1974). P. A. Lee, T. M. Rice, *Phys. Rev. B* **19**, 3970 (1979).
89. J. Bardeen, *Phys. Rev. Lett.* **42**, 1498 (1979); *ibid* **45** 1978 (1980); *Mol. Cryst. Liq. Cryst.* **81**, 719 (1981); G. Grüner, A. Zettl, W. G. Clark, J. Bardeen, *Phys. Rev. B* **24**, 7247 (1981).
90. L. P. Gor'kov, *Pisma Zh. Eksp. Teor. Fiz.* **38**, 76 (1983) [*Soviet Physics JETP Lett.* **38**, 87 (1983)].
91. J. Bardeen, E. Ben-Jacob, A. Zettl, G. Grüner, *Phys. Rev. Lett.* **49**, 493 (1982).
92. P. Monceau, J. Richard, M. Renard, *Phys. Rev. Lett.* **45**, 43 (1980); Z. Z. Wang, M. C. Saint-Lager, P. Monceau, M. Renard, P. Gressier, A. Meerschaut, L. Guemas, J. Rouxel, *Sol. St. Commun.* **46**, 325 (1983); Z. Z. Wang, P. Monceau, M. Renard, P. Gressier, L. Guemas, A. Meerschaut, *Sol. St. Commun.* **47**, 439 (1983); P. Monceau, M. Renard, J. Richard, M. C. Saint-Lager, H. Salva, Z. Z. Wang, *Phys. Rev. B* **28**, 1646 (1983).
93. M. Weger, G. Grüner, W. G. Clark, *Sol. St. Commun.* **35**, 243 (1980).
94. G. Grüner, *Mol. Cryst. Liq. Cryst.* **81**, 735 (1981).
95. M. J. Rice, *Phys. Rev. Lett.* **48**, 1640 (1982).
96. S. A. Brazovskii, I. E. Dzyaloshinskii, *Zh. Eksp. Teor. Fiz.* **71**, 2338 (1976) [*Soviet Physics JETP* **44** 1233 (1976)].
97. A. Kotani, *J. Phys. Soc. Japan* **42**, 408 and 416 (1977).
98. S. A. Brazovskii, *Pisma Zh. Eksp. Teor. Fiz.* **28**, 656 (1978); [*Soviet Physics JETP Lett.* **28**, 606 (1978)] and *Zh. Eksp. Teor. Fiz.* **78**, 677 (1980) [*Soviet Physics JETP* **51**, 342 (1980)]; S. A. Brazovskii, I. F. Dzyaloshinskii, N. N. Kirova *Zh. Eksp. Teor. Fiz.* **81**, 2279 (1981) [*Soviet Physics JETP* **54** 1209 (1981)].
99. H. Takayama, Y. R. Lin-Lin, K. Maki, *Phys. Rev. Lett.* **47**, 986 (1981).
100. J. Goldstone, F. Wilczek, *Phys. Rev. Lett.* **47**, 986 (1981).
101. I. Batistić, S. Barišić, in Ref. 5., p. 1543.
102. W. P. Su, J. R. Schrieffer, A. J. Heeger, *Phys. Rev. Lett.* **42**, 1698 (1979); *Phys. Rev. B* **22**, 2099 (1980); W. P. Su, *Phys. Rev. B* **27**, 370 (1983).
103. M. J. Rice, *Phys. Lett.* **71**A, 152 (1979).
104. Per Bak, V. L. Pokrovskii, *Phys. Rev. Lett.* **47**, 958 (1981) suggest how to include the effects of discreteness beyond Equation (33).
105. S. A. Brazovskii, I. E. Dzyaloshinskii, I. M. Krichever, *Zh. Eksp. Teor. Fiz.* **83**, 389 (1982) [*Soviet Physics JETP* **56**, 212 (1982)].
106. I. E. Dzyaloshinskii, I. M. Krichever, *Zh. Eksp. Teor. Fiz.* **83**, 1576 (1982) [*Soviet Physics JETP* **56**, 908 (1982)].
107. W. Dieterich, *Z. Physik* **270**, 239 (1974); *Adv. in Phys.* **25**, 615 (1976).
108. G. Toulouse, *N. Cimento* **23**B, 1974).
109. K. B. Efetov, A. J. Larkin, *Zh. Eksp. Teor. Fiz.* **66**, 2290 (1974) [*Soviet Physics JETP* **39** 1129 (1974)].
110. D. J. Scalapino, Y. Imry, P. Pincus, *Phys. Rev. B* **11**, 2024 (1975).
111. P. Manneville, *J. Physique* **36**, 701 (1975).

112. R. A. Klemm, H. Gutfreund, *Phys. Rev. B* **14**, 1086 (1976).
113. G. Theodorou, I. Batistić, S. Barišić, *Fizika (Yu)* **12**, 141 (1980).
114. I. Batistić, G. Theodorou, S. Barišić, *Sol. St. Commun.* **34**, 499 (1980); *J. Phys. C* **14**, 1905 (1980).
115. C. Rouceau, R. Ayroles, P. Monceau, L. Guemas, A. Meerschaut, J. Rouxel, *Phys. St. Sol. (a)* **62**, 483 (1980).
116. Z. Z. Wang, H. Salva, P. Monceau, M. Renard, C. Rouceau, R. Ayroles F. Levy, L. Guemas, A. Meerschaut, *J. Physique Lettres* **44**, 311 (1983); H. Salva, Z. Z. Wang, P. Monceau, J. Richard, M. Renard, *Phil. Mag.* **49**, 385 (1984).
117. K. Tsutsumi, T. Sambongi, S. Kagoshima, T. Ishiguro, *J. Phys. Soc. Japan* **44**, 1735 (1978).
118. A. W. Higgs, J. C. Gill, *Socl. St. Commun.* **47**, 737 (1983).
119. A. Bjeliš, S. Barišić, *Phys. Rev. Lett.* **48**, 684 (1982) and A. Bjeliš, personal communication.
120. A. Briggs, P. Monceau, M. Nunez-Regueiro, M. Ribault, J. Richard, *J. Physique* **42**, 1453 (1981).
121. W. W. Fuller, Ph.D. Thesis, Univ. of California at los Angeles 1980; W. W. Fuller, G. Grüner, P. M. Chaikin, N. P. Ong, *Chem. Scr.* **17**, 205 (1981); W. W. Fuller, P. M. Chaikin, N. P. Ong, *Sol. St. Commun.* **30**, 689 (1979).
122. K. Nishida, T. Sambongi, M. Ido, *Phys. Soc. Japan* **48**, 331 (1980).
123. V. J. Emery, R. Bruinsma, S. Barišić, *Phys. Rev. Lett.* **48**, 1039 (1982); V. J. Emery in Ref. 5.
124. J. Friedel, *J. Physique-Lettres* **36**, 279 (1975).
125. G. Bilboro, W. L. Mc Millan, *Phys. Rev. B***14**, 1887 (1976).
126. D. R. Tilley, *Proc. Phys. Soc. (London)* **85**, 1177 (1965); and **86**, 678 (1965).
127. S. Barišić, P. G. de Gennes, *Sol. St. Commun* **6**, 281 (1968).
128. P. W. Anderson, *J. Phys. Chem. Sol.* **11**, 26 (1959).
129. M. D. Whitmore, J. P. Carbotte, *J. Phys. F* **11**, 2585 (1981) and references therein.
130. H. Suhl, B. J. Matthias, L. R. Walker, *Phys. Rev. Lett.* **3**, 552 (1959).
131. D. R. Tilley, *Proc. Phys. Soc. (London)* **84**, 573 (1964).
132. V. K. Wong, C. C. Sung, *Phys. Rev. Lett.* **19**, 1236 (1967).
133. B. T. Geilikman, R. O. Zaitsev, V. Z. Kresin, *Sov. Phys. Sol. St.* **9**, 642 (1967).
134. A. A. Abrikosov, J. A. Ryzhkin, *Lecture Notes in Phys.* **65**, 187 (1977).
135. L. A. Turkevich, R. A. Klemm, *Phys. Rev. B* **19**, 2520 (1979).
136 W. A. Challender P. L. Richards, *Sol. St. Commun.* **52**, 117 (1984).

DYNAMICAL PROPERTIES OF QUASI-ONE-DIMENSIONAL CONDUCTORS: A PHASE HAMILTONIAN APPROACH

HIDETOSHI FUKUYAMA

Institute for Solid State Physics,
University of Tokyo,
Roppongi, Minatoku, Tokyo 106, Japan

and

HAJIME TAKAYAMA

Department of Physics,
Hokkaido University,
Sapporo 060, Japan

1. Introduction

Since the realization of quasi-one-dimensional conductors, there have been extensive investigations into the electronic properties of other such low-dimensional systems, both experimentally and theoretically [1–4]. Above all, systems having a Peierls transition [5] have been investigated in detail. The novelty of this Peierls–Fröhlich (PF) [6] state lies in the novel possibility of the transport phenomenon being associated with the collective degree of freedom, i.e. the charge density wave (CDW) [7]. In a CDW, electrons follow the periodic lattice distortion adiabatically, resulting in a periodic spatial variation of the self-consistent charge density. Lee, Rice and Anderson [8] have shown that the low-lying excitation of CDW is due to the sliding motion associated with the lattice distortion, which is described by the phase of the complex order parameter, i.e. the periodic lattice distortion. This phase is related to the choice of the origin of coordinates and then to the translational symmetry of the system. Hence this sliding motion, or sliding conductivity, is sensitive to the impurity scattering and the Umklapp scattering [8]. These scattering mechanisms result in impurity and commensurability pinning, respectively, whose various interesting properties have been revealed since that time. In these investigations the phase Hamiltonian, which is the effective Hamiltonian to describe the motion of the phase and is derived from the full Hamiltonian with electron–phonon interactions, has proved very useful [8–12]. Recently anomalous properties typically observed in $NbSe_3$ have also been discussed in a similar context [13, 14] and by Monceau in Part II of this volume. In these cases the dynamics of phase motion has been treated classically. This is due to the fact that the effective mass of CDW is high since electrons are moving together with lattice distortions.

Beside these investigations on the PF state there has been great progress in the last ten years in the understanding of the many-body problem in one-dimension, since the problem can been treated with relatively high accuracy (sometimes exactly) due to the limited dimensionality. (See e.g. the review paper by Solyom [15].) One of the

P. Monceau (ed.), Electronic Properties of Inorganic Quasi-One-Dimensional Materials, I, 41–104.

main reasons for the development of such theoretical research is the success of the bosonization, which was first introduced by Tomonaga [16] in 1950. This bosonization of the interacting fermion fields has lately been refined and extended by Luther and Peschel [17], Luther and Emery [18], Mattis [19] and Haldane [20, 21] to treat more complicated interaction processes. Through these investigations quantum phase variables, which represent the charge and spin degrees of freedom of electrons (and are different from the phase variable in the PF state), have been naturally introduced, and the phase Hamiltonian has then been derived [22–25], which is equivalent to the original Hamiltonian given in terms of fermions. In this case the quantum mechanical description is inevitable.

In realistic systems electrons are mutually interacting and are coupled to the lattice distortions. Such systems will then be described by phase variables representing electrons on one hand and lattice distortions with amplitude and phase on the other. This is the phase Hamiltonian approach to the whole system, and has been imployed in the examination of excitation spectra and the classification of solitons in the PF state [26–28]. Such an approach also turns out [29, 30] to be convenient in the discussion of the spin Peierls problem [31], which can be considered as a case of a strong repulsive interaction in the half-filled band.

The purpose of this review is to give a coherent description of the dynamical properties of electrons based on the phase Hamiltonian. In Section 2 the phase Hamiltonian is derived for interacting electrons, but without coupling to the lattice. In this section it is shown how conductivity and spin susceptibility are evaluated in terms of the phase variables. In Section 3 the PF state is discussed; especially the spin Peierls problem is examined in Section 3.7. In Section 4, the effects of impurities have been examined for the PF state on one hand and for the interacting electrons in the absence of PF state on the other. The latter is a new way to treat the Anderson localization [32–34] in the interacting systems. Based on the results obtained in the preceding sections the CDW conductivity in the PF state is examined in Section 5 and transport properties of MX_3 (M: transition metal, X: chalcogenide) will be discussed in Section 6 on a similar basis. A summary is given in Section 7.

We shall use units of $\hbar = k_B = 1$.

2. Phase Hamiltonian

2.1. Boson Representation and Phase Variables

We consider one-dimensional electrons represented by the following Hamiltonian, $\mathcal{H}$, [15]

$$\mathcal{H} = \mathcal{H}_0 + \mathcal{H}_{\text{int}} \tag{2.1}$$

$$\mathcal{H}_0 = \sum_{k>0} v_F(k-k_F)a_{k\sigma}^+ a_{k\sigma} + \sum_{k<0} v_F(-k-k_F)b_{k\sigma}^+ b_{k\sigma}, \tag{2.2}$$

$$\mathcal{H}_{\text{int}} = \frac{1}{L} \sum_{k_1, k_2, p} \sum_{\alpha,\beta} [(\gamma_{1/\!/}\delta_{\alpha,\beta} + \gamma_{1\perp}\delta_{\alpha,-\beta}) \times$$
$$\times\, a^{+}_{k_1\alpha} b^{+}_{k_2\beta} a_{k_2+2k_F+p\beta} b_{k_1-2k_F-p\alpha} +$$
$$+ (\gamma_{2/\!/}\delta_{\alpha,\beta} + \gamma_{2\perp}\delta_{\alpha,-\beta}) a^{+}_{k_1\alpha} b^{+}_{k_2\beta} b_{k_2+p\beta} a_{k_1-p\alpha} +$$
$$+ \frac{\gamma_{3\perp}}{2}\delta_{\alpha,-\beta}(a^{+}_{k_1\alpha} a^{+}_{k_2\beta} b_{k_2-2k_F+p\beta} b_{k_1+2k_F-p-G\alpha} + \text{h.c.}) +$$
$$+ \frac{\gamma_{4\perp}}{2}\delta_{\alpha,-\beta}(a^{+}_{k_1\alpha} a^{+}_{k_2\beta} a_{k_2+p\beta} a_{k_1-p\alpha} +$$
$$+ b^{+}_{k_1\alpha} b^{+}_{k_2\beta} b_{k_2+p\beta} b_{k_1-p\alpha})], \tag{2.3}$$

Here v_F and k_F are the Fermi velocity and the Fermi momentum, and $a_{k\sigma}(a^{+}_{k\sigma})$ and $b_{k\sigma}(b^{+}_{k\sigma})$ are the annihilation (creation) operators of electrons with $k > 0$ and $k < 0$, respectively. Equation (2.2) defines the kinetic energy with the linear spectrum as shown in Figure 1. In this case the density of states per spin, $N(0)$, is given by $N(0) = (\pi v_F)^{-1}$.

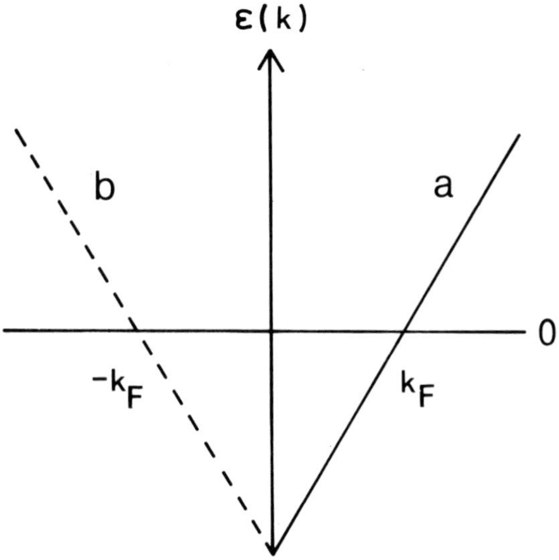

Fig. 1. Energy spectrum of one-dimensional electrons.

The interaction constants, $\gamma_{i/\!/}$ and $\gamma_{i\perp}$, in Equation (2.3) are those between parallel and antiparallel spins. We will introduce dimensionless coupling constants, g_i, given by $g_i = \gamma_i N(0)/2 = \gamma_i/2\pi v_F$. Processes characterized by each g_i (i = 1~4) are shown in Figure 2, where solid and broken lines represent electrons with $k > 0$ and $k < 0$, respectively. In this figure the suffices, $/\!/$ and $\perp$, are not shown. It should be noted that the momentum transfer, q, through g_2 and g_4 processes is small, i.e. $|q| \ll 2k_F$, whereas that in g_1 is large, $|q| \simeq 2k_F$ and then this process is called backward scattering. The g_3 represents the Umklapp process, only present if $4k_F \simeq G$, G being the reciprocal lattice vector. In Equation (2.3) all of the interactions are assumed to be instantaneous and local. This locality forbids processes corresponding to $g_{3/\!/}$ and $g_{4/\!/}$ due to the Pauli principle.

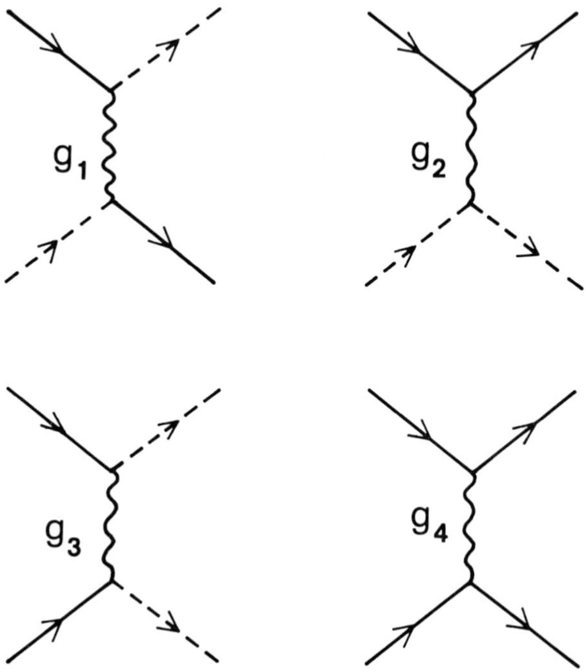

Fig. 2. Interaction processes, $g_1 \sim g_4$.

Some of the specific cases of $\mathcal{H}_{\text{int}}$ have been treated either rigorously or by the renormalization group. The example Tomonaga [16] first examined corresponds to taking $g_{2\parallel} \neq 0$ and $g_i = 0$ otherwise. Investigations with more generality have been performed later by Luther and Peschel [17], Luther and Emery [18] and Mattis [19], who found that the dynamical processes represented by electron field operators, $\psi_{is}(x)$ ($i = 1$ for $k > 0$ and $i = 2$ for $k < 0$), can be equivalently expressed by those of a boson field, $\phi_{is}(x)$, as follows;

$$\psi_{1s}(x) \to (2\pi\alpha)^{-1/2} \exp[-ik_F x - \phi_{1s}(x)] \equiv O_{1s}(x), \tag{2.4}$$

$$\psi_{2s}(x) \to (2\pi\alpha)^{-1/2} \exp[+ik_F x + \phi_{2s}(x)] \equiv O_{2s}(x), \tag{2.5}$$

where s is the spin index and α is the cut-off parameter to be set $\alpha \to 0$ whenever possible and ϕ_{is}'s are given by

$$\phi_{is}(x) = \sum_q A_q(x) \rho_{is}(q), \tag{2.6}$$

$$A_q(x) = \frac{2\pi}{Lq} \exp[-\tfrac{1}{2}\alpha|q| - iqx]. \tag{2.7}$$

Here L is the system size. In Equation (2.6) the density fluctuation operators, $\rho_{is}(q)$ ($q > 0$), are defined as follows,

$$\rho_{1s}(q) = \sum_{k>0} a^{\dagger}_{k+qs} a_{ks}, \tag{2.8a}$$

$$\rho_{1s}(-q) = \sum_{k>0} a^{\dagger}_{ks} a_{k+qs}, \tag{2.8b}$$

$$\rho_{2s}(q) = \sum_{k<-q} b^{\dagger}_{k+qs} b_{ks}, \tag{2.8c}$$

$$\rho_{2s}(-q) = \sum_{k<-q} b^{\dagger}_{ks} b_{k+qs}. \tag{2.8d}$$

As was found by Tomonaga [16], these operators, $\rho_{is}(q)$, satisfy following commutation relation,

$$\begin{aligned}[\rho_{1s}(-q), \rho_{1s'}(q')] &= [\rho_{2s}(q), \rho_{2s'}(-q')] \\ &= \frac{qL}{2\pi}\delta_{q,q'}\delta_{ss'},\end{aligned} \tag{2.9}$$

and commute otherwise.

The equivalence of ψ_{is} and 0_{is} is due to the fact that they yield same commutations with respect to $\rho_{is}(q)$; e.g. for $\rho_{1s}(q)$

$$[\psi_{1s}(x), \rho_{1s}(q)] = e^{-iqx}\,\psi_{1s}(x),$$

$$[0_{1s}(x), \rho_{1s}(q)] = e^{-iqx}\,0_{1s}(x),$$

and then for any operator, which is a functional of $\rho_{is}(q)$, $\psi_{1s}(x)$ and $0_{1s}(x)$ act equivalently. (The equivalence of ψ_{is} and 0_{is} can also be shown directly [21, 35].)

In terms of $\rho_{is}(q)$, $\mathcal{H}_0$ can be expressed as

$$\mathcal{H}_0{}^{\mathrm{B}} = \frac{2\pi v_F}{L}\sum_{q>0}\sum_{s}[\rho_{1s}(q)\rho_{1s}(-q) + \rho_{2s}(-q)\rho_{2s}(q)]. \tag{2.10}$$

This is due to the fact that the commutations of $\mathcal{H}_0$ and $\mathcal{H}_0{}^{\mathrm{B}}$ with respect to $\rho_{is}(q)$ are identical; e.g. for $\rho_{1s}(q)$

$$[\mathcal{H}_0, \rho_{1s}(q)] = v_F q\,\rho_{1s}(q), \tag{2.11a}$$

$$[\mathcal{H}_0{}^{\mathrm{B}}, \rho_{1s}(q)] = v_F q\,\rho_{1s}(q). \tag{2.11b}$$

Among various processes in $\mathcal{H}_{\mathrm{int}}$, Equation (2.3), those with γ_2 and γ_4 can be written in terms of the quadratic forms of $\rho_{is}(q)$,

$$\mathcal{H}_{\text{int}}^{(1)} = \frac{1}{L}\sum_{q>0}(\gamma_{2\parallel}+\gamma_{2\perp})\left[\rho_1(q)\rho_2(-q)+\rho_1(-q)\rho_2(q)\right] + $$
$$+\frac{1}{L}\sum_{q>0}(\gamma_{2\parallel}-\gamma_{2\perp})\left[\sigma_1(q)\sigma_2(-q)+\sigma_1(-q)\sigma_2(q)\right] + $$
$$+\frac{1}{L}\sum_{q>0}\gamma_{4\perp}\left[\rho_1(q)\rho_1(-q)+\rho_2(-q)\rho_2(q)\right] - $$
$$-\frac{1}{L}\sum_{q>0}\gamma_{4\perp}\left[\sigma_1(q)\sigma_1(-q)+\sigma_2(-q)\sigma_2(q)\right], \tag{2.12}$$

where $\rho_i(q)$ and $\sigma_i(q)$ are defined as follows,

$$\rho_i(q) = \frac{1}{\sqrt{2}}(\rho_{i\uparrow}(q)+\rho_{i\downarrow}(q)), \tag{2.13a}$$

$$\sigma_i(q) = \frac{1}{\sqrt{2}}(\rho_{i\uparrow}(q)-\rho_{i\downarrow}(q)). \tag{2.13b}$$

Thus $\mathcal{H}_0{}^B + \mathcal{H}_{\text{int}}^{(1)}$ is given by

$$\mathcal{H}_0{}^B + \mathcal{H}_{\text{int}}^{(1)} = \frac{2\pi v_F}{L}\sum_{q>0}\left[\left(1+\frac{g_{4\perp}}{2}\right)(\rho_1(q)\rho_1(-q)+\rho_2(-q)\rho_2(q)) + \right.$$
$$+\left(1-\frac{g_{4\perp}}{2}\right)(\sigma_1(q)\sigma_1(-q)+\sigma_2(-q)\sigma_2(q)) + $$
$$+\left(\frac{g_{2\parallel}+g_{2\perp}}{2}\right)(\rho_1(q)\rho_2(-q)+\rho_1(-q)\rho_2(q)) + $$
$$\left.+\left(\frac{g_{2\parallel}-g_{2\perp}}{2}\right)(\sigma_1(q)\sigma_2(-q)+\sigma_1(-q)\sigma_2(q))\right]. \tag{2.14}$$

It is obvious that Equation (2.14) can be exactly solved, since it is quadratic in $\rho_{is}(q)$ with commutators, Equation (2.9). The fact that Equation (2.14) can be solved is more clearly seen by the introduction of the quantum-mechanical phase variables;

$$\begin{aligned}\theta_\pm(x) &= \frac{i}{\sqrt{2}}\sum A_q(x)\left[\rho_1(q)\pm\rho_2(q)\right] \\ &= \frac{i}{2}\left[(\phi_{1\uparrow}+\phi_{1\downarrow})\pm(\phi_{2\uparrow}+\phi_{2\downarrow})\right],\end{aligned} \tag{2.15}$$

$$\begin{aligned}\phi_\pm(x) &= \frac{i}{\sqrt{2}}\sum A_q(x)\left[\sigma_1(q)\pm\sigma_2(q)\right] \\ &= \frac{i}{2}\left[(\phi_{1\uparrow}-\phi_{1\downarrow})\pm(\phi_{2\uparrow}-\phi_{2\downarrow})\right].\end{aligned} \tag{2.16}$$

By use of Equation (2.9), $\theta_\pm(x)$ and $\phi_\pm(x)$ are seen to satisfy the following commutation relations

$$[\theta_+(x), P(x')] = i\delta(x - x'), \tag{2.17}$$

$$[\phi_+(x), M(x')] = i\delta(x - x'), \tag{2.18}$$

where $P(x)$ and $M(x)$ are

$$P(x) = -\nabla\theta_-/2\pi, \tag{2.19}$$

$$M(x) = -\nabla\phi_-/2\pi. \tag{2.20}$$

Commutations other than Equations (2.17) and (2.18) vanish. By use of $\theta_\pm(x)$ and $\phi_\pm(x)$, $\mathcal{H}_0{}^B + \mathcal{H}_{\text{int}}^{(1)}$ can be written as

$$\mathcal{H}_0{}^B + \mathcal{H}_{\text{int}}^{(1)} = \int \mathrm{d}x\ [A_\rho'(\nabla\theta_+)^2 + C_\rho' P^2(x) + A_\sigma'(\nabla\phi_+)^2 + C_\sigma' M^2(x)], \tag{2.21}$$

where

$$A_\rho' = \frac{v_F}{4\pi}\left(1 + \frac{g_{2\parallel} + g_{2\perp} + g_{4\perp}}{2}\right), \tag{2.22a}$$

$$C_\rho' = \pi v_F \left(1 - \frac{g_{2\parallel} + g_{2\perp} - g_{4\perp}}{2}\right), \tag{2.22b}$$

$$A_\sigma' = \frac{v_F}{4\pi}\left(1 + \frac{g_{2\parallel} - g_{2\perp} - g_{4\perp}}{2}\right), \tag{2.22c}$$

$$C_\sigma' = \pi v_F \left(1 - \frac{g_{2\parallel} - g_{2\perp} + g_{4\perp}}{2}\right), \tag{2.22d}$$

Equation (2.21) is of a form of simple harmonic oscillators with excitation spectra linear in the momentum, $\omega_\rho(q)$ and $\omega_\sigma(q)$;

$$\omega_\rho(q) = 2\sqrt{A_\rho' C_\rho'}\ q, \tag{2.23}$$

$$\omega_\sigma(q) = 2\sqrt{A_\sigma' C_\sigma'}\ q. \tag{2.24}$$

The former (latter) represents the oscillation of $\theta_+(\phi_+)$. The physical implication of phase variables can be extracted by noting the following facts. By Equations (2.6) and (2.7) we see ($\alpha \to 0$)

$$\nabla\theta_+/\pi = \frac{\sqrt{2}}{L}\sum_q e^{-iqx}\,(\rho_1(q)+\rho_2(q))$$

$$\equiv n(x), \tag{2.25}$$

$$\nabla\phi_+/2\pi = \frac{1}{\sqrt{2L}}\sum e^{-iqx}\,(\sigma_1(q)+\sigma_2(q))$$

$$\equiv m(x), \tag{2.26}$$

where $n(x)$ and $m(x)$ can be interpreted as local density fluctuations of charge and spin with long wavelength. Equations (2.17) and (2.18) together with Equation (2.25) and (2.26) also imply

$$[n(x)/2, \theta_-(x')] = -i\delta\,(x-x'), \tag{2.27}$$

$$[m(x), \phi_-(x')] = -i\delta\,(x-x'). \tag{2.28}$$

Consequently θ_- and ϕ_- can be taken as quantum-mechanical representations of the local phase of superconductivity and of the azimuthal angle of the local spin density.

Other interaction processes in $\mathcal{H}_{\rm int}$, i.e. γ_1 and γ_3 can be treated as follows. First the contribution with $\gamma_{1/\!/}$ can be written as

$$-\frac{\gamma_{1/\!/}}{L}\sum_{k_1k_2p}\sum_s a^\dagger_{k_1s}a_{k_2+2k_F+ps}b^\dagger_{k_2s}b_{k_1-2k_F-ps} \tag{2.29a}$$

$$= -\frac{\gamma_{1/\!/}}{L}\sum_q\sum_s \rho_{1s}(q)\,\rho_{2s}(-q) \tag{2.29b}$$

$$= -\frac{\gamma_{1/\!/}}{L}\sum_q [\rho_1(q)\,\rho_2(-q)+\sigma_1(q)\,\sigma_2(-q)]\,, \tag{2.29c}$$

where we noted $k_1 \sim k_F$ and $k_2 \sim -k_F$ in Equation (2.29a). Equation (2.29c) is similar to Equation (2.14) and then this contribution is represented as Equation (2.21) with the slight modification of A_ρ', A_σ', C_ρ' and C_σ'.

On the other hand the $\gamma_{1\perp}$-process has a different feature. By use of Equations (2.4) and (2.5) we obtain, as follows:

$$\frac{\gamma_{1\perp}}{2}\sum_s\int dx\,[\psi^\dagger_{1-s}(x)\,\psi^\dagger_{2s}(x)\,\psi_{1s}(x)\,\psi_{2-s}(x)+\text{h.c.}] \tag{2.30a}$$

$$= \frac{\gamma_{1\perp}}{2(2\pi\alpha)^2}\sum_s\int dx\,[\exp\{\phi_{1-s}-\phi_{2s}-\phi_{1s}+\phi_{2-s}\}+\text{h.c.}]\,, \tag{2.30b}$$

$$= \frac{\gamma_{1\perp}}{2\pi^2\alpha^2}\int dx\,\cos 2\phi_+(x), \tag{2.30c}$$

where $\phi_+(x)$ is defined by Equation (2.16).

Finally the γ_3-process will be transformed as follows.

$$\frac{\gamma_{3\perp}}{2}\sum_s \int dx\,(\psi_{2s}^+\psi_{2-s}^+\psi_{1-s}\psi_{1s}e^{iGx} + \text{h.c.}), \tag{2.31a}$$

$$= \frac{\gamma_{3\perp}}{2\,(2\pi\alpha)^2}\sum_s \int dx\,[\exp\{i(G-4k_F)x - \sigma_{1s} - \phi_{1-s} - \phi_{2s} - \phi_{2-s}\} + \text{+ h.c.}], \tag{2.31b}$$

$$= \frac{\gamma_{3\perp}}{2\pi^2\alpha^2}\int dx\,\cos\,[(G-4k_F)x + 2\theta_+]. \tag{2.31c}$$

Consequently the total Hamiltonian, $\mathcal{H}_0 + \mathcal{H}_{\text{int}}$ Equation (2.1), is transformed to the following phase Hamiltonian.

$$\mathcal{H} = \mathcal{H}_\rho + \mathcal{H}_\sigma, \tag{2.32}$$

$$\mathcal{H}_\rho = \int dx\,[A_\rho\,(\nabla\theta)^2 + B_\rho\cos\,[(G-4k_F)x + 2\theta)] + C_\rho P^2], \tag{2.33}$$

$$\mathcal{H}_\sigma = \int dx\,[A_\sigma\,(\nabla\phi)^2 + B_\sigma\cos 2\phi + C_\sigma M^2], \tag{2.34}$$

$$A_\rho = \frac{v_F}{4\pi}\left(1 - \frac{g_{1/\!/} - g_{2/\!/} - g_{2\perp} - g_{4\perp}}{2}\right), \tag{2.35}$$

$$B_\rho = v_F g_{3\perp}/2\pi\alpha^2, \tag{2.36}$$

$$C_\rho = \pi v_F\left(1 + \frac{g_{1/\!/} - g_{2/\!/} - g_{2\perp} + g_{4\perp}}{2}\right), \tag{2.37}$$

$$A_\sigma = \frac{v_F}{4\pi}\left(1 - \frac{g_{1/\!/} - g_{2/\!/} + g_{2\perp} + g_{4\perp}}{2}\right), \tag{2.38}$$

$$B_\sigma = v_F g_{1\perp}/2\pi\alpha^2, \tag{2.39}$$

$$C_\sigma = \pi v_F\left(1 + \frac{g_{1/\!/} - g_{2/\!/} + g_{2\perp} - g_{4\perp}}{2}\right). \tag{2.40}$$

Here $\theta(x) = \theta_+(x)$ and $\phi(x) = \phi_+(x)$.

It is to be noted that in the present interaction processes the charge and spin degrees of freedom are separated as $\mathcal{H}_\rho$ and $\mathcal{H}_\sigma$, respectively [18]. Unless $g_{3\perp}$ and $g_{1\perp}$ processes, which introduce non-linearity, are present, $\mathcal{H}$, Equation (2.32), can be solved exactly, since the resulting Hamiltonian is simply harmonic. The excitation spectra in this case are linear in q as is schematically shown in Figure 3, where the velocities, v_ρ and v_σ, are given as follows

$$v_\rho = 2\sqrt{A_\rho C_\rho}\,, \tag{2.41}$$

$$v_\sigma = 2\sqrt{A_\sigma C_\sigma}\,. \tag{2.42}$$

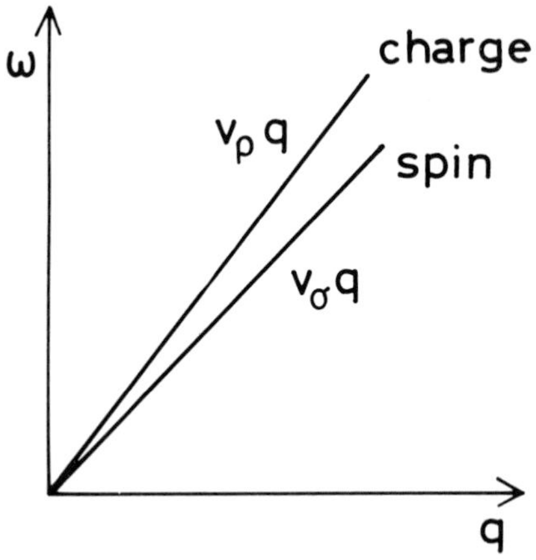

Fig. 3. Excitation spectra in the absence of the nonlinearity.

The present model for the interaction processes is very general and various limiting cases correspond to known models, which are as follows.

(i) *Tomonaga–Luttinger (LT) model* [16, 36]

$g_{2/\!/} \neq 0, \quad g_i = 0 \quad \text{otherwise}$

(ii) *Spin-Dependent Tomonaga (SDT) model*

$g_{2/\!/} \neq 0, \quad g_{2\perp} \neq 0, \quad g_i = 0 \quad \text{otherwise}$

(iii) *Bychkov–Gorkov–Dzyaloshinsky (BDG) model* [37]

$g_{1/\!/} = g_{1\perp} \equiv g_1 \neq 0,$

$g_{2/\!/} = g_{2\perp} \equiv g_2 \neq 0,$

$g_i = 0 \quad \text{otherwise}$

(iv) *Luther–Emery (LE) model* [18]

$g_{1/\!/} \neq 0, \quad g_{1\perp} \neq 0,$

$g_{2/\!/} \neq 0, \quad g_{2\perp} \neq 0,$

$g_i = 0 \quad \text{otherwise.}$

The derivation of the phase Hamiltonian of the type of Equation (2.34) (for spin fluctuations) was first performed by Luther [24]. Efetov and Larkin [23] also noted the existence of the phase Hamiltonian for the charge fluctuations, and they used this in the calculation of the response functions. Suzumura [25] evaluated the various response functions based on $\mathcal{H}_\rho$ and $\mathcal{H}_\sigma$, Equations (2.33) and (2.34).

We have seen that the Hamiltonian of interacting fermions is given in terms of the phase variables. The physical quantities can also be expressed by these variables. In Equations (2.25) and (2.26) the local density fluctuations varying slowly in space are given in terms of $\nabla\theta$ and $\nabla\phi$. On the other hand the electron density operator with wave number around $\pm 2k_F$ is given by

$$\rho_{\pm 2k_F,s}(x) = \frac{1}{2\pi\alpha} \exp[\pm i(2k_F x + \theta + s\phi)], \tag{2.43}$$

where $s = \pm 1$ is the spin index. Consequently, the total density operators of charge and spin, $N(x)$ and $M(x)$, are as follows

$$N(x) = \nabla\theta/\pi + \frac{2}{\pi\alpha} \cos(2k_F x + \theta) \cos\phi, \tag{2.44}$$

$$M(x) = \nabla\phi/2\pi - \frac{1}{\pi\alpha} \sin(2k_F x + \theta) \sin\phi. \tag{2.45}$$

2.2. THE g-OLOGY

There does not exist any phase transition at finite temperatures, T, in one-dimension [38]. However, various response functions have divergent singularities as $T \to 0$ which indicate the types of possible ordered states at $T = 0$ [15]. The phase diagram of the types of the ordered state in the plane of coupling constants is called the g-ology. For the g-ology we consider the following four operators [25];

$$O^+_{\text{CDW}}(x) = \psi^+_{1\uparrow}\psi_{2\uparrow} = (2\pi\alpha)^{-1} \exp[-i\phi_+ + i(2k_F x - \theta_+)], \tag{2.46a}$$

$$O^+_{\text{SDW}}(x) = \psi^+_{1\uparrow}\psi_{2\downarrow} = (2\pi\alpha)^{-1} \exp[-i\phi_- + i(2k_F x - \theta_+)], \tag{2.46b}$$

$$O^+_{\text{SS}}(x) = \psi^+_{1\uparrow}\psi^+_{2\downarrow} = (2\pi\alpha)^{-1} \exp[-i\phi_+ - i\theta_-], \tag{2.46c}$$

$$O^+_{\text{TS}}(x) = \psi^+_{1\uparrow}\psi^+_{2\uparrow} = (2\pi\alpha)^{-1} \exp[-i\phi_- - i\theta_-]. \tag{2.46d}$$

These represent fluctuations of the charge density wave (CDW), the spin density wave (SDW), the singlet superconductivity (SS) and the triplet superconductivity (TS), respectively. The g-ology will be investigated by evaluating the response function

$$\chi_A(q, i\omega_\lambda) = \int_{-\infty}^{\infty} \mathrm{d}x \int_{-\beta}^{\beta} \mathrm{d}\tau \exp(-iqx - i\omega_\lambda\tau) \times$$
$$\times \langle T_\tau O_A(x,\tau) O_A{}^+(0,0)\rangle, \tag{2.47}$$

where $\omega_\lambda = 2\pi\lambda T$, $\beta = T^{-1}$ and A stands for CDW, SDW, SS and TS and $O_A(x, \tau)$ is defined by

$$O_A(x,\tau) = e^{\mathcal{H}\tau} O_A(x) e^{-\mathcal{H}\tau}. \tag{2.48}$$

In Equation (2.47), T_τ is the chronological operator and $\langle\ \rangle$ is meant to take thermal average with respect to $\mathcal{H}$, Equation (2.32). The separability of the Hamiltonian into the charge and the spin density fluctuations, $\mathcal{H}_\rho$ and $\mathcal{H}_\sigma$, allows the separate evaluations of $\langle T_\tau O_A(x, \tau) O_A{}^+(0, 0,)\rangle$ with respect to $\theta_\pm$ and $\phi_\pm$.

As the temperature is lowered, $\chi_A(q, 0)$ can have divergent singularities

$$\chi_A(q, 0) \propto T^{-\gamma_A}, \tag{2.49}$$

where $q = 2k_F$ for CDW and SDW and $q = 0$ for SS and TS.

We will explicitly examine $\chi_A(q, i\omega_\lambda)$ in the SDT model, which has exact solutions. In this case the Hamiltonian is harmonic and we then manipulate as follows, e.g., for the charge fluctuations and for large $|x|$,

$$\langle T_\tau \exp - i[\theta(x, \tau) - \theta(0, 0)]\rangle \tag{2.50a}$$

$$= \exp[-\tfrac{1}{2}\langle T_\tau(\theta(x, \tau) - \theta(0, 0))^2\rangle] \tag{2.50b}$$

$$= (2\pi T/\omega_F)^{\eta_1}\ |\exp(|x|/\xi_1) - \exp(-|x|/\xi_1 - i2\pi T\tau)|^{-\eta_1}, \tag{2.50c}$$

where $\omega_F = v_\rho \alpha^{-1}$ and

$$\eta_1 = \left[\left(1 - \frac{g_{2\parallel} + g_{2\perp}}{2}\right)\bigg/\left(1 + \frac{g_{2\parallel} + g_{2\perp}}{2}\right)\right]^{1/2}, \tag{2.51}$$

$$\xi_1 = v_F\left[1 - \left(\frac{g_{2\parallel} + g_{2\perp}}{2}\right)^2\right]^{1/2}\bigg/\ \pi T \equiv v_\rho/\pi T. \tag{2.52}$$

In deriving Equation (2.50c) from Equation (2.50b) we note the following as $|x| \to \infty$;

$$\langle T_\tau(\theta(x, \tau) - \theta(0, 0))^2\rangle$$

$$= \frac{2C}{\pi}\ \rho T \sum_{\omega_l} \int \mathrm{d}k\ \frac{1}{\omega_l{}^2 + \omega_\rho(k)^2}\ [1 - \exp(ikx - i\omega_l\tau)],$$

$$= \frac{2C}{\pi}\ \rho T\left[\int \mathrm{d}k\ \frac{1 - \exp ikx}{\omega_\rho(k)^2} + \sum_{\omega_l \neq 0} \int \mathrm{d}k\ \frac{1}{\omega_l{}^2 + \omega_\rho(k)^2} - \right.$$

$$\left. - T \sum_{\omega_l \neq 0} \int \mathrm{d}k\ \frac{\exp[i(kx - \omega_l\tau)]}{\omega_l{}^2 + \omega_\rho(k)^2}\right],$$

$$= \frac{1}{\pi}\sqrt{\frac{C_\rho}{A_\rho}}\left[\frac{|x|}{\xi_1} + \ln \omega_F/2\pi T + \ln\left|1 - \exp\left(-\frac{2|x|}{\xi_1} - i2\pi T\tau\right)\right|\right]. \tag{2.53}$$

By use of Equation (2.50c) and the similar equation for the spin fluctuations, we obtain following γ_A defined in Equation (2.49).

$$\gamma_{\text{CDW}} = 2 - \gamma_\rho - \gamma_\sigma, \tag{2.54a}$$

$$\gamma_{\text{SDW}} = 2 - \gamma_\rho - \frac{1}{\gamma_\sigma}, \tag{2.54b}$$

$$\gamma_{\text{SS}} = 2 - \frac{1}{\gamma_\rho} - \gamma_\sigma, \tag{2.54c}$$

$$\gamma_{\text{TS}} = 2 - \frac{1}{\gamma_\rho} - \frac{1}{\gamma_\sigma}, \tag{2.54d}$$

where

$$\gamma_\rho = \left[\left(1 - \frac{g_{2\parallel} + g_{2\perp}}{2}\right) \Big/ \left(1 + \frac{g_{2\parallel} + g_{2\perp}}{2}\right)\right]^{1/2}, \tag{2.55}$$

$$\gamma_\sigma = \left[\left(1 - \frac{g_{2\parallel} - g_{2\perp}}{2}\right) \Big/ \left(1 + \frac{g_{2\parallel} - g_{2\perp}}{2}\right)\right]^{1/2}. \tag{2.56}$$

In Figure 4 the types of response functions with the largest γ_A are shown in the plane of $g_{2\perp}$ and $g_{2\parallel}$. This is the g-ology for the SDT model.

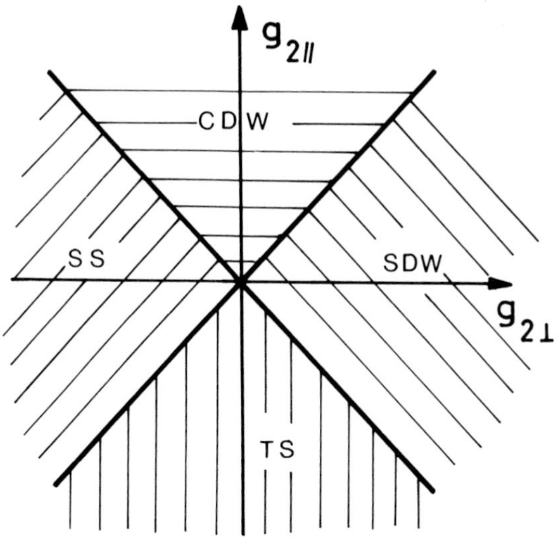

Fig. 4. The g-ology of the spin-dependent Tomonaga (SDT) model.

In the presence of B_ρ and/or B_σ, there do not exist exact solutions in general and some kind of approximations is inevitable. The self-consistent harmonic approximation (SCHA), which is proposed by Dashen *et al.* [39] and by Coleman [40] in the present context, is a simple one, and will be explained in the following. We consider the model quantum sine-Gordon Hamiltonian which typifies either $\mathcal{H}_\rho$ or $\mathcal{H}_\sigma$;

$$\mathcal{H} = \int dx \, [A(\nabla\theta)^2 + CP^2 - B \cos 2\theta]\,, \tag{2.57}$$

where B is assumed to be positive. In SCHA, $\cos 2\theta$ is approximated as

$$\cos 2\theta \to \exp[-2\langle\theta^2\rangle]\,[1 - 2(\theta^2 - \langle\theta^2\rangle)]\,, \tag{2.58}$$

where $\langle\theta^2\rangle$ is the average to be taken with respect to the approximated Hamiltonian. By Equation (2.58) the excitation spectrum of Equation (2.57) is given by

$$\omega(q) = 2\,[C(Aq^2 + 2\bar{B})]^{1/2}\,, \tag{2.59}$$

where $\bar{B} = B\exp[-2\langle\theta^2\rangle]$, which will be evaluated as follows, at $T = 0$, by

$$\langle\theta^2\rangle = \frac{C}{L}\sum_q \frac{1}{\omega(q)} = \frac{1}{2\pi}\sqrt{\frac{C}{A}}\ln\frac{2\pi}{\alpha q_0}\,. \tag{2.60}$$

Here the integration over q in Equation (2.60) is cut off by α^{-1} and q_0 is defined by

$$q_0{}^2 = 2\bar{B}/A = (2B/A)\exp[-2\langle\theta^2\rangle]\,. \tag{2.61}$$

Equations (2.60) and (2.61) constitute self-consistent equations for q_0, i.e.

$$q_0{}^2 = (2B/A)(\alpha q_0/2\pi)^{\frac{1}{\pi}\sqrt{\frac{C}{A}}}\,. \tag{2.62}$$

This equation always has a solution $q_0 = 0$, in which case the excitation spectrum does not have a gap. On the other hand it can have a physical non zero solution if the following inequality is satisfied;

$$(C/A)^{1/2} < 2\pi. \tag{2.63}$$

In the presence of finite q_0 the excitation spectrum has a gap, $\omega_0 \equiv 2\sqrt{AC'}q_0$, as schematically shown in Figure 5. In this case there exists finite restoring force of the oscillations around $\theta = 0$ (i.e. the phase is pinned) but the force is reduced by $\bar{B}/B$ due to the smearing of the potential cause by the quantum oscillations. Unless Equation (2.63) is satisfied, $\omega_0 = 0$ even if $B \neq 0$ due to the overwhelming quantum fluctuations.

In the BGD model there exists such a type of nonlinearity in $\mathcal{H}_\sigma$, even though $\mathcal{H}_\rho$ is harmonic, i.e.

$$A_\sigma = \frac{v_F}{4\pi}\left(1 - \frac{g_1}{2}\right), \tag{2.64a}$$

$$C_\sigma = \pi v_F\left(1 + \frac{g_1}{2}\right), \tag{2.64b}$$

$$B_\sigma = v_F g_1/2\pi\alpha^2. \tag{2.64c}$$

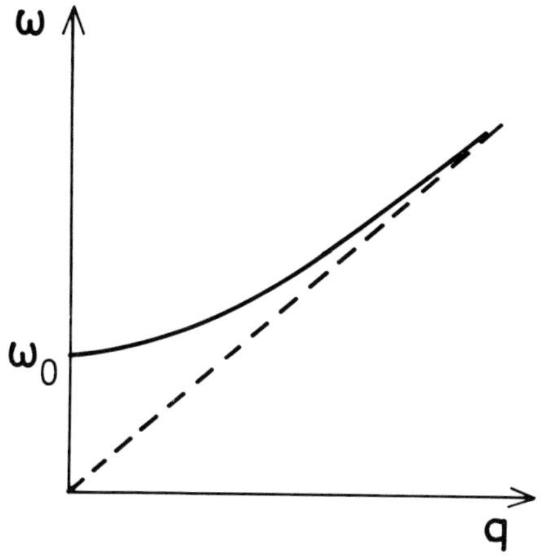

Fig. 5. Excitation spectrum for Equation (2.57).

From Equation (2.63) we see that the excitation spectrum is gapless if $g_1 > 0$ and that $\omega_0 \neq 0$ for $g_1 < 0$. The former (latter) corresponds to the weak (strong) coupling regime. The fact that $g_1 = 0$ separates these two regimes is in agreement with those obtained by the renormalization group treatment [15, 41, 42]. However it remains an unresolved difficulty to obtain, in terms of $\mathcal{H}_\rho$ and $\mathcal{H}_\sigma$, the correct exponents of the response functions even if $g_1 > 0$.

2.3. Conductivity and Spin-Susceptibility

The Kubo formula for the conductivity, σ, is given by

$$\sigma(\omega) = i \int_0^\infty \mathrm{d}t\, e^{-i\omega t} \langle [J(t), P] \rangle, \tag{2.65}$$

where J and P are total electric current and dipole moment, $\dot{P} \equiv (\mathrm{d}/\mathrm{d}t)P = J$. The former is given by the phase variable as

$$J \equiv \int \mathrm{d}x\, j(x) = \frac{e}{\pi} \int \mathrm{d}x\, \dot{\theta} = \dot{P}\,, \tag{2.66}$$

where $-e$ is the electronic charge. This is due to the continuity equation $j - e\, \partial n/\partial t = 0$, with the local number density, n, Equation (2.25). Thus $\sigma(\omega)$, Equation (2.65), is rewritten in terms of the Green function, $D(q, q'; i\omega_n)$, as follows

$$\sigma(\omega) = \frac{i\omega}{2} \left(\frac{e}{\pi}\right)^2 L D(0, 0; i\omega_n) \big|_{i\omega_n \to \omega - i0^+}, \tag{2.67}$$

where

$$D(q, q'; i\omega_n) = \int_{-\beta}^{\beta} \mathrm{d}\tau\, e^{i\omega_n \tau} \langle T_\tau \theta_q(\tau) \theta_{-q'}(0) \rangle, \tag{2.68}$$

and

$$\theta(x) = \frac{1}{\sqrt{L}} \sum_q e^{iqx} \theta_q. \tag{2.69}$$

In the case of quantum sine-Gordon Hamiltonian, Equation (2.57), treated in the self-consistent harmonic approximation, $D(q, q', i\omega_n)$ is given as

$$D(q, q', i\omega_n) = \delta_{q,q'} \frac{4C}{\omega_n{}^2 + \omega(q)^2}, \tag{2.70}$$

where $\omega(q)$ is defined by Equation (2.59) and then the frequency dependence of $\sigma(\omega)$ is

$$\sigma(\omega) = \frac{2e^2}{\pi^2} CL \frac{i\omega}{\omega_0{}^2 - \omega^2}, \tag{2.71}$$

with $\omega_0 \equiv \omega(q = 0)$. If $\omega_0 \neq 0$, i.e. in the presence of pinning of the phase, the static conductivity is vanishing, while if $\omega_0 = 0$,

$$\sigma(\omega) = \frac{2e^2}{\pi^2} CL \frac{1}{i\omega} = \frac{Ne^2}{m^*} \frac{1}{i\omega}, \tag{2.72}$$

where $N = 2k_F L/\pi$ is the total electron number and $m^* = \pi k_F/C$ is the effective mass. The divergence of $\sigma(\omega)$, in this case, Equation (2.72), is due to the sliding of the whole electronic system (sliding mode) [8].

The spin susceptibility, χ, per unit length is defined in terms of the magnetization, M, as

$$\chi = M/HL, \tag{2.73}$$

where H is the uniform magnetic field applied in the z-direction. By use of Equation (2.26), M is given by

$$M = -g\mu_B \sum_i \langle S_{iz} \rangle = -\frac{g\mu_B}{2\pi} \int \mathrm{d}x \, \langle \nabla\phi \rangle, \tag{2.74}$$

where g and $\mu_B = e/2mc$ are the g-factor and the Bohr magneton, respectively, and S_i is the spin matrix. On the other hand, the Zeeman energy, $\mathcal{H}_z$, is written as follows

$$\mathcal{H}_z = h \int \mathrm{d}x \, \nabla\phi, \tag{2.75}$$

$$h = g\mu_B H/2\pi. \tag{2.76}$$

Thus the first term of the Hamiltonian for the spin part, $\mathcal{H}_\sigma$, Equation (2.34) is modified by the magnetic field as follows (except some H-dependent constant),

$$A_\sigma(\nabla\phi)^2 \to A_\sigma(\nabla\phi + q_0)^2 \tag{2.77}$$

$$q_0 = h/2A_\sigma. \tag{2.78}$$

If $B = 0$ in Equation (2.34), the spatial variation of ϕ in the ground state is given so as to minimize Equation (2.77) i.e. $\langle\nabla\phi\rangle = -q_0$ so that χ is given by

$$\chi = (g\mu_B)^2/8\pi^2 A_\sigma. \tag{2.79}$$

This is the Pauli paramagnetism modified by interactions.

In the presence of B_σ the classical part of $\mathcal{H}_\sigma$ has the following form (through redefinition of 2ϕ by ϕ and the renormalization and change of sign of numerical factors)

$$E = \int \mathrm{d}x\ [A(\nabla\phi - q_0)^2 - B\cos\phi], \tag{2.80}$$

which has the typical energies to describe the commensurate-incommensurate (C–IC) transition [43, 44]. (In Equation (2.80) we assume $A > 0, B > 0, q_0 > 0$.)

The spatial variation of ϕ in the ground state is governed by the variational equation

$$2A\nabla^2\phi - B\sin\phi = 0, \tag{2.81a}$$

or

$$A(\nabla\phi)^2 + B\cos\phi = \text{constant} \equiv C. \tag{2.81b}$$

The constant, C, is to be determined so as to minimize E, Equation (2.80). If $C < |B|$, the variation of ϕ is bounded. On the other hand ϕ is steadily increasing as a function of x in the case of $C > |B|$ and the spatial extent, $2\pi/q$, for ϕ to vary 2π is given by

$$\frac{2\pi}{q} = \int_0^{2\pi} \mathrm{d}\phi\ \frac{\sqrt{A}}{\sqrt{C - B\cos\phi}}. \tag{2.82}$$

The ground state energy per unit length, ϵ, is

$$\epsilon = \frac{q}{2\pi}\int_0^{2\pi/q} \mathrm{d}x\ [A(\nabla\phi - q_0)^2 - B\cos\phi], \tag{2.83a}$$

$$= Aq_0^2\left[1 - \sqrt{2\pi}\ \frac{\sqrt{s}}{kK(k)} + s\left\{4\ \frac{E(K)}{k^2K(k)} + 1 - \frac{2}{k^2}\right\}\right], \tag{2.83b}$$

$$\equiv Aq_0^2\,[1 - \sqrt{2\pi}\,\sqrt{s}\,f(k,s)], \tag{2.83c}$$

where use is made of Equation (2.82) to derive Equation (2.83b) from Equation (2.83a) and $s = B/Aq_0{}^2$. In Equation (2.83b) $E(k)$ and $K(k)$ are defined by

$$E(k) = \int_0^{\pi/2} d\theta \sqrt{1 - k^2 \sin^2 \theta},$$

$$K(k) = \int_0^{\pi/2} d\theta \frac{1}{\sqrt{1 - k^2 \sin^2 \theta}},$$

and $k^2 = 2B/(C + B)$. The constant, C, or equivalently, k, is determined by the condition $\partial\epsilon/\partial k = 0$ or $\partial f/\partial k = 0$, $f(k, y)$ being defined in Equation (2.83c). The C–IC transition occurs when $\partial f/\partial k = 0$ is satisfied at $k = 1$, i.e. at $q_0 = Q_c$ where

$$Q_c = \left(\frac{8B}{A\pi^2}\right)^{1/2}. \tag{2.84}$$

The y $(\equiv Aq_0{}^2/B)$ dependence of q is schematically shown in Figure 6. The spatial variation of ϕ is given by [45, 46]

$$\sin\{\tfrac{1}{2}(2\phi - \pi)\} = \mathrm{sn}(x/dk; k), \tag{2.85}$$

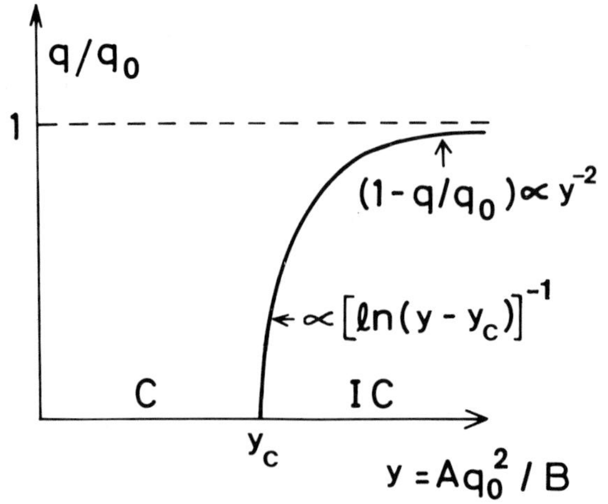

Fig. 6. The dependence of q defined by Equation (2.82) on $y = Aq_0{}^2/B$.

where $d = (2A/|B|)^{1/2}$ and $\mathrm{sn}(z; k)$ is a Jacobian elliptic function with the parameter k. Its explicit spatial variation in the case $q_0 \gtrsim Q_c$ is shown in Figure 7. As seen the spatial variation is not a straight line but has structures known as discommensurations (DC) [47], or a soliton lattice (see Section 3.6).

Since the magnetization, M, is given by q as $M/L = g\mu_B q/4\pi^2$, the strong nonlinearity

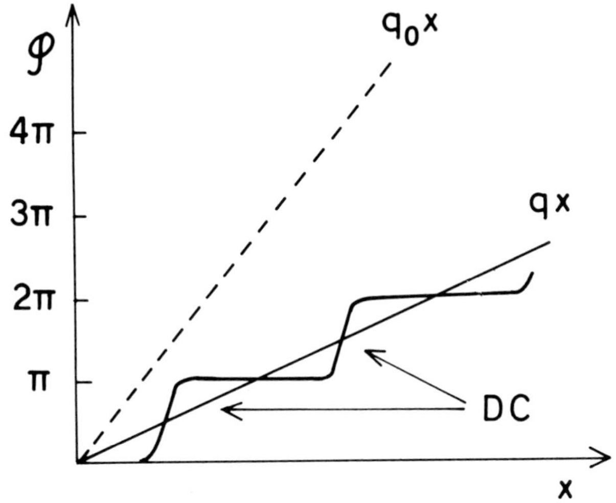

Fig. 7. Spatial variation of ϕ described by Equation (2.85).

of the spin susceptibility is expected. Detailed theoretical examinations on such non-linearity have been given [48] in the context of the experimental finding in $(TMTSF)_2PF_6$ [49, 50].

2.4. Three-Dimensionality and Phase Transition

So far we have discussed one-dimensional systems where there does not exist any phase transition at $T \neq 0$. In the presence of such three-dimensionality as interchain electron transfer or/and interchain interactions one can except a phase transition at finite temperature. In the present scheme the onset of true long range order can be viewed as the pinning of the phase variables to characterize each ordered state (see (2.46)). For the description of the onset of long range order the scheme originally proposed by Scalapino, Imry and Pincus [51] and by Efetov and Larkin [52] is most suited. In this scheme the large fluctuations intrinsic to one-dimensional systems are treated rigorously but the weak three-dimensionality is taken into account in the mean field approximation. This kind of approximation will be valid if the critical temperature, T_c, of the onset of long range order is much lower than the mean field temperature, T_M, below which the short range order (SRO) is present along each chain. This situation is schematically shown in Figure 8.

In order to illustrate what this scheme implies in the phase Hamiltonian approach we consider the case where the three dimensionality is due to the Coulomb interactions between electrons on different chains forming a square lattice, R_i, (Figure 9). This interchain interaction can be written as

$$\mathcal{H}' = \tfrac{1}{2} \sum_{i \neq j} \int dx \int dx' \, \rho_i(x) \rho_j(x') v_{ij}(x - x'),$$

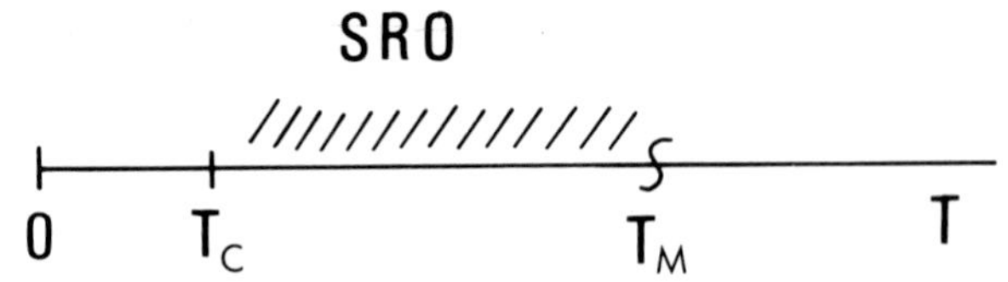

Fig. 8. Schematic representation of various regions in the phase transition of quasi-one-dimensional systems; T_C, the true three-dimensional transition temperature, and T_M, the mean field temperature.

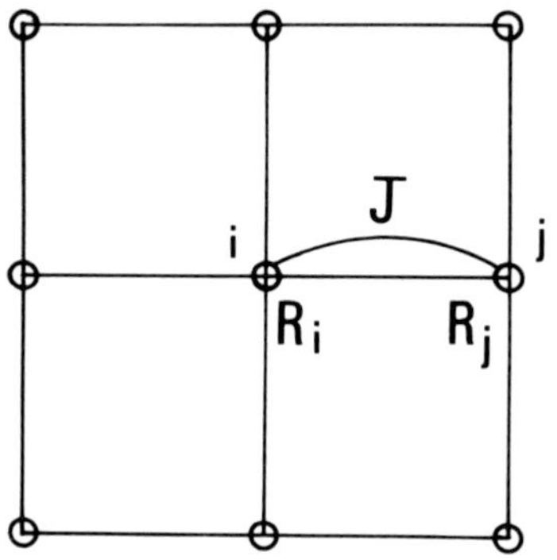

Fig. 9. Model lattice structure in the plane perpendicular to the chain axis.

where $\rho_i(x)$ is the electronic density on the i th chain and $v_{ij}(x - x')$ is given by

$$v_{ij} = \frac{e^2}{\epsilon_0} \frac{1}{\sqrt{(R_i - R_j)^2 + (x - x')^2}},$$

where ϵ_0 is the dielectric constant. By use of Equation (2.44) for each chain we obtain

$$\mathcal{H}' = \tfrac{1}{2} \sum_{i \neq j} J_{ij} \int_{-\infty}^{\infty} dx \cos \phi_i(x) \cos \phi_j(x) \cos (\theta_i(x) - \theta_j(x)), \tag{2.86}$$

where θ_i and ϕ_i are phase variables on the i-th chain and J_{ij} is given by

$$\begin{aligned} J_{ij} &= \frac{2e^2}{\epsilon_0 (\pi\alpha)^2} \int_{-\infty}^{\infty} dx \cos 2k_F x \frac{1}{\sqrt{(R_i - R_j)^2 + x^2}} \\ &= \frac{4e^2}{\epsilon_0 (\pi\alpha)^2} K_0 (2k_F |R_i - R_j|). \end{aligned} \tag{2.87}$$

where $K_0(z)$ is the modified Bessel function. In the evaluation of Equation (2.86) we ignore the contribution from the first term in Equation (2.44), since this term does not lead to the pinning of the phase variables. In deriving Equation (2.87) we assumed that the spatial variation of $\theta_i(x)$ is relatively weak in the sense of $|\nabla\theta_i|/2k_F \ll 1$.

The interchain coupling of the type of Equation (2.86) but without the spin degree of freedom, ϕ_i, has been employed in the discussion of the phase transition in the Peierls systems [53–63], where the quantum fluctuations can also be ignored due to the large effective mass (see Section 3). In this case the effective Hamiltonian is reduced to

$$E = \int \mathrm{d}x \left[\sum_i A(\nabla\theta_i)^2 + J \sum_{(ij)} \cos(\theta_i - \theta_j) \right], \tag{2.88}$$

where J_{ij} only between the nearest neighbor chains, (i, j), is retained due to the strong z-dependence of $K_0(z)$ once $z \gtrsim 1$. In the case of a square lattice of our interest, we can change J to $-J$ in Equation (2.88) since the origin of phase variables on one kind of sublattice (Figure 9) can be shifted by π. (Such a transformation is not possible on a triangular lattice, which needs a different treatment.)

In the mean field approximation to the interchain coupling, Equation (2.88) is written as

$$E = \sum_i \int \mathrm{d}x\, [A(\nabla\theta_i)^2 - \Delta \cos\theta_i], \tag{2.89}$$

where the order parameter Δ is given by

$$\Delta = J \sum_j{}' \langle\cos\theta_j\rangle = zJ\langle\cos\theta_i\rangle, \tag{2.90}$$

and z is the number of nearest neighbour chains ($z = 4$ for Figure 9). Equations (2.89) and (2.90) are self-consistent equations. The critical temperature, T_c, of the onset of long range order is obtained by linearizing the r.h.s. of Equation (2.90) in terms of Δ;

$$1 = (zJ/T_c) \int_{-\infty}^{\infty} \mathrm{d}x'\, K(x, x'), \tag{2.91}$$

$$K(x, x') = \int \mathrm{d}\theta\, e^{-\beta_c E_0} \cos\theta(x) \cos\theta(x') \Big/ \int \mathrm{d}\theta\, e^{-\beta_c E_0}, \tag{2.92}$$

$$E_0 = A \int \mathrm{d}x\, (\nabla\theta)^2, \tag{2.93}$$

where $\beta_c = 1/T_c$. and the suffix i of the chain is omitted for brevity. The evaluation of $K(x, x')$, Equation (2.92), can be made by the Fourier transformation of $\theta(x)$ as in Equation (2.69),

$$E_0 = \sum_k Ak^2 \theta_k \theta_{-k}, \tag{2.94}$$

$$\cos\theta(x) \cos\theta(x') \to \tfrac{1}{2} \operatorname{Re} \exp i[\theta(x) - \theta(x')]$$

$$= \tfrac{1}{2} \operatorname{Re} \exp\left[i \frac{1}{L} \sum_k \theta_k (e^{ikx} - e^{ikx'}) \right]. \tag{2.95}$$

In Equation (2.95) we ignored the term $\operatorname{Re} \exp i[\theta(x) + \theta(x')]$ since it results in vanishing contribution. Now the integration over the functional space of θ is performed over those of each Fourier component of θ_k, which is a complex variable,

$$\int d\theta \to \prod_{k>0} \int d\theta'_k \, d\theta''_k,$$

where θ'_k and θ''_k are the real and imaginary parts of θ_k. The resulting Gaussian integration can be performed exactly and we obtain

$$\begin{aligned} K(x, x') &= \tfrac{1}{2} \exp\left[-\frac{T}{L} \sum_{k>0} \frac{1 - \cos k(x - x')}{Ak^2}\right], \\ &= \tfrac{1}{2} \exp[-|x - x'|/\xi], \end{aligned} \tag{2.96}$$

where ξ is the correlation length given by

$$\xi = 4A/T. \tag{2.97}$$

Consequently Equation (2.91) implies $1 = zJ\xi/T_c$ or [57, 59]

$$T_c = (4AzJ)^{1/2}. \tag{2.98}$$

The physical reason for Equation (2.98) is simple: ξ is the characteristic length for the short range order in a chain and then the energy gain due to phase coherence between chain is of the order of $zJ\xi$, which should compete with the thermal fluctuation, T. The true long-range order will be established once $zJ\xi \sim T$.

Below T_c, Δ of Equation (2.90) is finite and is proportional to $(T_c - T)^{1/2}$ near T_c, which is the typical dependence in the mean field approximation. As the temperature is lowered the quantum fluctuations neglected in Equation (2.88) become important once T is comparable to the zone-boundary frequencies of phason; i.e. $T \lesssim v_\rho \alpha^{-1}$ or $v_\sigma \alpha^{-1}$. In this case the Hamiltonian is reduced to that of a quantum sine-Gordon equation of the type of Equation (2.57).

3. Phase Hamiltonian for the Peierls–Fröhlich State

3.1. PEIERLS–FRÖHLICH STATE

In the argument described in the preceding section we have implicitly replaced an underlying lattice by a rigid uniform background of positive charge. Here we examine explicitly its degrees of freedom and their coupling to electrons. Peierls [5] first pointed out that in one-dimensional conductors a metallic state is unstable in comparison to the insulating state, in which the lattice is distorted periodically with the wave number $2k_F$ associated with the CDW condensation with the same wave number, whereby a gap opens at the Fermi level in the electronic spectrum. This is because the gain in the electronic energy due to the gap formation overcomes the expense of the elastic lattice energy due to the distortion. Fröhlich [6] then argued that the CDW condensate can slide through a system without expenditure of energy, which was claimed at that time to be a possible origin of the superconductivity. It is now well-established that Fröhlich's superconducting mechanism does not work in actual conductors because the CDW condensate is pinned by impurity potential [8] and/or underlying lattice potential [8], and because its motion, even after depinning, suffers from frictions. Many interesting phenomena, however, associated with the Peierls–Fröhlich (PF) state have been observed in quasi-one-dimensional conductors, and they are extensively discussed in this volume.

Below, we first derive the phase Hamiltonian for the PF state in a clean one-dimensional conductor with k_F incommensurate to the reciprocal lattice vector. We then examine the effects of the underlying lattice (or commensurability) potential. Effects of the impurity potential will be explained in the next section in the context of Anderson localization. In this section we mostly ignore the mutual interactions, except in subsections 3.5 and 3.7.

3.2. Dynamics of Phonon Condensate

A model Hamiltonian describing electrons in a linear chain coupled to phonons is given by

$$\mathcal{H} = \sum_{k,s} \epsilon_k c_{ks}^{+} c_{ks} + \sum_{q} \omega_q \beta_q^{+} \beta_q + $$

$$+ \hat{g} \sum_{k,q,s} \frac{1}{\sqrt{2N\omega_q}} c_{k+qs}^{+} c_{ks} (\beta_q + \beta_{-q}^{+}), \tag{3.1}$$

where c_{ks}^{+} and β_q^{+} are creation operators for the electron and longitudinal phonon, respectively, $\hat{g}$ is the electron–phonon coupling constant. Here we neglect both interactions between electrons and the coupling between different chains. Because of a perfect nesting of the Fermi surface in one dimension, static lattice distortion with wave number $2k_F \equiv Q$ exist in the ground state, associated with the CDW condensation with the same wave number. Therefore we shall be interested in electrons with momenta $k \simeq \pm k_F$ and in phonons with wave number $q \sim \pm Q$, and rewrite Equation (3.1) as

$$\mathcal{H} = \mathcal{H}_0 + \omega_Q \sum_{|q|<<Q} \{\beta_{Q+q}^{+} \beta_{Q+q} + \beta_{-(Q+q)}^{+} \beta_{-(Q+q)}\} + $$

$$+ \frac{\hat{g}}{\sqrt{2N\omega_Q}} \left[\sum_{\substack{k>0,s \\ |q|<<Q}} a_{k-Q-q,s}^{+} a_{k,s} (\beta_{-(Q+q)} + \beta_{Q+q}^{+}) + \text{h.c.} \right], \tag{3.2}$$

where $\mathcal{H}_0$ is given by Equation (2.2) and we put $q \to Q + q$ and $\omega_{\pm(Q+q)} \simeq \omega_Q$. In the space-coordinate representation Equation (3.2) becomes

$$\mathcal{H} = \mathcal{H}_e(\phi_{ph}) + \mathcal{H}_{ph}, \tag{3.3}$$

$$\mathcal{H}_e(\phi_{ph}) = \sum_s \int dx\ \psi_s^+(x) \left\{ -iv_F \sigma_3 \frac{\partial}{\partial x} + \right.$$
$$\left. + \hat{g}\left[\sigma_- \phi_{ph}^*(x) e^{iQx} + \sigma_+ \phi_{ph}(x) e^{-iQx} \right] \right\} \psi_s(x), \tag{3.4}$$

$$\mathcal{H}_{ph} = \int dx\ \{\omega_Q{}^2\ |\phi_{ph}(x)|^2 + |\pi_{ph}(x)|^2\}. \tag{3.5}$$

Here we have introduced the spinor representation $\psi_s{}^+ = (\psi_{1s}^+, \psi_{2s}^+)$ with Pauli matrices $\sigma_i(\sigma_\pm = \sigma_1 \pm i\sigma_2)$. The phonon field $\phi_{ph}(x)$ and its conjugate momentum $\pi_{ph}(x)$ are defined by

$$\phi_{ph}(x) = \frac{1}{\sqrt{2\omega_Q N}} \sum_q\ e^{-iqx}(\beta_{Q+q}^+ + \beta_{-Q-q}), \tag{3.6}$$

$$\pi_{ph}(x) = i\sqrt{\frac{\omega_Q}{2N}} \sum_q\ e^{iqx}(\beta_{-Q-q}^+ - \beta_{Q+q}). \tag{3.7}$$

Note that the actual lattice distortion $d(x)$ is given by

$$d(x) = 2\ \mathrm{Re}\,\{\phi_{ph}(x) e^{-iQx}\}, \tag{3.8}$$

i.e. the phonon field ϕ_{ph} here represents the slowly varying part of the lattice distortion. The PF state is characterized by a macroscopic condensation either of the $\pm Q$ phonon mode ($\langle\phi_{ph}\rangle \neq 0$, i.e. a periodic lattice distortion) or by that of the CDW ($\langle O_{\mathrm{CDW}}^+\rangle \neq 0$, O_{CDW}^+ being defined in Equation (2.46a)). In equilibrium both are related by the condition

$$\left\langle \frac{\delta\mathcal{H}}{\delta\phi_{ph}^*} \right\rangle = \omega_Q{}^2 \langle\phi_{ph}\rangle + \hat{g} e^{iQx} \langle O_{\mathrm{CDW}}^+\rangle = 0. \tag{3.9}$$

Dynamical properties of the PF state are also characterized by dynamics either of the phonon condensate or of the CDW condensate. We will explain these two approaches separately and show that both approaches yield identical results in case where the coupling between lattice distortion and CDW is adiabatic.

Dynamics of the phonon condensate in the PF state was first discussed by Lee, Rice and Anderson [8]. Here we derive an effective Lagrangian $\mathcal{L}(\phi_{ph})$ following the mean field type analysis by Brazovskii and Dzyaloshinskii [64]. The method is to write the partition function, Z, of the system in the functional form of $\phi(x, \tau) \equiv e^{\mathcal{H}\tau} \phi_{ph}(x) e^{-\mathcal{H}\tau}$, thereby eliminating the electronic degrees of freedom. In other words, we derive the action for ϕ_{ph} in the Euclidian space. Neglecting a certain overall normalization constant we write

$$Z = \int D\phi_{ph}(x, \tau) \exp\left[-\int_0^\beta d\tau \, \mathscr{L}^{(0)}(\phi_{ph})\right] Z_e[\phi_{ph}(x, \tau)]$$

$$\equiv \int D\phi_{ph}(x, \tau) \exp\left[-\int_0^\beta d\tau \, \mathscr{L}(\phi_{ph})\right], \tag{3.10}$$

where $\mathscr{L}^{(0)}(\phi_{ph})$ represents the free phonon part

$$\mathscr{L}^{(0)} = \int dx \, \phi_{ph}^* \left\{-\frac{\partial^2}{\partial\tau^2} + \omega_Q{}^2\right\} \phi_{ph}, \tag{3.11}$$

and $Z_e[\phi_{ph}]$ is the electronic contribution to Z

$$Z_e[\phi_{ph}] = \mathrm{Tr}\left(T_\tau \exp\left[-\int_0^\beta d\tau \, \mathcal{H}_e(\phi_{ph})\right]\right), \tag{3.12}$$

with $\mathcal{H}_e(\phi_{ph})$ being given by Equation (3.4). In Equation (3.12) Tr is over the electronic states in the presence of the potential $\phi_{ph}(x, \tau)$. This formulation is therefore based on the adiabatic approximation: electrons respond instantaneously to the phonon field.

Now we introduce the equilibrium value and the amplitude and phase fluctuations of the field ϕ_{ph}, i.e. Δ, δ and χ, respectively, by

$$g\phi_{ph}(x, \tau) = \{\Delta + \delta(x, \tau)\}e^{i\chi(x,\tau)}. \tag{3.13}$$

Then by a rotation of the spinor field defined by

$$\tilde{\Psi}_s = U^+ \Psi_s U,$$

$$U = \exp\left(\frac{i}{2}\chi\sigma_3\right),$$

where $\tilde{\Psi}_s = (e^{ik_F x}\psi_{1s}, e^{-ik_F x}\psi_{2s})$. Equation (3.4) becomes

$$\mathcal{H}_e = \mathcal{H}_e^0 + \mathcal{H}_e{}^{\text{fluc}}, \tag{3.14}$$

$$\mathcal{H}_e^0 = \sum_s \int dx \, \tilde{\Psi}_s^+(x) \left\{-iv_F\sigma_3 \frac{\partial}{\partial x} + \Delta\sigma_1\right\} \tilde{\Psi}_s(x), \tag{3.15}$$

$$\mathcal{H}_e{}^{\text{fluc}} = \sum_s \int dx \, \tilde{\Psi}_s^+(x) \left\{\sigma(x, \tau)\sigma_1 + \frac{v_F}{2}\frac{\partial\chi}{\partial x}\sigma_0 + \frac{i}{2}\frac{\partial\chi}{\partial\tau}\sigma_3\right\} \tilde{\Psi}_s(x). \tag{3.16}$$

Without the fluctuation part, $\mathcal{H}_e{}^{\text{fluc}}$, Equation (3.12) is readily evaluated, and the self-consistent equation for Δ derived by the condition $\partial\mathscr{L}/\partial\Delta = 0$ becomes

$$\frac{\omega_Q^2}{\hat{g}^2} = 2T \sum_{\epsilon_n} \sum_{k>0} \frac{1}{\epsilon_n^2 + (v_F k)^2 + \Delta^2}, \tag{3.17}$$

which is a BCS-type gap equation ($\epsilon_n = (2n + 1)\pi T$). It results in a phase transition temperature T_{c0} and Δ at $T = 0$ given by

$$T_{c0} = 1.13\epsilon_F e^{-1/\lambda}, \tag{3.18a}$$

$$\Delta = 2\epsilon_F e^{-1/\lambda}, \tag{3.18b}$$

where

$$\lambda = \frac{\hat{g}^2}{\pi v_F \omega_Q^2} \tag{3.19}$$

is the dimensionless electron–phonon coupling constant.

If fluctuation effects are fully taken into account, the phase transition at a finite temperature disappears in one dimension. But the T_{c0} defined by Equation (3.18a) has the following physical significance: below around $0.25T_{c0}$ a short-range order of ϕ_{ph} is developed [65], so that we can practically regard the amplitude of ϕ_{ph} as being constant. The dynamics of the (quasi) phonon condensate are then analyzed by evaluating contributions of $\mathcal{H}_e^{\text{fluc}}$ perturbationally, regarding δ, $\partial\chi/\partial x$ and $\partial\chi/\partial\tau$ (but not necessarily χ itself) as small parameters. Their explicit evaluation is straightforward, by means of the thermal Green function technique [66], and the resulting Lagrangian becomes (after the analytical continuation $\tau \to it$)

$$\mathcal{L}(\delta, \chi) = \frac{1}{\pi v_F \lambda \omega_Q^2} \int \mathrm{d}x \left\{ -\left(\frac{\partial\delta}{\partial t}\right)^2 + a_T \lambda \omega_Q^2 \delta^2 + u_a^2 \left(\frac{\partial\delta}{\partial x}\right)^2 + \right.$$
$$\left. + \Delta^2 \left[-\left(\frac{\partial\chi}{\partial t}\right)^2 + u_p^2 \left(\frac{\partial\chi}{\partial x}\right)^2 \right] \right\}, \tag{3.20}$$

where the velocities of the phase and amplitude modes, u_p and u_a, are given by

$$u_p^2 = \frac{\lambda \omega_Q^2}{4\Delta^2} a_T v_F^2, \tag{3.21}$$

$$u_a^2 = \frac{b_T}{3a_T} u_p^2, \tag{3.22}$$

with

$$a_T = \pi T \sum_{\epsilon_n} \frac{\Delta^2}{(\epsilon_n^2 + \Delta^2)^{3/2}}, \tag{3.23a}$$

$$b_T = 3\pi T \Delta^2 \sum_{\epsilon_n} \frac{\epsilon_n^2}{(\epsilon_n^2 + \Delta^2)^{5/2}}. \tag{3.23b}$$

The above coefficients, a_T and b_T, are unity at $T = 0$ and weakly dependent on T at low temperatures. Note that the time-derivative terms in $\mathscr{L}^{(0)}$, Equation (3.11), appear directly in Equation (3.20), while the space-derivative terms are modified by the electron–phonon interaction (so that we have put $\omega_{Q+q} \simeq \omega_Q$ in Equation (3.2)). The electron–phonon interaction also reduces a gap in the amplitude mode ($\omega_Q \to \lambda^{1/2}\,\omega_Q$). The effective phase Hamiltonian $\mathcal{H}^0_{\text{eff}}$ is thus given by

$$\mathcal{H}^0_{\text{eff}} = \pi v' \int \mathrm{d}x \left\{ p^2 + \frac{1}{4\pi^2} \left(\frac{u_p}{v'} \right)^2 \left(\frac{\partial \chi}{\partial x} \right)^2 \right\}, \tag{3.24}$$

where $p = \partial \mathscr{L}/\partial \dot{\chi}$ and $v' = u_p{}^2/v_F$. The phase Hamiltonian Equation (3.24) has been derived by Pietronero *et al.* [9] and Fukuyama [10] by an intuitive argument.

The excitation spectra of the amplitude and phase modes derived from Equation (3.20) agree with those obtained by Lee, Rice and Anderson [8]. (We note that there is a difference of numerical factors between Equation (3.20) and the corresponding Lagrangian originally derived Brazovskii and Dzyaloshinskii [64], who do not consider spin.) Kurihara [67] and Takano [68] have also derived a similar Lagrangian to Equation (3.20), microscopically, but by separating $\hat{g}\phi_{ph}$ into its real and imaginary parts in the evaluation of the electronic contribution, instead of using the direct introduction of the amplitude and phase of $\hat{g}\phi_{ph}$ (Equation (3.13)). They have also calculated such higher-order terms in $\mathscr{L}(\delta, \chi)$ as δ^3, $\delta\chi^2$ and $\delta^2\chi^2$.

3.3. Dynamics of CDW Condensate

We examine here dynamical properties of the PF state by a complementary approach, namely, the boson representation of the electronic state discussed in Section 2.1. By this approach, interactions between electrons are easily incorporated, as we will see, and $\mathcal{H}_e(\phi_{ph})$ of Equation (3.4) is written as

$$\mathcal{H}_e(\phi_{ph}) = \mathcal{H}_e + \mathcal{H}_{e-p}, \tag{3.25}$$

$$\mathcal{H}_e = \int \mathrm{d}x\ \{A_\rho(\nabla\theta)^2 + C_\rho p^2 + A_\sigma(\nabla\phi)^2 + C_\sigma M^2\}, \tag{3.26}$$

$$\mathcal{H}_{e-p} = \tilde{g} \int \mathrm{d}x \cos\phi\{e^{-i\theta}\,\phi_{ph}(x) + e^{i\theta}\,\phi^*_{ph}(x)\}, \tag{3.27a}$$

$$= -\frac{2}{\pi\alpha} \int \mathrm{d}x\ \Delta \cos\phi \cos(\theta - \chi), \tag{3.27b}$$

Hamiltonian (3.26) is just Equation (2.32), but without the B_ρ and B_σ terms (neglected for the moment). We made use of Equation (2.4) and (2.5) to derive Equation (3.27a) where $\tilde{g} = \hat{g}/\pi\alpha$ and then we put

$$\hat{g}\phi_{ph} = -\Delta\, e^{i\chi} \tag{3.28}$$

in Equation (3.27a) instead of Equation (3.13).

To discuss the physical significance of the term $\cos(\theta - \chi)$ in Equation (3.27b) it is very important to recognize whether the underlying lattice distortion is static (rigid) or moving with the CDW [26–28]. In case of the static distortion we fix Δ and χ. For simplicity let us consider the case with a uniform Δ and $\chi = 0$. Then Equation (3.25) describes a coupled sine-Gordon system of θ and ϕ, and both modes have an energy gap. At first glance, the gap may involve the cut-off paramater α, but we can show that in the SCHA scheme described in Section 2.2 the renormalized gap in the non-interacting limit coincides with the PF gap 2Δ. That is to say, the excitation spectrum, which corresponds to Equation (2.59), reduces to

$$\omega^2(q) = v_F{}^2 q^2 + 4\Delta^2. \tag{3.29}$$

for both θ and ϕ modes in the present problem. Thus in case of rigid lattice distortion θ and ϕ represent electron-hole pair excitations across the PF gap. It is worth pointing out here that the ground state energy evaluated by the SCHA scheme does not yield correct results. It appears to be due to failure of the SCHA scheme in evaluating shifts in individual energy levels of electrons caused by the lattice distortion, which are supposed to overcompensate the increase of the lattice elastic energy. We emphasize, however, that SCHA is in general a powerful scheme in discussing excitation spectra.

To investigate the case of moving lattice distortion we write down the equations of motion for θ, ϕ and χ deduced from the total Hamiltonian (3.3) as

$$\ddot{\theta} - 4C_\rho A_\rho \nabla^2 \theta + \frac{4C_\rho}{\pi\alpha} \Delta \cos\phi \sin(\theta - \chi) = 0, \tag{3.30}$$

$$\ddot{\phi} - 4C_\sigma A_\sigma \nabla^2 \phi + \frac{4C_\sigma}{\pi\alpha} \Delta \sin\phi \cos(\theta - \chi) = 0, \tag{3.31}$$

$$\Delta\ddot{\chi} - \frac{\hat{g}^2}{\pi\alpha} \cos\phi \sin(\theta - \chi) = 0. \tag{3.32}$$

Here we consider a situation where the amplitude Δ is static. The adiabatic motion of the CDW condensate coupled to the lattice distortion is represented by the condition

$$\theta = \chi + \chi_0, \tag{3.33}$$

where χ_0 is a certain constant. Then from Equations (3.30) and (3.32) we obtain the equation of motion for θ (or χ) as

$$\left(1 + \frac{4C_\rho \Delta^2}{\hat{g}^2}\right)\ddot{\theta} - 4C_\rho A_\rho \nabla^2 \theta = 0. \tag{3.34}$$

In the weak coupling limit ($\lambda \ll 1$) of non-interacting electrons, Equation (3.34) becomes identical to the equation of motion derived from Equation (3.24). Thus we conclude that the phase Hamiltonian, Equation (3.24), with $\chi \sim \theta$ properly describes the dynamics of the CDW condensate adiabatically coupled to the lattice distortion [27, 69–72].

Note that in most of the phenomena of interest (low frequency response and not very close to the PF transition temperature), the dynamics of the CDW amplitude can be neglected.

3.4. COMMENSURABILITY ENERGY

If the system is nearly commensurate with the periodicity of the lattice with degree M, (i.e. $G/2k_F \cong M$) there exists an extra energy called the commensurability energy, E_c, as has been noted by Lee, Rice and Anderson [8]. This energy is due to the $2k_F$ scattering of electrons by lattice distortion successively through the states whose momenta are near $(2l-1)k_F (1 \leq l \leq M)$ combined with the Umklapp scattering. In order to evaluate E_c explicitly we consider the tight-binding model (instead of the continuum model of Section 3.2) [70]

$$\mathcal{H}_0 = -\sum_{n,s} t_{n+1,n}(c^+_{n+1,s}c_{ns} + \text{h.c.}), \tag{3.35}$$

with

$$t_{n+1,n} = t_0 - \hat{\gamma}(u_{n+1} - u_n), \tag{3.36}$$

where $c_{n,s}$ and u_n are related to the corresponding quantities in the continuum approximation as follows:

$$c_{n,s}a^{-1/2} = \psi_{1s}(x) + \psi_{2s}(x), \tag{3.37a}$$

$$u_n = \frac{1}{2i}\, u(x)\exp(i\chi - iQna) + \text{c.c.} \tag{3.37b}$$

Here a is the lattice constant, $Q = 2k_F$ and $u(x)\exp(i\chi)$ $(= \phi_{ph}(x)/2i)$ is the slowly varying part of lattice distortion. The coupling constant, $\hat{\gamma}$, in this tight-binding model is related to $\hat{g}$ in the continuum model by

$$\hat{g} = 2i\hat{\gamma}\,[\sin k_F a - \sin(k_F + Q)a]\,.$$

Note that the factor i connecting the corresponding quantities in the two models is irrelevant, since it simply introduces a constant shift in the phase.

In the present model E_c is given as follows, [28],

$$E_c = -\frac{a}{2}\, g_M \sum_s \int \mathrm{d}x\; [u^{M-1}\exp[i(M-1)(\chi - Qx) + iGx]\, \psi^+_{2s}\, \psi_{1s} + \text{h.c.}], \tag{3.38}$$

where g_M is

$$g_M = \frac{2}{a}\,(-2\hat{\gamma}\sin k_F a)^{M-1}\cos 2k_F a \times$$
$$\times \prod_{1<l<M} \cos 2lk_F a / |\epsilon_{k_F} - \epsilon_{(2l-1)k_F}|. \tag{3.39}$$

Here we noted that the energy band of electrons, $\epsilon(k)$, in this case is given by $\epsilon(k) = -2t_0 \cos ka$.

In terms of the phase variables, θ and ϕ, Equation (3.38) in the commensurate system ($MQ = G$) reduces to

$$E_c = -\tilde{g}_M \int \mathrm{d}x\, u^{M-1} \cos(\theta + (M-1)\chi) \cos\phi, \tag{3.40}$$

where $\tilde{g}_M = g_M/\pi\alpha$.

3.5. Solitons

The total Hamiltonian describing a one-dimensional electron-lattice system with M-commensurability (i.e. $G/2k_F = M$) consists of Equations (2.32), (3.5), (3.27a) and (3.40). In terms of our phase variables representing the electronic state, θ and ϕ, and the amplitude and phase of the slowly varying part of the lattice distortion, u and χ, this Hamiltonian is explicitly given by

$$\begin{aligned}\mathcal{H} = &\int \mathrm{d}x\, [A_\rho(\nabla\theta)^2 + C_\rho P^2 + \delta_{M,2} B_\rho \cos 2\theta + \\ &+ A_\sigma(\nabla\phi)^2 + C_\sigma M^2 + B_\sigma \cos 2\phi + \\ &+ \{\hat{g}u \cos(\theta - \chi) - \tilde{g}_M u^{M-1} \cos(\theta + (M-1)\chi)\} \cos\phi + K' u^2], \end{aligned} \tag{3.41a}$$

$$\equiv E[\theta, \phi, u, \chi] + \int \mathrm{d}x\, [C_\rho P^2 + C_\sigma M^2], \tag{3.41b}$$

where $E[\theta, \phi, u, \chi] \equiv E_T$ is the classical potential energy and $K' \simeq (2\omega_Q)^2$.

Before a systematic analysis of the full Hamiltonian (3.41a), let us consider the simplest situation, namely $\phi(=0)$ and $u(=u_0)$ are fixed, $B_\rho = 0$, and with the adiabatic condition Equation (3.33) in the classical approximation. Then Equation (3.41a) reduces to

$$\mathcal{H}' = \frac{A}{2} \int \mathrm{d}x\, \{\dot{\theta}^2 + c_0{}^2 (\nabla\tilde{\theta})^2 + 2\omega_M{}^2 (1 - \cos\tilde{\theta})\} + \text{const.} \tag{3.42}$$

where $\tilde{\theta} = M\theta$, $A = (2M^2 C_\rho)^{-1}$, $c_0{}^2 = 4C_\rho A_\rho$, $\omega_M{}^2 = 2M^2 C_\rho \tilde{g}_M u^{M-1}$, and $\dot{\theta} = 2C_\rho P$ is used. (It should be noted that the non-linear term similar to that in Equation (3.42) results also from the interchain coupling in the IC system treated in the mean field approximation as shown in Equation (2.89). However this approximation to the three-dimensionality is valid for low lying excitations only if the system consist of different kinds of chains.) Equation (3.42) is the well-known sine-Gordon Hamiltonian which bears nonlinear solitary waves, i.e., solitons [73]. Solitons are of current interest in various fields of condensed-matter physics [74]; in particular, in connection with the transport properties of the CDW condensate which is discussed elsewhere in this volume and below in Sections 5 and 6.

The Hamiltonian, Equation (3.42), has degenerate, uniform ground states $\tilde{\theta}(x) = 2\pi n$, n being an integer. The linear (small amplitude) excitation spectrum has a gap ω_0, in contrast to the phase mode (phason) in the incommensurate PF state as seen from Equation (3.24). Solitons are nonlinear (large amplitude) excitations connecting these discrete ground states. The soliton in the classical approximation is explicitly given by

$$\tilde{\theta}_s(x, t) = 4 \tan^{-1} \{\pm\exp(\gamma_R \omega_0 (x - vt)/d)\}, \tag{3.43}$$

where v is the velocity, $\gamma_R = (1 - v^2/c_0{}^2)^{-1/2}$, and + (–) indicates soliton (antisoliton). The soliton is localized in space within a range of the order $2d$, where $d = c_0/\omega_0$. Its formation (excitation) energy E_s at $v = 0$ is given by

$$E_s = 8A\omega_0 c_0. \tag{3.44}$$

Now let us return to the full Hamiltonian (3.41). In the classical approximation the ground states are given as configurations to minimize E_T with respect to all variables, θ, ϕ, u and χ. In the present model of Equation (3.41) the variational equations for u and χ result in equations relating u and χ with θ and ϕ (as we will see explicitly later), i.e., $E[\theta, \phi, u(\theta, \phi), \chi(\theta, \phi)] \equiv V[\theta, \phi]$. Thus the ground states can be specified in the plane of θ and ϕ. Due to the trigonometric dependence on θ and ϕ of the energy, E_T, there exists a discrete degeneracy in the ground state as in the simpler case of Equation (3.42). The possible types of solitons, which are topological excitations connecting between these degenerate ground states, can be described as trajectories in the plane of these phase variables as x is varied from $-\infty$ to ∞. These trajectories can be given by the potential, $V[\theta, \phi]$, since solitons are variational solutions. Besides its formation energy, such as Equation (3.44), a soliton possesses its own charge and spin, Q and S, which are given by [26]

$$Q = -\frac{e}{\pi} [\theta(\infty) - \theta(-\infty)], \tag{3.45}$$

$$S = \frac{1}{2\pi} [\phi(\infty) - \phi(-\infty)], \tag{3.46}$$

These equations naturally result from Equations (2.25) and (2.26). In a quantum-mechanical treatment Equations (3.45) and (3.46) still hold if the phase, $\theta(x)$ and $\phi(x)$, are understood as those of the classical parts, $\theta_{cl}(x)$ and $\phi_{cl}(x)$ around which quantum fluctuations, $\hat{\theta}(x)$ and $\hat{\phi}(x)$ are present, i.e. $\theta(x) = \theta_{cl}(x) + \hat{\theta}(x)$ and $\phi(x) = \phi_{cl}(x) + \hat{\phi}(x)$. Ho [75] and Horovitz [76] derived equations similar to Equation (3.45) and (3.46) after the introduction of the order parameters of CDW and SDW. In their treatment θ and ϕ correspond to the phases of these order parameters.

In the following we will discuss the incommensurate, dimerized and trimerized cases separately. [26, 28]

(i) *Incommensureate case*

In this case we can ignore the Umklapp energy in E_T and take $u(x) > 0$. In the ground state θ always follows χ, which is arbitrary. Consequently θ can not be pinned. There exists, however, a discrete degeneracy in ϕ; i.e. $\phi = m\pi (m = 0, \pm 1, \pm 2, \ldots)$ for $B_\sigma < \tilde{g}^2/8K'$ and $\phi = \pi/2 + m\pi$ for $B_\sigma > \tilde{g}^2/8K'$. This is due to the fact that the variation of E_T with respect to u results in

$$2K'u + \tilde{g} \cos \phi = 0, \tag{3.47}$$

and the substitution of this relation into E_T yields

$$V[\theta, \phi] = \int dx \left[A_\rho (\nabla\theta)^2 + A_\sigma (\nabla\phi)^2 + B \cos 2\phi - \frac{g^2 \cos^2 \phi}{4K'} \right]. \tag{3.48}$$

Thus possible solitons in this case are [77]

$$Q = 0, \qquad S = \pm\tfrac{1}{2}. \tag{3.49}$$

(ii) *Dimerized Case* ($M = 2$)

In the case of half-filled band, $E_T \equiv E_2$ is written as follows

$$E_2 = \int dx\, [A_\rho (\nabla\psi)^2 + A_\sigma (\nabla\phi)^2 - B_\sigma \cos 2\psi + B_\sigma \cos 2\phi + \\ + K'u^2 - 2\tilde{g}u \cos \psi \cos \phi], \tag{3.50}$$

where we put $\chi = \pi/2$ so that $u(x)$ represents an amplitude of the staggered displacement u_n, and $\psi = \theta + \pi/2$, and $\tilde{g}_2 \equiv \tilde{g}$. Varying with respect to $u(x)$, we obtain

$$u(x) = \frac{\tilde{g}}{K'} \cos \psi \cos \phi. \tag{3.51}$$

Thus the effective Hamiltonian for ϕ and ψ is given as follows

$$V_2 [\psi, \phi] = \int dx\, [A_\rho (\nabla\psi)^2 + A_\sigma (\nabla\phi)^2 + U_2 (\psi, \phi)], \tag{3.52}$$

$$U_2 (\psi, \phi) = -\frac{\tilde{g}^2}{K'} \cos^2 \psi \cos^2 \phi - B_\rho \cos 2\psi + B_\sigma \cos 2\phi. \tag{3.53}$$

It is clear that ψ and ϕ in the ground state are spatially constant. They are given by the minimum points of $U_2 (\psi, \phi)$. It is seen that the ground state is dimerized under the following conditions,

$$\frac{2K'B_\sigma}{\tilde{g}^2} < 1, \quad \frac{2K'B_\rho}{\tilde{g}^2} > -1, \quad \text{and} \quad \frac{2K'}{\tilde{g}^2} (B_\sigma - B_\rho) < 1. \tag{3.54}$$

The magnitude of the bond alternation u_0 in the ground state and the energy E_0 per unit length are given by

$$u_0 = \tilde{g}/K', \tag{3.55}$$

$$E_0 = B_\sigma - B_\rho - \tilde{g}^2/K'. \tag{3.56}$$

Possible soliton excitations from this ground state are described as solutions of the following simultaneous equations.

$$A_\rho \nabla^2 \psi - B_\rho \sin 2\psi - \frac{\tilde{g}^2}{K'} \sin \psi \cos \psi \cos^2 \phi = 0, \tag{3.57}$$

$$A_\sigma \nabla^2 \phi + B_\sigma \sin 2\phi - \frac{\tilde{g}^2}{K'} \sin \phi \cos \phi \cos^2 \psi = 0. \tag{3.58}$$

Various types of solitons are characterized by the trajectories in the $\theta - \phi$ plane as x is varied from $-\infty$ to ∞, which is schematically shown in Figure 10. In this figure dots

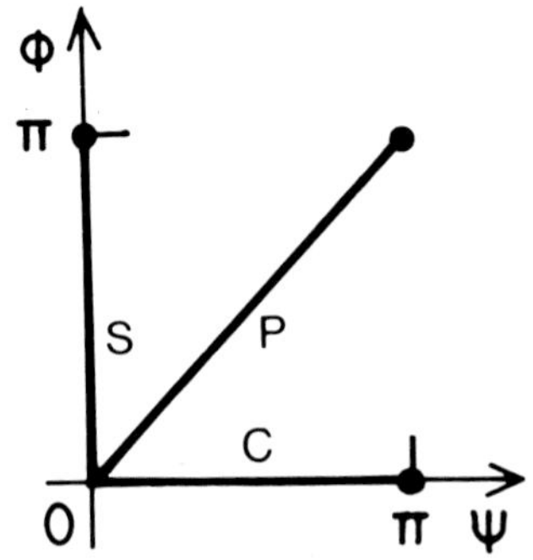

Fig. 10. Classifications of solitons in the dimerized system.

correspond to the configuration in the ground state. From this figure it is evident that the fundamental types of solitons are those denoted as C, S, and P;

C(Charge Soliton)	:	$Q = \pm e$,	$S = 0$,
S(Spin Soliton)	:	$Q = 0$,	$S = \pm 1/2$,
P(Polaron)	:	$Q = \pm e$,	$S = \pm 1/2$.

These charge and spin solitons were discussed by Su, Schrieffer and Heeger [78. 79] and independently by Rice [80] based on the tight-binding model of Equations (3.35) and (3.36) to interpret various interesting properties of polyacetylene. Later Takayama, Lin-Liu and Maki [81] and Horovitz [82] found the exact solutions for these solitons in a continuum version of their model. Furthermore, the polaron has been observed in dynamical computations by Su and Schrieffer [83], and been analytically examined in a continuum limit by Campbell and Bishop [84] and by Brazovskii [85, 86]. In these investigations, however, the mutual Coulomb interaction between electrons is not explicitly considered. By the phase Hamiltonian approach described above [28], on the other hand, we could easily take into account such interaction effects in the most general form, and can at the same time obtain a unified description for various solitons. In this respect it is to be noted that, though the net Q and/or S is not changed, the trajectories in θ–ϕ plane are not necessarily straight in the presence of mutual interactions. For some choice of interaction paramaters the solitons of the type shown in Figure 11 can also be expected. These are termed charge (solid line) and spin (broken line) bisolitons.

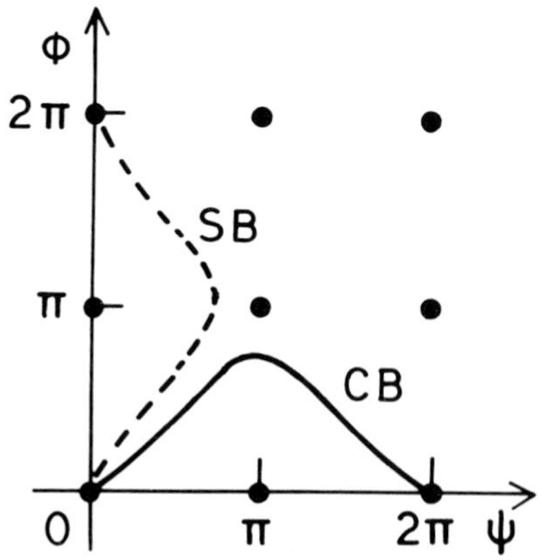

Fig. 11. Trajectories of charge and spin bisolitons.

(iii) *Trimerized case* ($M = 3$)

For a third-filled band, $E_T \equiv E_3$ is written as follows.

$$E_3 = \int dx \; [A_\rho(\nabla\theta)^2 + A_\sigma(\nabla\phi)^2 + B_\sigma \cos 2\phi + K'u^2 + + \{\tilde{g}u \cos(\theta - \phi) - \tilde{g}_3 u^2 \cos(\theta + 2\chi)\} \cos\phi] . \tag{3.59}$$

The local minimum condition for E_3 with respect to u and χ yields

$$v \equiv u \sin\chi = \frac{\tilde{g}}{2K'} \cos\phi \sin\theta \; [2\kappa \cos\theta \cos\phi - 1] , \tag{3.60}$$

$$w \equiv u \cos\chi = -\frac{\tilde{g}}{2K'} \cos\phi \; [\cos\theta + \kappa \cos\phi \cos 2\theta] , \tag{3.61}$$

where v and w are approximated by keeping terms up to first order of small parameter $\kappa = \tilde{g}_3/K'$. Substituting v and w into E_3, we obtain effective Hamiltonian for θ and ϕ as

$$V_3\,[\theta, \phi] = \int dx \; [A_\rho(\nabla\theta)^2 + A_\sigma(\nabla\phi)^2 + U_3(\theta, \phi)] , \tag{3.62}$$

$$U_3(\theta, \phi) = -\frac{\tilde{g}^2}{4K'} \cos^2\phi \; [1 + \kappa \cos\phi \cos 3\theta] + B_\sigma \cos 2\phi . \tag{3.63}$$

The ground states are trimerized in the following region

$$\frac{8K'B_\sigma}{\tilde{g}^2} < (1 + \kappa). \tag{3.64}$$

As in the case of the dimerized system, various possible types of soliton excitations can be expected and they are schematically shown in Figure 12. The shapes of solitons are determined by the following equations.

$$A_\rho \nabla^2 \theta - \frac{3}{8} \frac{\tilde{g}^2}{K'} \kappa \cos^3 \phi \sin 3\,\theta = 0, \tag{3.65}$$

$$A_\sigma \nabla^2 \phi + B_\sigma \sin 2\phi - \frac{\tilde{g}^2}{16K'} \sin 2\,\phi\,(2 + 3\kappa \cos \phi \cos 3\theta) = 0. \tag{3.66}$$

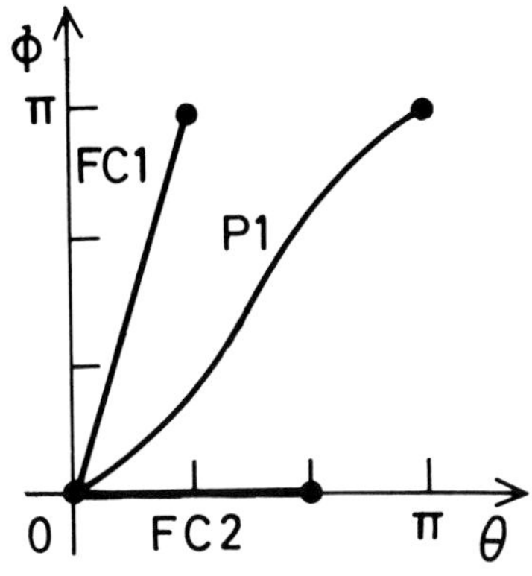

Fig. 12. Classification of solitons in the trimerized system.

The fundamental types of solitons classified typically as $FC1$, $FC2$, P in Figure 12 have following features;

$FC1$(Fractional Charge Soliton) : $Q = \pm e/3$, $S = \pm\frac{1}{2}$,

$FC2$(Fractional Charge Soliton) : $Q = \pm 2e/3$, $S = 0$,

P(Polaron) : $Q = \pm e$, $S = \pm\frac{1}{2}$.

These solitons with fractional charges in trimerized state without interactions between electrons have been discussed based on the tight-binding model by Su and Schrieffer [78, 87] and on its continuum version by Gammel and Krumhansl [88].

3.6. Soliton Lattice

Next we examine the incommensurate PF state which is nearly commensurate, i.e. $2k_F \cong G/M$. For simplicity we consider here only the charge degree of freedom in the presence of static lattice distortion. The effective phase Hamiltonian derived from Equations (3.5), (3.26), (3.38) becomes [89]

$$\mathcal{H}^c_{\text{eff}} = \pi v' \int \mathrm{d}x \left[p^2 + \frac{1}{4\pi^2 v'^2} \times \right.$$

$$\left. \times \left\{ u_p{}^2 (\nabla\theta - q_0)^2 + \frac{2\omega_M{}^2}{M^2} (1 - \cos M\theta) \right\} \right], \tag{3.67}$$

instead of $\mathcal{H}^0_{\text{eff}}$ of Equation (3.24). Here $q_0 = 2k_F - G/M$ and ω_M, already appeared in Equation (3.42), is given in this case by $\omega_M{}^2 = \omega_Q{}^2 (\Delta/\epsilon_F)^{M-2}$. The ground state is determined by the classical part of Equation (3.67). The latter is essentially identical to Equation (2.80), which has already been discussed in Section 2.3. The stationary solution of Equation (3.67) is given by Equation (2.85) with $2\phi = M\theta$ and $d = u_p/\omega_M$ in this case. For a small q_0 this solution, as shown schematically in Figure 7, is regarded as a periodic array of solitons, i.e. a soliton lattice. Its period l is given by Equation (2.82), i.e. $l = 2\pi/q = 2dkK(k)$, where $K(k)$ is the complete elliptic integral with a parameter k.

It is worth noting here the difference between q_0 and q. In the susceptibility problem discussed in Section 2.3, q_0 is a given parameter proportional to an external magnetic field (Equation (2.78)), while q, or equivalently a parameter k (or a constant C in Equation (2.81b)), is determined so as to minimize the energy of Equation (2.80). In the present case the incommensurability of $2k_F$ to G, i.e., a nonvanishing q_0, is determined by the charge transfer between the band which undergoes the PF condensation and the remaining bands (which are often called as a charge reservoir). The degree of the charge transfer is in turn determined so as to minimize the total energy involving also the energy of the charge reservoir and then the problem is not determined within the band which undergoes the PF transition. What is observed directly in experiments on the incommensurate PF state is q. The nonvanishing parameter q_0 simply indicates that the charge transfer yields an incommensurate k_F. The situation is more clearly seen in the problem of a doped polyacetylene. Since there is no charge reservoir in this system, q is directly related to the excess charge introduced by doping through Equation (2.44) and (2.82) [69].

3.7 SPIN-PEIERLS STATE

In Section 3.5 we have emphasized that by the present phase Hamiltonian approach we can easily take into account the mutual Coulomb interaction. In the argument there, however, the mutual interaction is assumed to be weak and then the charge and spin degrees of freedom are treated on an equal footing. In the case of very strong repulsive interactions, however, the charge fluctuations are usually suppressed. Especially, if the band is half-filled, the low lying excitations are exclusively of spin origin. This is seen most easily in the Hubbard model

$$\mathcal{H} = \mathcal{H}_0 + U \sum_i n_{is} n_{i-s}, \tag{3.68}$$

where $\mathcal{H}_0$ is given by Equation (3.35) and $n_{is} = c_{is}^+ c_{is}$. If $U \gg t_0$ in the half-filled band, the low lying excitations of Equation (3.68) are described by the following Hamiltonian of Anderson's kinetic exchange interactions;

$$\mathcal{H} = \sum_n J_{n+1,n} S_{n+1} S_n, \tag{3.69}$$

where $J_{n+1,n} = 4t_{n+1,n}^2/U$, and the problem is reduced to that of a spin-Peierls transition [31]. Below we explain an alternative phase Hamiltonian approach to this problem [29, 30].

(a) *Undistorted lattice*

If the lattice is undistorted, $J_{n+1,n}$ is uniform in space, $J_{n+1,n} = J$. The Hamiltonian is conveniently transformed into that of the spinless fermions by use of the Jordan–Wigner transformation [31];

$$S_l^+ = \exp\left[-i\pi \sum_j^{l-1} a_j^+ a_j\right] a_l^+ = (S_l^-)^+, \tag{3.70a}$$

$$S_l^z = -\tfrac{1}{2} + a_l^+ a_l, \tag{3.70b}$$

$$S_l^+ S_{l+1}^- = a_l^+ a_{l+1}. \tag{3.70c}$$

The resulting Hamiltonian is

$$\mathcal{H} = \sum \epsilon(k) a_k^+ a_k + \frac{1}{N} \sum V(q) a_{k+q}^+ a_{k'-q}^+ a_{k'} a_k, \tag{3.71}$$

where

$$\epsilon(k) = J(\cos ka - 1), \tag{3.72a}$$

$$V(q) = J \cos qa. \tag{3.72b}$$

Here a is the lattice spacing and the energy band of this spinless fermion is half-filled, i.e. $k_F a = \pi/2$. The second term of the r.h.s. of Equation (3.71) can be treated as in Section 2 by taking the scattering near or across the Fermi surface into account; i.e. $q \simeq 0$ or $|q| \simeq Q$ and $|q| \simeq 2Q$ (Umklapp scattering).

By repeating the argument of Section 2, we immediately see that Equation (3.71) can be transformed to the phase Hamiltonian, and the result is [29, 30]

$$\mathcal{H} = \int \mathrm{d}x\ [A(\nabla\phi)^2 + CM^2 - D\cos 2\phi], \tag{3.73}$$

$$A = \frac{Ja}{8}, \tag{3.74}$$

$$C = \frac{\pi^2}{2} Ja, \tag{3.75}$$

$$D = \frac{\pi^2 J}{8a}. \tag{3.76}$$

Here the cut-off parameter, α, is chosen as a/π. The last term in r.h.s. of Equation (3.73) is due to the Umklapp scattering [91]. The results for A and C, Equation (3.74) and (3.75), already take into account of the corrections caused by the linearization of the energy spectrum, $\epsilon(k)$, Equation (3.72a), and have been adjusted along the procedures proposed by Cross and Fisher [92] so that Equation (3.73) yields the known exact results. The phase variable, $\phi(x)$, is related to the spin operator, $S_z(x)$, by

$$S_z(x) = \frac{1}{a} \cos(2k_F x + \phi(x)) + \frac{1}{2\pi} \nabla\phi(x). \tag{3.77}$$

The Umklapp term will be treated in SCHA, introduced in Section 2.2, and the result of this approximation is that for the values of A and C of Equation (3.74) and (3.75), $\bar{D} \equiv D \exp[-2 \langle \phi^2 \rangle] = 0$, i.e. the quantum fluctuations destroy the last term in Equation (3.73) and then there is no preference for any particular value of phase, ϕ, resulting in degeneracy with respect to its value, Reflecting this degeneracy, the spectrum of the excitation, which is the spin wave, does not have a gap.

It is to be noted that $\phi = 0$ corresponds to the Néel state, as may be seen from Equation (3.77). (On each lattice site, R_n, $2k_F R_n$ is a multiple of π.) On the other hand, $\phi = \pi/2$ corresponds to the non-magnetic state. Then the degeneracy in ϕ implies that the ground state of a spin 1/2 Heisenberg antiferromagnet can have both aspects of the Néel state and the non-magnetic (singlet) state, as stressed by Anderson [93]. The present phase Hamiltonian approach reveals this fact very clearly [30].

(b) *Distorted lattice*

In the presence of the coupling between spin and lattice distortion, we write $J_{n+1,n}$ in Equation (3.69) as

$$J_{n+1,n} = J\left[1 + \lambda \frac{u_{n+1} - u_n}{2a}\right], \tag{3.78a}$$

$$u_n = (-)^n u. \tag{3.78b}$$

The modulation of the exchange constant introduces an extra energy to Equation (3.73), which is

$$\mathcal{H}' = \int \mathrm{d}x \left[-B \sin\phi + \frac{2Ku^2}{a}\right], \tag{3.79}$$

where

$$B = J\lambda u/a^2, \tag{3.80}$$

and K is the elastic constant. The B-term in Equation (3.79) prefers the state with $\phi = \pi/2$. The total Hamiltonian, $\mathcal{H}_0 + \mathcal{H}'$, can also be treated in SCHA, which leads to the conclusion that the energy gain of a spin system due to the coupling to the lattice distortion, i.e., the B-term, is proportional to $u^{4/3}$. This 4/3 power, which is in contrast to $u^2 ln^2 u$ obtained in the mean field approximation, has been derived by Cross and Fisher [92] and is consistent with the result of numerical calculations [94, 95] and the renormalization group treatment [96–98]. The presence of such a large energy gain is due to the fact, mentioned previously, that the ground state contains wave functions of a singlet character, $\phi = \pi/2$.

It is to be noted that the present phase Hamiltonian is capable of describing such a subtle problem [30] as well as nonlinear excitations, i.e. solitons in this system [29], in a simple way.

4. Impurity Pinning and Anderson Localization

4.1. Impurity Pinning of CDW

So far only clean systems without imperfections have been considered. In actual systems there exist various kinds of randomness. In this section, the effects of impurities will be considered as a typical example of randomness.

The interaction energy between electrons and impurities is given by

$$\mathcal{H}_{\text{imp}} = \sum_i \int \mathrm{d}x\, \rho(x) v(x - R_i), \tag{4.1}$$

where $\rho(x)$ is the local electron density and the summation is over impurity sites, R_i. The impurity potential, $v(x)$, is assumed to be of short range, $v(x) = \tilde{V}_0 \delta(x)$. By use of Equation (2.44) we write Equation (4.1) as

$$\mathcal{H}_{\text{imp}} = \tilde{V}_0 \sum_i \nabla\theta(R_i)/\pi + V_0 \sum \cos\phi(R_i) \cos(QR_i + \theta(R_i))$$

$$= \mathcal{H}^{(1)}_{\text{imp}} + \mathcal{H}^{(2)}_{\text{imp}}, \tag{4.2}$$

where $Q = 2k_F$ and $V_0 = 2\tilde{V}_0/\pi\alpha$. The first term on the r.h.s. of Equation (4.2) represents the coupling of long wave fluctuations to the impurities, whereas the second term corresponds to the backward scattering.

In the case of the PF state, which is considered in this sub-section, the spin degree of freedom is not essential, and needs not to be treated explicitly. Hence we ignore the factor $\cos \phi(R_i)$ by incorporating the possible effects into V_0. Bak and Brazovskii [59] indicated that $\mathcal{H}^{(1)}_{\text{imp}}$, through which only $\nabla\theta$ is coupled to the random potential, does not necessarily destroy the long range order. On the other hand it has been shown by a general argument [99–101] that the perturbation, which directly couples θ to impurities, destroys the long range order, and $\mathcal{H}^{(2)}_{\text{imp}}$ can be considered as an example of such a kind of perturbation [10, 11, 102, 103]. By this reason $\mathcal{H}^{(2)}_{\text{imp}}$ is more important in the pinning problem and then we will mainly be concerned with this term. Consequently, by use of Equation (3.24), the ground state of the incommensurate CDW in the adiabatic approximation is described by the following Hamiltonian, with $A = v_F/4\pi$,

$$E = \int dx \left[A(\nabla\theta)^2 + V_0 \sum_i \delta(x - R_i) \cos(Qx + \theta) \right]. \tag{4.3}$$

The problem characterized by Equation (4.3) is complicated because of the existence of two competing energies, the elastic energy (the first term) which prefers the spatially uniform phase, and the impurity energy (the second term) which introduces the distortion of the phase so that $\cos(QR_i + \theta(R_i))$ can be as small ($V_0 > 0$ is assumed) as possible at each impurity. The former energy prefers the translationally invariant state, while the latter prefers the distorted phase fixed in space, reflecting the random configurations of impurities. This is the impurity pinning of the phase variable. As is easily understood, the problem is highly nonlinear and ordinary perturbation theory is not appropriate.

The characteristic parameter in the present problem is

$$\epsilon = V_0/n_i v_F, \tag{4.4}$$

where n_i is the average density of impurities. This is seen as follows; if the impurity potential dominates, then the phase will adjust itself so that $\Theta(R_i) = \pi$ with the modulus of 2π at each impurity site, where $\theta(x)$ is the total phase $\Theta(x) = Qx + \theta(x)$. The system will gain the impurity potential energy of $n_i V_0$, while paying the elastic energy of the order of $A(1/n_i)n_i^2 = O(n_i v_F)$ per impurity. Thus the ratio of these, which is given by Equation (4.4), is expected to characterize the impurity pinning mechanism in each system. In fact, in the strong pinning limit $\epsilon \gg 1$, the preferable configuration should satisfy the condition $\cos\theta = -1$ at each R_i as shown schematically in Figure. 13. In the

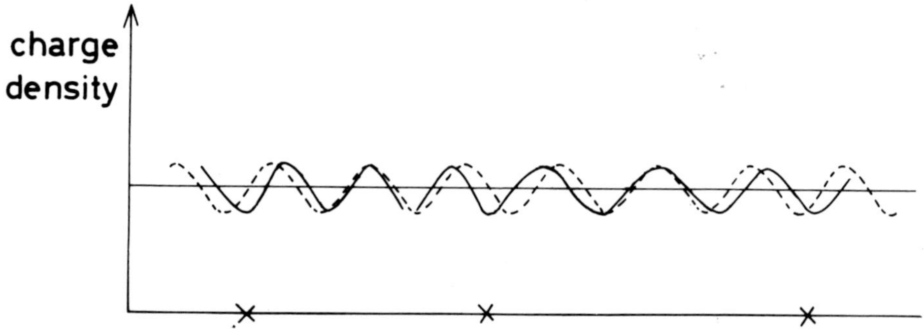

Fig. 13. Schematic representation of the case of the strong pinning, $\epsilon \gg 1$, where crosses represent impurities.

weak pinning limit $\epsilon \ll 1$, on the other hand, $\theta(x)$ is varying slowly from one impurity site to the next and $\Theta(R_i)$ are distributed almost randomly. But they are not completely random and the impurity pinning mechanism does exist, even in this limit. This is due to the following fact.

If $\epsilon \ll 1$, the characteristic length, L_0, can be considered, which is much longer than the average distance between impurities, n_i^{-1}, and over which θ varies of the order of π. This L_0 is to be determined variationally. For any point on a chain, x, the phase is taken as common in the interval $(x - L_0/2, x + L_0/2)$, and we treat the impurity potential as [11, 104]

$$V_0 \sum_i{}' \cos(Qx_i + \theta(x_i))$$

$$= V_0 \operatorname{Re} e^{i\theta(x)} \sum_i{}' e^{iQx_i}, \tag{4.5a}$$

$$= V_0 \sqrt{n_i L_0} \cos(\alpha(x) + \theta(x)), \tag{4.5b}$$

where the summation is over impurities in the interval $(x - L_0/2, x + L_0/2)$ and $\alpha(x)$ is as defined in Figure 14. In Equation (4.5a) it is assumed that Qx_i is completely random

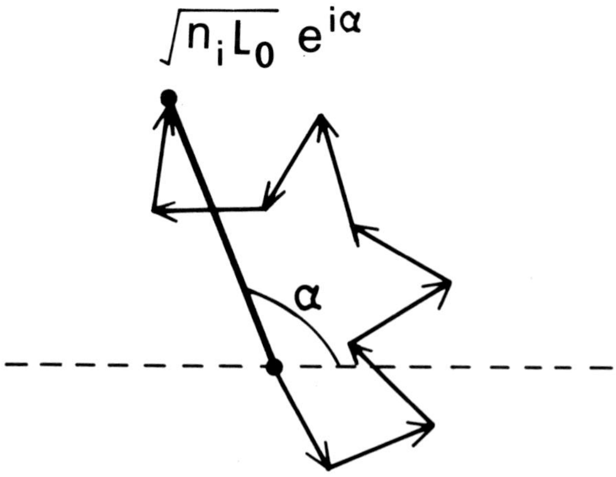

Fig.14. The determination of $\alpha(x)$ of Equation (4.5b) in the case of weak pinning, $\epsilon \ll 1$.

and $n_i L_0 \gg 1$. The fluctuation of a number of impurities in the interval L_0 is ignored. By definition $|\nabla\alpha| \simeq L_0^{-1}$. If $\theta(x)$ follows $\alpha(x)$, i.e. $\cos(\alpha(x) + \theta(x)) = -1$, there exists an energy gain from the impurity potential, $-\sqrt{n_i/L_0}\, V_0$, per unit length. Then E, Equation (4.3), is estimated as

$$\frac{E}{L} = A \frac{1}{bL_0^2} - \sqrt{\frac{n_i}{L_0}}\, V_0, \tag{4.6}$$

where b is a numerical factor of the order of unity; $b = 3/\pi^2$ if the phase difference from section to section is random between $-\pi$ and π and if the true solution smoothly interpolates between them.

The variation of Equation (4.6) with respect to L_0 results in

$$L_0^{-1} = [(b\pi V_0/v_F)^2 n_i]^{1/3}, \tag{4.7a}$$

or in terms of ϵ, Equation (4.4),

$$(n_i L_0)^{-1} = (b\pi\epsilon)^{2/3}. \tag{4.7b}$$

This is the result for the case of weak pinning. On the other hand, $L_0^{-1} \sim n_i$, in the case of strong pinning, i.e. $\epsilon \gg 1$.

The existence of finite L_0, in either case, results in a vanishing static conductivity at $T = 0$, since the pinning frequency ω_0 in Equation (2.71) will be typically of the order of $\omega(q)$ at $q = L_0^{-1}$. Analytical calculations of $\sigma(\omega)$ at $T = 0$ in the limits of weak and strong pinning have been done by Fukuyama and Lee [11], and the result is schematically shown in Figure 15. In contrast to the case with one definite ω_0, Re

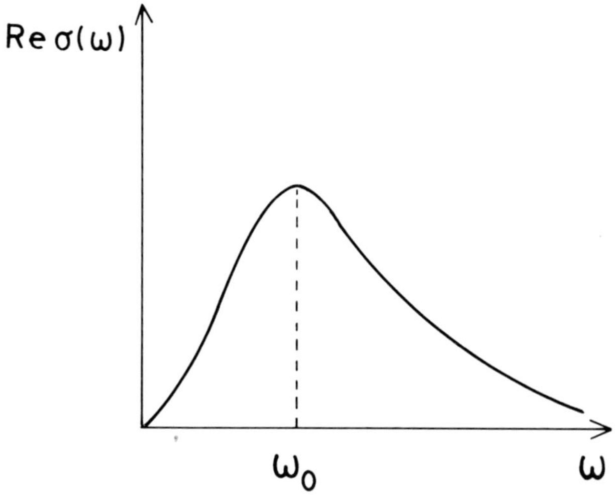

Fig.15. Schematic representation of Re $\sigma(\omega)$.

$\sigma(\omega)$ has a broad spectrum reflecting the random configuration of impurities. More detailed analytical investigation of this problem has been performed by Feigelman [105], who employed the transfer matrix and the distribution function for random systems and obtained results essentially similar to the previous discussions.

There exist also numerical calculations. Ternaishi and Kubo [106] solved the following equation for a fixed configuration of impurities derived by the variation of Equation (4.3) with respect to θ;

$$2A\nabla^2\theta + V_0 \sum_i \delta(x - R_i)\sin(Qx + \theta) = 0. \tag{4.8}$$

Their results for $\sigma(\omega)$ and the analytical results, are in close agreement, and the existence of two characteristic regime of weak and strong pinning has been confirmed. Similar results have also been obtained by molecular-dynamics studies by Koehler and Lee [107].

Weisz, Sokoloff and Sacco [108] extended the model to two and three dimensions on a discrete lattice and found that the low lying excitations are localized, i.e. pinned by impurities. Lee and Fukuyama [109] have shown the existence of finite L_0 even for the three dimensional systems with the long range Coulomb interactions which play an important role.

In the presence of finite electric field, E, one should take the following energy, E_e, into account;

$$E_e = PE = \frac{e}{\pi}\int \mathrm{d}x\, \theta(x)E, \tag{4.9}$$

where P is the electric dipole moment defined in Equation (2.66). In the presence of E_e, Equation (4.9), one could expect a strong non-Ohmicity, which is actually demonstrated in the numerical calculations by Teranishi and Kubo [106]. Especially they found that at $T \neq 0$ the static conductivity is of activation type, the activation energy of which decreases linearly in proportion to the electric field. In this non-Ohmic conduction process, and also for the Ohmic conduction at $T \neq 0$, the large local changes of θ overcoming the potential barriers are essential (see also [144]).

As the field gets stronger the CDW is expected to be depinned [110, 111]. Lee and Rice [112] estimated such depinning field, E_T, in quasi-one-dimensional systems by balancing the energy gain by impurity pinning with that by the electric field when the phase is advanced by 2π. They concluded that $E_T \propto n_i$ and n_i^2 for strong and weak pinning, respectively. (Quite recently Klemm and Schrieffer [113] obtained $E_T \propto n_i^2$ for weak pinning by different method.) Sokoloff [114] performed molecular-dynamical studies of the depinning by extending the calculations of Teranishi and Kubo [106] to longer chains and observed the existence of finite E_T (see also [215]).

In the commensurate or near-commensurate systems the energy to be considered is, instead of Equation (4.3), typically given by

$$E = \int \mathrm{d}x\, [A(\nabla\theta - q_0)^2 - B\cos M\theta + V_0 \sum_i \delta(x - R_i)\cos(Qx + \theta)]. \tag{4.10}$$

In the case of $q_0 = 0$ and $M = 1$ [104] it is found that the long range order is found to be stable at absolute zero as far as the impurity potential is not too strong, while the glassy state (charge density glass (CDG)) where the phase is pinned by impurities is realized for strong impurity potentials. On the other hand, Nakanishi [115] examined the case of $q_0 = 0$ and $M = 3$, who found that the long range order is unstable for an arbitrary weak impurity potential in spite of the existence of the commensurability potential. Littlewood and Rice [116] examined the case where $q_0 \neq 0$ and $B = 0$ by Monte-Carlo calculations in the context of the pressure-induced hysteresis in Cr and found that the wave vector of CDW lags behind the equilibrium value, and that changes by addition or removal of phase solitons result in the characteristic feature of the glassy state: the metastability.

Similar problems but with $\mathcal{H}_{\mathrm{imp}}^{(1)}$ instead of $\mathcal{H}_{\mathrm{imp}}^{(2)}$ in Equation (4.2) have been considered. Baeriswyl and Bishop [117] examined the static correlation function and the dynamic structure factor in the case of $q_0 = 0$ and $M = 1$ and Rice *et al.* [118] considered the impurity pinning of the discommensurations for $q_0 = 0$ and arbitrary M. The dynamics of sine-Gordon solitons have also been examined by Fogel *et al.* [119] in the presence of an impurity.

The complete pinning at $T = 0$ so far explained is affected by thermal fluctuations and by possible thermal depinning, as the case may be [118, 120–122]. Above the critical temperature, T_P, of the Peierls transition the fluctuations affect the conductivity; in the temperature regime very close to T_P and then if L_0, Equation (4.7a), is shorter than the coherence length ξ, the impurity will reduce the conductivity [123] as ξ gets longer, while at higher temperatures where $\xi \ll L_0$ conductivity will increase as ξ gets longer since impurity scattering will be irrelevant [124, 125, 125a].

As has been discussed in detail so far, the important feature of the impurity pinning lies in the spatial distortions of the phase in the ground state, reflecting the random configurations of impurities. However, if the quantity of interest is not so sensitive to such spatial variations, the pinning may conveniently be described by an averaged potential deduced from Equation (4.5b), i.e.

$$V_0 \sum_i \cos(Qx_i + \theta(x_i)) \rightarrow W \int \mathrm{d}x \cos\theta(x), \tag{4.11}$$

where $W \sim AL_0{}^{-2}$. In Section 5 we will see consequences of such simplifying approximation.

4.2 ANDERSON LOCALIZATION

In the preceeding section we discussed the impurity pinning of the phase of the charge density wave in Peierls systems, and found that impurities play an essential role. Such impurity pinning in the Peierls system can be viewed as one aspect of Anderson localization, as we will see explicitly. In usual cases, without the Peierls distortion, however, the spin degree of freedom, ϕ, has to be taken into account. This will be examined in this section for one-dimensional systems.

Since the scaling theory by Abrahams *et al.* [32] there has been substantial progress nowadays in the microscopic understanding of Anderson localization [33, 34] and it is realized that the mutual interactions have profound effects on localization [34, 128–130]. In 1D systems such combined effects of interactions and localization [131–133] will be more clearly seen than in other dimension since the randomness alone always results in localization irrespective of the location of the Fermi energy, ϵ_F [134], while the interactions alone lead to the divergences of the various response functions, as we saw in Section 2. The phase Hamiltonian gives a simple description to the examination of such a problem [135]. For the sake of clarity we confine ourselves to the SDT model for the mutual interactions.

Our model Hamiltonian is then

$$\mathcal{H} = \int \mathrm{d}x\, [A_\rho(\nabla\theta)^2 + C_\rho P^2 + A_\sigma(\nabla\phi)^2 + C_\sigma M^2 + \sum_i V_0 \cos\phi(x) \cos(2k_F x + \theta(x))\, \delta(x - R_i)], \tag{4.12}$$

where the last term is due to the coupling to impurities distributed randomly over R_i. The impurity potential is assumed to be of short range and only the backward scattering is taken into account. The similarity of the Anderson localization and the impurity pinning problem of the Peierls phase is now clear when Equation (4.3) and (4.12) are compared. The presence of $\cos\phi$ in Equation (4.12) indicates the possible fluctuations of spin degree of freedom in the present interacting fermion system in contrast to the Peierls state. Another difference of the present problem from the Peierls phase is the importance of the quantum fluctuations represented by terms including P^2 and M^2 in Equation (4.12). This quantum effect is taken into account by writing $\theta = \theta_{\text{cl}} + \hat{\theta}$ and $\phi = \phi_{\text{cl}} + \hat{\phi}$ where θ_{cl} and ϕ_{cl} are slowly varying classical parts and $\hat{\theta}$ and $\hat{\phi}$ are quantum fluctuations around θ_{cl} and ϕ_{cl}. The latter, $\hat{\theta}$ and $\hat{\phi}$, are treated in the SCHA (see Section 2.3),

$$\begin{aligned}&\cos\phi(x)\cos(2k_F x + \theta(x))\\ &\quad\to \gamma[1 - \tfrac{1}{2}(\hat{\theta}(x)^2 - \langle\hat{\theta}^2\rangle + \hat{\phi}(x)^2 - \langle\hat{\phi}^2\rangle)] \times\\ &\quad\times \cos\phi_{\text{cl}}(x)\cos(2k_F x + \theta_{\text{cl}}(x)),\end{aligned} \tag{4.13}$$

$$\gamma = \exp[-\tfrac{1}{2}(\langle\hat{\theta}^2\rangle + \langle\hat{\phi}^2\rangle)]. \tag{4.14}$$

In Equation (4.13) and (4.14), $\langle\hat{\theta}^2\rangle$ and $\langle\hat{\phi}^2\rangle$ are the averages to be determined self-consistently. We ignore the spatial variations of these quantities. By this approximation the problem reduces essentially (but not exactly) to that of a classical system where the effective impurity potential contains the quantum effect through γ, Equation (4.14).

It is seen from Equation (4.13) that ϕ_{cl} is not directly influenced by the random distribution of impurities and then that ϕ_{cl} can be fixed so that $|\cos\phi_{\text{cl}}|$ takes a maximum value. The characteristic length, L_0, of the distortion of θ_{cl} to be determined can be treated by the same method as in Section 4.1. In the discussion of localization–delocalization the weak pinning is relevant and θ_{cl} is considered to be constant over distance, L_0, much longer than the average spacing of impurities and the energy gain due to such distortion of θ_{cl} is given as in Equation (4.6)

$$\frac{1}{L_0}\sum_i{}' \cos(2k_F R_i + \theta_{\text{cl}}(R_i)) = -(n_i/L_0)^{1/2}. \tag{4.15}$$

By Equations (4.13) and (4.15) the last term in the r.h.s. of Equation (4.12) is written as

$$-\left(\frac{n_i}{L_0}\right)^{1/2} V_0\,\gamma\int_0^L \mathrm{d}x\left[1 - \frac{1}{2}(\hat{\theta}^2 - \langle\theta^2\rangle + \hat{\phi}^2 - \langle\phi^2\rangle)\right]. \tag{4.16}$$

Then the self-consistency equation for γ, (4.14), is given by

$$\gamma = \exp\left[-\frac{1}{8\pi}\int_{-\alpha^{-1}}^{\alpha^{-1}} \mathrm{d}q\, T\sum_m \sum_{\nu=\rho,\sigma} \frac{1}{\omega_m{}^2/4C_\nu + A_\nu(q^2 + q_{c\nu}{}^2)}\right], \tag{4.17}$$

where

$$A_\nu q_{c\nu}^2 = \frac{\sqrt{n_i L_0}}{2L_0} V_0 \gamma, \qquad \nu = \rho, \sigma, \tag{4.18}$$

and $\omega_m = 2\pi m T$, m being the integer. Equation (4.17) is evaluated as

$$\gamma = C(V_0/n_i v_F)^{\eta/(4-\eta)} (n_i L_0)^{(-\eta/2)/(4-\eta)}, \tag{4.19}$$

where

$$C = \left[\left(\frac{\pi/2}{1+\tilde{\alpha}}\right)^{\eta_\rho/4} \left(\frac{\pi/2}{1+(g_{2\parallel} - g_{2\perp})/2}\right)^{\eta_\sigma/4} (n_i \alpha)^{\eta/2}\right]^{4/(4-\eta)}. \tag{4.20}$$

and $\tilde{\alpha} = (g_{2\parallel} + g_{2\perp})/2$. Quantities η_ρ, η_σ and η are defined by

$$\eta_\rho = \left[\frac{1-(g_{2\parallel} + g_{2\perp})/2}{1+(g_{2\parallel} + g_{2\perp})/2}\right]^{1/2}, \tag{4.21}$$

$$\eta_\sigma = \left[\frac{1-(g_{2\parallel} - g_{2\perp})/2}{1+(g_{2\parallel} - g_{2\perp})/2}\right]^{1/2}, \tag{4.22}$$

and $\eta = \eta_\rho + \eta_\sigma$. Thus the excess energy ΔE due to the distortion of θ_{cl} is given by

$$\begin{aligned} \frac{\Delta E}{L} &= A_\rho/bL_0^2 - L_0^{-1}\sqrt{n_i L_0}\, V_0 \gamma \left(1 + \frac{\langle\theta^2\rangle}{2} + \frac{\langle\rho^2\rangle}{2}\right) + \\ &\quad + \frac{1}{2L} \sum_{\nu=\rho,\sigma} \sum_q \left(\frac{b_\nu}{a_\nu}\right)^{1/2} [(q^2 + q_{c\nu}^2)^{1/2} - |q|] \\ &= n_i^2 v_F \left[-\frac{(1-\eta/4)\gamma V_0/n_i v_F}{\sqrt{n_i L_0}} + \frac{1+\tilde{\alpha}}{4\pi b (n_i L_0)^2}\right], \end{aligned} \tag{4.23}$$

where b is same as in Equation (4.6). The minimization of Equation (4.23) with respect to L_0 results in the characteristic length given by

$$n_i L_0 = \left(\frac{\pi b C}{1+\tilde{\alpha}}\right)^{-(4-\eta)/2/(3-\eta)} \times \left(\frac{V_0}{n_i v_F}\right)^{-2/(3-\eta)} \tag{4.24}$$

Consequently $\eta = 3$ separates the region of localization and delocalization, i.e. $L_0 = \infty$. This is shown in Figure 16 in the plane of $g_{2\parallel}$ and $g_{2\perp}$, where the broken lines are boundaries obtained in Figure 4.

Such a localization–delocalization transition induced by interactions has been indicated originally by Chui and Bray [131], Apel [132], Apel and Rice [133] by different methods.

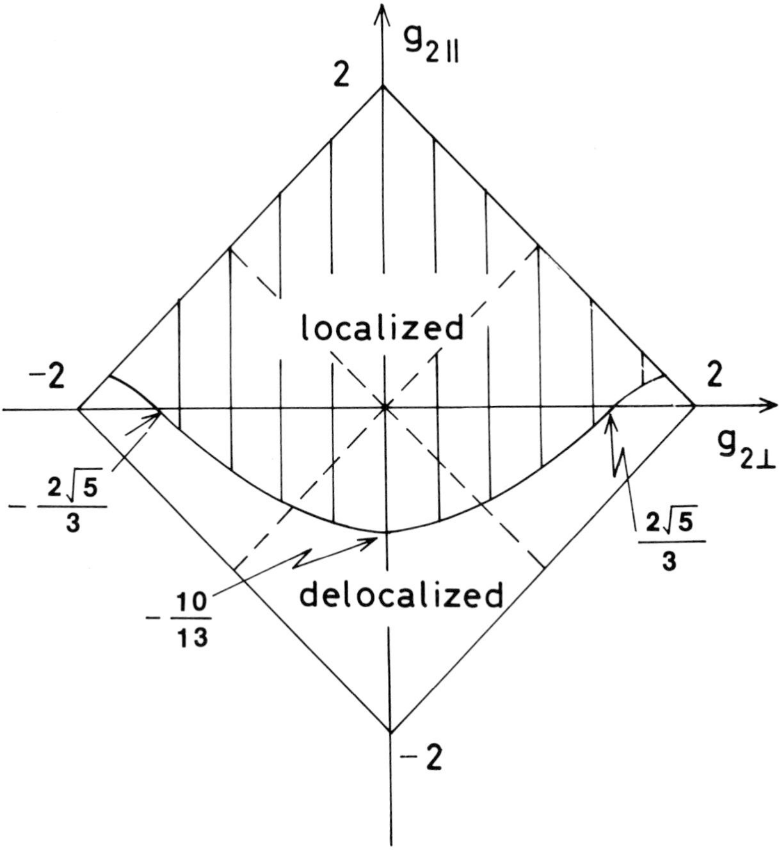

Fig. 16. Phase diagram of localized and delocalized regimes for SDT model (cf. Fig. 4).

In the impurity pinning problem treated in this and the preceeding subsections the dynamical properties (such as Ohmic conductivity) are determined by the fluctuations of phases $\hat{\theta}$ and $\hat{\phi}$ around those distorted ones θ_{cl} and ϕ_{cl} in the ground state. Such distortions of θ_{cl} and ϕ_{cl} are essential in the pinning and the simple perturbative treatment with respect to the impurity potential ignoring θ_{cl} and ϕ_{cl} is not generally justified though apparently acceptable results are obtained in some cases [10, 126, 127].

5. CDW Conductivity in the Peierls-Fröhlich State

5.1 Langevin Equation Approach

Based on the phase Hamiltonian derived in the preceeding sections, we discuss here the dynamical nature of the CDW condensate and the resulting electrical conductivity in the PF state in the context of various interesting phenomena observed in quasi-one-dimensional conductors $NbSe_3$, TaS_3 and so on (for the experimental details see the article by Monceau in Part II of this work). For this purpose we assume that the mutual Coulomb interaction between electrons can be neglected and the phase variable, ϕ, responsible for the spin density wave is discarded. Also we only consider the case that the CDW condensate coupled to the lattice distortion adiabatically, and the CDW phase θ is treated as classical variable. The effective Hamiltonian, which describes the dynamical properties of θ, is thus written as

$$\mathcal{H}_{\text{eff}} = \pi v' \int \mathrm{d}x \left[p^2 + \frac{1}{4\pi^2 v'^2} \{u_p{}^2 (\nabla\theta)^2 + 2\omega_F{}^2 V(\theta)\} - \frac{e}{\pi^2 v'} E\theta \right], \tag{5.1}$$

where $V(\theta)$ represents either the commensurability pinning potential in Equation (3.67), i.e., $\omega_F = \omega_M$ and $V(\theta) = (1 - \cos M\theta)/M^2$, or the averaged impurity pinning potential of Equation (4.11), i.e., $\omega_F = \omega_{\text{imp}} \equiv [2\pi v' W]^{1/2}$ and $V(\theta + 2\pi) = V(\theta)$ is assumed. The last term in Equation (5.1) is the coupling energy to an external electric field E, Equation (4.9).

For the proper description of transport properties associated with CDW dynamics we have to know its damping mechanism, which is not given by Equation (5.1). It arises partly from interactions between CDW and other modes, such as phonons not participating the CDW condensate and quasiparticles thermally excited (or those on other parts of the Fermi surface surviving even in the PF state). Besides the averaged pinning potential, Equation (4.11), impurities also cause damping of a CDW motion. Up to now, however, microscopic analysis of these damping mechanisms has been very limited [111, 136, 137, 137a]. We therefore introduce the following phenomenological argument: all the effects which give rise to the damping of a CDW motion are represented by random forces $F(x, t)$ on the CDW phase $\theta(x, t)$ and by the corresponding frictional force $-\eta p$, η being the frictional constant. Consequently the equation of motion for θ is given by

$$\dot{p} = -\frac{\delta \mathcal{H}_{\text{eff}}}{\delta\theta} - \eta p + F. \tag{5.2}$$

Substituting Equation (5.1) into Equation (5.2), we obtain the following Langevin equation

$$\frac{\partial^2\theta}{\partial t^2} - u_p{}^2 \frac{\partial^2\theta}{\partial x^2} + \eta \frac{\partial\theta}{\partial t} + \omega_F{}^2 \frac{\partial V}{\partial\theta} = 2v'eE + 2\pi v' F. \tag{5.3}$$

Usually the random force $F(x, t)$ is assumed to satisfy the conditions $\langle F(x, t)\rangle = 0$ and

$$\langle F(x, t) F(x', t)\rangle = \eta \frac{T}{\pi v'} \delta(x - x')\delta(t - t'), \tag{5.4}$$

where T is the temperature. In the following subsections we discuss CDW conductivity derived from various limits of Equation (5.3).

At this point, we briefly mention three-dimensionality (3-D) effects on the CDW dynamics. Various interesting phenomena are in fact observed in the 3-D ordered PF state of quasi-1-D conductors. Because of the coupling between CDW's on different chains, they move coherently. Since the resulting CDW motion is still predominantly in the chain direction [138–140], the 1-D description of Equation (5.3) will be appropriate as a simplest approximation, but the degree of the spatial extent of the coherence is an important problem when experimental results are to be analyzed quantitatively [112, 141, 142].

5.2. CLASSICAL PARTICLE MODEL

The motion of a rigid CDW condensate (i.e. sliding motion of CDW) is described by Equation (5.3) without the spatial gradient term. In this case the effective momentum, charge and force are reinterpreted as those per unit length. In this case we obtain $p = M^*\dot{\theta}$ and the equation of motion

$$\frac{\partial^2 \theta}{\partial t^2} + \eta \frac{\partial \theta}{\partial t} + \tilde{\omega}_F^2 \frac{\partial V}{\partial \theta} = \left(\frac{n_s eE}{\pi} + \tilde{F}\right)\Big/ M^*. \tag{5.5}$$

Here $\tilde{\omega}_F{}^2 = \omega_F{}^2/n_s$ and $M^* = n_s/2\pi v' \equiv n_s m^*$ plays a role of the effective mass of the CDW, where n_s is the density of electrons participating a sliding motion of CDW. The microscopic evaluation of n_s will be discussed later in Section 6. Similarly the current associated with the sliding CDW is given by

$$J_{\mathrm{CDW}} = \frac{n_s e}{\pi} \frac{\partial \theta}{\partial t}\,. \tag{5.6}$$

Equation (5.5) describes a Brownian particle model in a periodic potential $V(\theta)$, and has been extensively studied in connection with the Josephson effect [143, 144] as well as the superionic conductivity [145, 146]. In the former case the potential $V(\theta)$ is undoubtedly a sinusoidal one, 1-cos θ, arising from the Josephson relation [147]. Usually the (Gaussian) random force $\tilde{F}$, which now satisfies $\langle \tilde{F}(t)\tilde{F}(t')\rangle = 2\eta M^* T\delta(t - t')$, is incorporated by a Fokker–Planck-type equation for the distribution function $P(p, \theta; t)$ derived from Equation (5.5). For a system with strong damping, or, more explicitly, if the condition

$$\left| \frac{\tilde{\omega}_F{}^2}{\eta^2}\left(\frac{\partial^2 \bar{V}}{\partial \theta^2}\right)\right| \ll 1, \tag{5.7}$$

is satisfied, the Fokker–Planck-type equation can be reduced to the Smoluchowski equation for the distribution function of θ, $S(\theta, t)$ [143, 148]. If the effective mass M^* is large enough in addition to the condition of Equation (5.7), i.e.

$$\gamma \equiv \frac{2M^*\tilde{\omega}_F^2}{T} \gg 1, \tag{5.8}$$

a non-Ohmic conductivity with a sharp threshold field E_T is obtained. Explicitly we obtain the following expression for a dc current (we assume $V(\theta) \geq V(0) = V(2\pi) = 0$)

$$J(E) = \begin{cases} O(e^{-\gamma}), & E < E_T, \\ J_c \left[\int_0^{2\pi} \frac{d\theta}{(E/E_0) - (\partial V/\partial \theta)} + O(\gamma^{-1}) \right]^{-1}, & E > E_T, \end{cases} \tag{5.9}$$

where $J_c = 2\pi E_0 J_\infty / E$ with $J_\infty = (n_s e/\pi)^2 E/\eta M^*$ and

$$E_0 = \frac{\pi M^* \tilde{\omega}_F{}^2}{n_s e} = \frac{\tilde{\omega}_F{}^2}{2v' e}. \tag{5.10}$$

Notice that the electric force and the pinning force with $dV/d\theta = 1$ balance at $E = E_0$, and then E_T is given by $E_T = E_0 \max (dV/d\theta)$. In the limit $\gamma \to \infty$, J vanishes when $E < E_T$, increases rather sharply at $E \gtrsim E_T$, and becomes proportional to E for $E \gg E_T$. The explicit dependences of J on E for $E > E_T$ are $J/J_\infty = (\epsilon^2 - 1)^{1/2}$ (A), $[\ln \{(\epsilon + 1)/(\epsilon - 1)\}]^{-1}$ (B) and $\epsilon - 1/\epsilon$ (C) with $V(\theta) = 1 - \cos\theta$ (A), and $(1/2)\theta^2$ (B) and $|\theta|$ (C) (in $|\theta| \leq \pi$), respectively, where $\epsilon = E/E_T$. These non-Ohmic conductivities $\sigma(E) = J/E$ are shown in Figure 17, where we have also shown for comparison the result of Bardeen's tunneling model [149]. Notice that the non-Ohmicity at $E \gg E_T$ derived from Equation (5.9) has a common behavior, $\sigma(\infty) - \sigma(E) \propto E^{-2}$, as long as $V(\theta)$ is continuous. On the other hand, J at $E \gtrsim E_T$ is proportional to $(E - E_T)^{1/2}$ if $dV/d\theta \simeq a - b(\theta - \theta_m)^2$ near its maximum ($\theta = \theta_m$), while $J \propto (E - E_T)$ for the potential (C).

The above results, in the limit of strong damping and large effective mass, are directly obtained from a simpler equation of motion

$$\eta \frac{\partial \theta}{\partial t} + \tilde{\omega}_F{}^2 \frac{\partial V}{\partial \theta} = e n_s E/\pi M^*, \tag{5.11}$$

which was first introduced by Grüner *et al.* [150] and discussed further by Monceau *et al.* [141, 151]. In the case of superconducting Josephson effects Equation (5.11) with a sinusoidal potential $V(\theta)$ is called a resistively shunted junction model and has been extensively studied [144]. When $V(\theta) = 1 - \cos\theta$, Equation (5.11) is solved for a static field E larger than E_T, and $J_{\mathrm{CDW}}(t) = (n_s e/\pi)(\partial\theta/\partial t)$ is obtained as [144, 151]

$$J_{\mathrm{CDW}}(t) = J_\infty \sqrt{\epsilon^2 - 1} \left\{ 1 + 2 \sum_{n=1} K^n \cos n(\Omega t + \theta_0) \right\}, \tag{5.12}$$

where $K = \epsilon - \sqrt{\epsilon^2 - 1}$, θ_0 is an integration constant, and

$$\Omega = \frac{\tilde{\omega}_F{}^2}{\eta} \sqrt{\epsilon^2 - 1} = \frac{\pi}{e\eta} J_{\mathrm{CDW}}^{\mathrm{DC}}. \tag{5.13}$$

Here $J_{\mathrm{CDW}}^{\mathrm{DC}}$ is the first term of Equation (5.12), i.e. J given by Equation (5.9). The

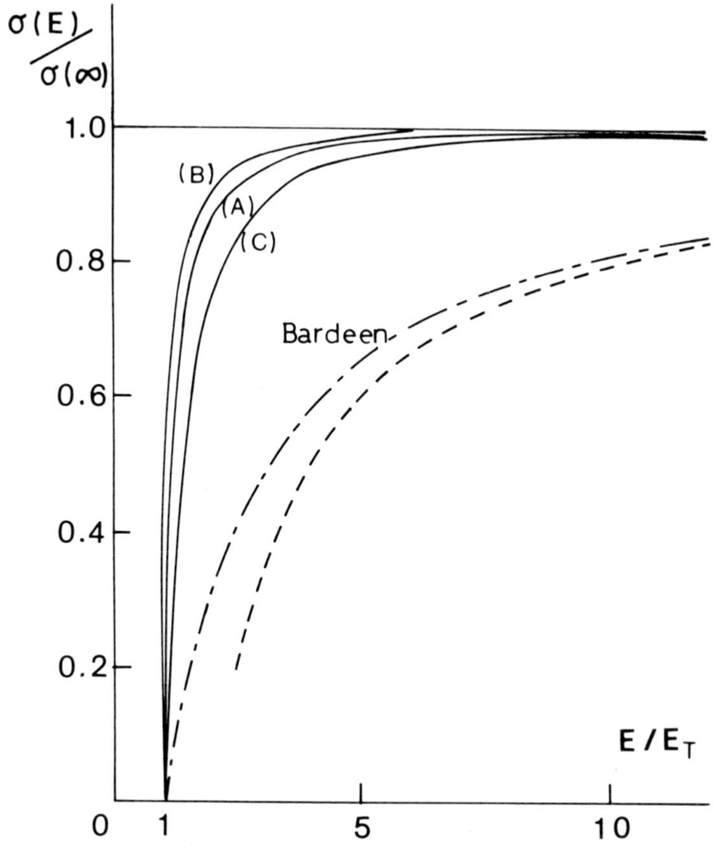

Fig. 17. The electric field dependences of the conductivity for various models. The broken line is explained in Section 6.1.

periodic variation of J_{CDW} reflects a sliding motion of CDW in the tilted (by E) periodic potential, suffering from a strong frictional force. The period is the time required for θ to change by 2π, or in other words, the time required for CDW to traverse a wavelength $\lambda = 2\pi/Q$ in a real space. The fundamental frequency Ω of Equation (5.13), which is proportional to $J^{\mathrm{DC}}_{\mathrm{CDW}}$, can be interpreted as that of a narrow band noise, first observed by Fleming and Grimes [152]. Equation (5.12) also implies that the amplitudes of its higher harmonics are reduced by the factor K^n.

When only an ac external field with small amplitude is applied, overdamped oscillation around the bottom of the pinning potential ($\theta = 0$) takes place. The ac conductivity in this case is simply given by

$$\sigma(\omega) = \sigma_\infty \frac{i\omega\tau_{\mathrm{eff}}}{1 + i\omega\tau_{\mathrm{eff}}}, \tag{5.14}$$

where $\sigma_\infty = n_s e^2/\pi^2 \eta M^*$ and $\tau_{\mathrm{eff}} = \eta/\tilde{\omega}_F{}^2 V''(0)$.

More interesting is a case when both ac and dc fields are applied at the same time. One of the important problems in this case is to understand how the external ac field couples with the periodic variation of J_{CDW}, Equation (5.12), when CDW is forced to slide by the dc field. Such synchronization phenomena are discussed within the present model by Richard *et al.* [151] to explain sharp peaks in their data on $\mathrm{d}V/\mathrm{d}I$ of $NbSe_3$ (see also [144]). Another important problem is an occurrence of non-Ohmicity under the combined ac and dc fields. When a dc field less than E_T is applied, the particle locates at the minimum of the potential. If the amplitude of an added ac field is so large that the particle sways out from the confining potential, dc current attributable to sliding CDW appears and the non-Ohmicity results. From this point of view Grüner *et al.* [153] have explained their data on ac-induced dc conductivity of $NbSe_3$.

5.3. SOLITON AND SOLITON LATTICE

Next we will take the possible spatial variations into account and consider dynamical properties associated with soliton lattices. The systematic part of Equation (5.3) is a nonlinear 'relativistic' wave equation:

$$\frac{\partial^2\theta}{\partial t^2} - u_p{}^2\,\frac{\partial^2\theta}{\partial x^2} + \omega_F{}^2\,\frac{\partial V}{\partial\theta} = 0. \tag{5.15}$$

There exists a class of exact nonlinear solutions, i.e. solitons, for a potential $V(\theta)$ which has degenerate minima and is only implicity dependent of x and t. The sine-Gordon soliton discussed in Section 3.5 is a typical example. The solution of Equation (5.15), $\theta_\pm(x-ut)$, and its excitation energy $E_\pm(u)$ are given by [154]

$$\pm\gamma_R(x-ut) = \frac{d}{\sqrt{2}}\int_{\theta_\pm(0)}^{\theta_\pm(x-ut)}\frac{\mathrm{d}\theta}{\sqrt{V(\theta)}}, \tag{5.16}$$

$$E_\pm(u) = \frac{\gamma_R u_p{}^2}{\sqrt{2}\pi v' d}\int_0^{2\pi}\mathrm{d}\theta\sqrt{V(\theta)}, \tag{5.17}$$

where $\gamma_R = (1 - u^2/u_p{}^2)^{1/2}$ and $d = u_p/\omega_F$. These are the generalization of Equation (3.43) and (3.44), respectively. The dynamical aspects of the soliton-bearing system described by Equation (5.15) has been extensively studied [154–156], although details of the dynamical processes, such as the creation or anihilation of solitons, have not yet fully been established.

Concerned with transport property of the PF state, Rice *et al.* [12] have first emphasized that such solitons are a new type of current-carrying excitations in a weakly pinned CDW condensate. An explicit evaluation of the static non-Ohmic conductivity, which is based on their idea, has been performed by Trullinger *et al.* [157]. They solved a discretized (in space) version of Equation (5.3) with $V(\theta) = 1 - \cos\theta$. This is a model of coupled pendulums subject to random forces. Restricting to a strong damping case as in Section 5.2, they derived a Smoluchowski equation for $S(\{\theta_i\},t)$, where the number of variables θ_i is now macroscopic and coupled with each other. They introduce a factorization ansatz (the Hartree approximation) into the part of $S(\{\theta_i\},t)$ which describes the response to an external force. When the external field is weak, the resulting conductivity at low temperature is Ohmic and proportional to E_s/T times the equilibrium density of excited solitons which is proportional to $(E_s/T)^{1/2}$ $\exp(-E_s/T)$, where $E_s = E_\pm(0)$ of Equation (3.44). When the field exceeds a certain characteristic magnitude E_c, a non-Ohmicity develops rapidly, and then the conductivity saturates (again Ohmic) at high field limits. These qualitative features are in accordance with the experimental observations, though quantitative comparisons of E_c are difficult.

The Fokker–Planck-type equation for the same problem, but not restricted to the strong damping limit, has been examined by Imada [158] in the linear response regime involving an ac external field. He reproduces a central peak in the dynamical structure factor, which is attributable to solitons and is observed also in the computor simulations [159, 160]. The dynamical conductivity turns out to be not so sensitive to solitons, and the frequency dependence of $\sigma(\omega)$ is dominated by oscillations of the pinned CDW. Solitons give rise to significant, but relatively small contributions to $\sigma(\omega)$ only for $\omega \simeq 0$.

Büttiker and Landauer [161] have performed detailed analysis of the origin of the non-Ohmic conduction based on the same model discussed by Trullinger *et al.* [157]. They attributed it to combined processes of the nucleation of soliton and antisoliton pairs, their subsequent separation by the external force, and the eventual annihilation by collision with other solitons. Since only the overdamped case is of interest, the inertia term $\partial^2\theta/\partial t^2$ in Equation (5.3) has been neglected from the beginning, but the spatial correlation effect is fully taken into account, in contrast to the Hartree approximation by Trullinger *et al.* In particular the above-mentioned nucleation rate is evaluated by use of the classical energy of the saddle-point configuration $\theta_N(x)$ and the eigenvalues of the normal modes around $\theta_N(x)$. At $x = \pm\infty$, this configuration $\theta_N(x)$ approaches one of the local minima of the tilted sinusoidal potential θ_s, and moves into the next minimum in between. In case $(\Delta\Omega)^2 \equiv \omega_F{}^2 - 2ev'E \gtrsim 0$ they find a small-amplitude nucleus given by $\theta_N(x) = \theta_s + \Delta\theta_m \operatorname{sech}^2(x/2\xi)$, for which all the evaluations mentioned above can be carried out analytically, where $\Delta\theta_m = 3\sqrt{2}\Delta\Omega/\omega_F$ and $\xi = d/\sqrt{\cos\theta}$. They obtain a nonlinear conductivity proportional to $(\Delta E_N/T)^{1/2} \exp(-\Delta E_N/T)$ in the limit $\Delta\Omega \to 0$ and at low temperatures $\Delta E_N \gg T$, where $\Delta E_N = (3\sqrt{2}/5)E_s\,[\Delta\Omega/\omega_F]^{5/2}$. We also note here that Maki [162] has discussed a similar nucleation process of soliton and antisoliton pairs but through quantum mechanical tunneling. An extension of his theory to finite temperatures has been proposed [163], where a finite number of weakly bound soliton-antisoliton pairs already exist.

Solitons discussed so far are local excitations from a uniform commensurate ground state, as was explained in Section 3.5. When the wave number of CDW is nearly commensurate, the ground state configuration is a soliton lattice explained in Section 3.6. Now we briefly discuss a motion of the soliton lattice, in connection with the problem of the non-Ohmic conduction. A uniformly moving soliton lattice is described by Equation (2.85) with x replaced by $\gamma_R(x - ut)$. It is a solution of Equation (5.3) with vanishing E, F and η. With finite E and η, Equation (5.3) also describes a resistive Josephson line subject to a constant current density, and it has been investigated by Marcus and Imry [164]. The system exhibits non-Ohmic behavior very similar to that obtained by

Trullinger *et al.* [157]. When u reaches the critical velocity u_p, the soliton lattice cannot respond to the external field, instead a whole CDW starts to move. Weger and Horovitz [165] have argued that the non-Ohmicity appears through these two steps of different origins in an underdamped case, while in an overdamped case the Ohmic conduction due to a motion of the soliton lattice is followed directly by the non-Ohmic conduction of a sliding CDW.

Based on a soliton lattice picture, Bak [166] has adopted a little different point of view. The soliton lattice is considered to be pinned by impurity potential as a whole CDW considered before. The non-Ohmicity appears, then, because of the depinning of the soliton lattice but not of a whole CDW. This idea leads to the following results. We see from Equation (3.46) that each soliton in this soliton lattice carries effective charge $2e/M$, M being the degree of the commensurability. The unit of charge is therefore fractional for $M \geq 3$ ($M = 4$ for $NbSe_3$) [78]. Also the fundamental noise frequency expected in the non-Ohmic regime has to be determined by the time required for CDW to traverse a period of the lattice. It is estimated in case of $NbSe_3$ to be ten times smaller than given by Equation (5.13).

6. Transport Properties of $NbSe_3$ and TaS_3

Drastic non-Ohmic conductivity in the PF state of $NbSe_3$ was first observed by Monceau and co-workers [167, 168]. Extensive study since then [13, 14] has now established that, as described by Monceau in Part II, the various anomalous transport properties of $NbSe_3$ are attributable to dynamical nature of the CDW condensate, whose wave number is incommensurate but close to the $M = 4$ commensurability [169]. These anomalies include a very small characteristic field E_c (several mV/cm) which scales behavior of the non-Ohmicity [167], a sudden onset of the non-Ohmic conductivity at a well-defined threshold field E_T which is comparable to E_c [152, 170], an ac linear conductivity very sensitive to the frequency [171], an occurrence of narrow band noises superimposed by a broad band noise in the non-Ohmic region [152, 170], certain hysteresis, such as 'overshooting' [172, 173] and 'switching' [174, 175], associated with a dynamical onset of the non-Ohmic conduction. Almost all of such peculiar transport properties are observed also in TaS_3 [176, 176a], where the properties of this material, even its crystal structure, depend sensitively on the sample preparation [177]. Quite recently similar transport properties have also been found in materials, which belong to a different class of quasi-1-D conductors from $NbSe_3$, namely in a blue bronze $K_{0.30}MoO_3$ [178] and a novel inorganic compound $(TaSe_4)_2I$ [179]. In this section we discuss these interesting transport properties based on the theories described in Section 5, i.e. theories derived from phase Hamiltonian approach to the CDW dynamics.

6.1. DC NON-OHMIC CONDUCTIVITY

The experimental fact that the non-Ohmic behavior is scaled by a small characteristic field E_c excludes an idea that the phenomenon is due to quasi-particles excited above the PF gap. Further the existence of a well-defined threshold field E_T denies a possibility that nonlinear excitations, such as solitons described in Section 5.3, could be responsible for the non-Ohmic conductivity. Actually the large excitation energy of solitons, though reduced by an applied field, is already incompatible with the very small characteristic field E_c. (This does not necessarily mean that conduction of excited solitons does not exist in any circumstances [140, 180].) The existence of a well-defined, but small, E_T implies that some large entities carrying electric charge start to move as a whole at this field. Based on this fact two models have been proposed up to now: the sliding of a whole CDW and that of a soliton lattice on the CDW mentioned in Section 5.3. The latter proposal, i.e. the sliding soliton lattice, due to Bak [166] is based on the fact that $NbSe_3$ [169] (and monoclinic TaS_3 [177]) has an incommensurate CDW close to the $M = 4$ commensurability. This model, though not yet examined in detail [181], yields essentially the same results as the former models and the differences are in the unit of elementary charge and in the narrow-band noise frequency, which are not yet conclusive experimentally [166, 175].

As regards the model of the sliding CDW there exist two points of view. The one is a classical particle model, first proposed by Grüner, Zawadowski and Chaikin [150], and explained in Section 5.2, and the other is Bardeen's model, in which the depinning of CDW is interpreted as a (macroscopic) quantum tunneling process [149, 182]. The latter model is explained by Bardeen himself in this volume. Both of these models can explain some, but not all, of the transport phenomena mentioned above, as we will see in the following. Here we discuss some details of the stationary non-Ohmic conductivity based on the classical particle model.

The experimental data of the (excess) non-Ohmic conductivity above the threshold field E_T in $NbSe_3$ are well fitted with the following empirical formula

$$\frac{\sigma(E)}{\sigma(\infty)} = (1 - 1/y) \exp(-1/y), \tag{6.1}$$

where $y = E/E_T$ [183, 184]. The classical particle model with the periodic model potentials discussed in Section 5.2 does not reproduce Equation (6.1), in particular, its asymptotic behavior at large y as mentioned below Equation (5.10).

One of the possible modifications of the model is to introduce distribution of CDW domains [141, 151, 185]. Actually by assuming an exponential distribution of the spatial extent of coherent regions Portis [185] has derived almost the same expression as Equation (6.1) for the excess current

$$J(E) = \sigma_b (E - E_T) \exp(-E_c/(E - E_T)), \qquad (E > E_T), \tag{6.2}$$

which was proposed empirically by Fleming [170]. In this treatment the depinned and then mobile region increases as the field is increased. Consequently the non-Ohmicity in this model results from the field dependence of the effective carrier number, which appears not to be the case experimentally [13]; i.e. the mobility, not the carrier number, varies once the sliding commences. We believe this is a very important point to be seriously checked by experiments. If the non-Ohmicity is entirely due to the change in the mobility, then Bardeen's tunneling theory, which reproduces Equation (6.1), has also to be reconsidered. By his theory [149], the prefactor $(1 - 1/y)$ in Equation (6.1), which predominantly governs the increase of J_{CDW} just above E_T, is due to the increase of the number of carriers. Another difficulty of this domain-distribution model is in the distribution of E_T, which results in the existence of sliding domains among those still pinned [141].

In order to incorporate, within the classical particle model, further possible change in the mobility than that due to a simple tilting of the potential by the field E, i.e. $V(\theta) - E\theta/E_c$, we have to introduce a certain field dependence of the effective potential $V(\theta)$ itself. An underlying assumption of this idea is that the pinning mechanism is predominantly due to impurities (or defects) which are randomly distributed. Stationary configurations of the CDW phase, which are determined by the two competing energies as explained in Section 4.1, are expected to depend sensitively on the field E, so that the effective potential $V(\theta)$ is, too. This picture, which is similar to that of spin glasses in the presence of the magnetic field [186–188], may be compatible also with the hysteretic phenomena mentioned above. In connection with this variable potential model we note [189] that the empirical formula Equation (6.1) might be interpreted as an interpolation of a function which is proportional to $(1 - 1/y)$ near $y = 1$ and $(1 - 2/y)$ at large y. The triangular potential is shown in Section 5.2 to reproduce the same feature near $y = 1$. On the other hand, the sawtooth potential, which is defined by $V(\theta) = a\ (\theta - 2n\pi)$ in $2n\pi \leq \theta < 2(n + 1)\pi$ and then has discontinuities, yields $\sigma(E)/\sigma(\infty) = 1 - a/y$, where $y = E/E_0$, with E_0 being given by Equation (5.10). Thus a variable potential $V(\theta)$ from the triangular, (C) of Section 5.2 with $E_T = E_0$, to a sawtooth-type with $a \simeq 2$ is expected to reproduce $\sigma(E)$ qualitatively similar to that given by Equation (6.1) (see Figure 17 where the broken line represents $\sigma(E)$ of the sawtooth potential with a fixed $a = 2$).

Although this example of the variable potential is purely phenomenological, it indicates the important roles played by internal degrees of freedom of the CDW condensate in the presence of an external field. A theory, in which the coupling of the internal modes with a sliding CDW itself is taken into account, has been proposed by Sneddon *et al.* [190]. They predicted $E^{-1/2}$ dependence of the non-Ohmic conductivity at large E, which seems not to be observed in the experiments by Oda and Ido [191] with E as large as 50 E_T. If their data are replotted as in [190], a distinct deviation from the $E^{-1/2}$ dependence is seen at $E \geq 15\ E_T$ [191]. On the other hand $J(E)$ convex downwards at $E \gtrsim E_T$ has been recently derived by Fisher [190a] for a model impurity potential whose amplitude is also random. (See also note added in proof.)

6.2. AC LINEAR CONDUCTIVITY

The remarkable ω-dependence of the ac conductivity even at low frequencies ω is another feature associated with the CDW condensate [171, 192]. In the case of a strong damping in the classical particle model, the dynamics is an overdamped oscillation yielding the ac conductivity of Equation (5.14). This picture is, of course, oversimplified. As mentioned above, internal degrees of freedom of the CDW condensate, which are expected to have their own dynamics, certainly have to be taken into account. Cooperative modes involving many degrees of freedom cannot follow the external field if it changes too rapidly. Even for low frequency we may rather expect that some portions of the CDW, not necessarily as a whole, follow the field. This means that the potential $V(\theta)$, which determines τ_{eff} ($\propto V''(O)^{-1}$) in Equation (5.14), is not necessarily identical to $V(\theta)$, which governs motions of a whole CDW [174]. The above considerations have to be taken into account also in analysis of the ac-induced dc conductivity, which is briefly mentioned at the end of Section 5.2. The effect is reduced at higher frequencies [153]. This property is expected from the above argument, but is opposite to that expected from the tunneling theory. On the other hand it is not clear for the moment that the scaling relation between the dc non-Ohmic conductivity $\sigma(E)$ and the ac linear conductivity $\sigma(\omega)$, which has been predicted by the tunneling theory [149, 193] and has been observed in $NbSe_3$ [194], can be accounted for by the impurity pinning [216].

Recently Horovitz and Trullinger [195] have analyzed the ac conductivity $\sigma(\omega)$ of a nearly commensurate case. Their discussions are based on Equation (5.3) but with $F = 0$ in contrast to those by Imada [158] for the commensurate case with $F \neq 0$. They emphasize the role of solitons, or more explicitly, the modification of the continuum modes (phasons) by the presence of solitons, and claim that the ω-dependence becomes sharper as a CDW is closer to the commensurability.

6.3. Noise Associated with the Non-Ohmic Conductivity

Another dramatic phenomenon associated with the CDW dynamics is the occurence of narrow band noises superimposed by a broad band noise, which are observed only in the non-Ohmic region [152]. The narrow band noise spectrum consists of a few or a single [196] fundamental harmonics and their subharmonics [151, 152]. There have been some proposals that the phenomena may be attributed to a chaotic motion of the CDW condensate [197–200]. Since, however, the fundamental frequency Ω_0 of narrow band noise obeys Equation (5.13) well, as already mentioned [151], it is now thought that the narrow band noise is tightly connected to a uniform motion of a whole CDW (or a soliton lattice). A problem remains concerning the relationship between Ω_0 and $J_{\text{CDW}} = en_s v$ with the velocity v; in other words, concerning a characteristic period l_0 in space defined by $\Omega_0 = 2\pi v/l_0$. As pointed out in Section 5.3, $l_0 = l$, the period of the lattice, in the sliding soliton-lattice model. In the sliding CDW (as a whole) model, l_0 depends on the pinning potential: $l_0 = \lambda$ ($\simeq 4a$ for $NbSe_3$), the wave-length of CDW, for the impurity pinning, while $l_0 = a$, the period of the underlying lattice, for the commensurability pinning. This will be answered once an accurate value of n_s is determined experimentally [201–203]. Other mechanisms of the narrow band noise have also been proposed by Portis [185] and Barnes and Zawadowski [204]. We also note that Ong *et al.* [205, 206] have recently claimed that the noise phenomena are not a bulk effect, but they are observed only in the inhomogeneous region in a sample near the contact to leads.

The power spectrum of the large broad band noise, which appears also in the non-Ohmic regime, follows a $f^{-\alpha}$ dependence with $\alpha = 0.8$ in $NbSe_3$ [207] and $\alpha = 0.4$ in TaS_3 [208]. The mechanism of this broad band noise has not yet been well established [207–209].

6.4. HYSTERESIS

Recently observations on dynamical process of the CDW depinning reveal further interesting aspects of the sliding CDW. When the pulsed electric field applied to $NbSe_3$ were alternative in sign, the voltage developed across the specimen rises beyond its dc value, towards which it subsequently decays with a time constant of the order of several $10\mu s$. This is an 'overshooting' phenomenon first observed by Gill [172]. Grüner and Zettl [175] reported a 'switching' phenomenon; i.e. a jump or 'switch' observed in the intermediate region of Ohmic and non-Ohmic conductivity in dc I–V traces for $NbSe_3$ (see also [141, 174]). These hysteresis, which become more distinct at lower temperatures, indicate that some large entities carrying electrical charge are by no means rigid. They are hardly explained by the commensurability potential in Equation (3.67) alone or by the simplified impurity potential of Equation (4.11). We certainly need to consider a number of metastable configurations under the original random potential of Equation (4.5a), and dynamical transfer of the system between such metastable states [210, 211]. Such hysteresis phenomenon of CDW in the presence of randomness has been noted before by Littlewood and Rice [116] in another context. As mentioned before, these hysteresis also remind us of remanent phenomena in spin glasses [186–188], although their characteristic time scales differ greatly in two systems. We think the idea of charge-density-glass (CDG) [104] has to be further investigated.

6.5. THEORETICAL PROBLEMS

Compared with the huge accumulation of experimental data, theoretical development is yet far from satisfactory. Various experimental observations described above certainly require theories involving dynamics of the internal degrees of freedom of the CDW condensate. Even if only the simple commensurability potential is taken into account, or if the impurity effect is represented by the simplified potential of Equation (4.11), full analysis of the Langevin type equation (5.3), derived from our phase Hamiltonian approach, is not easy because of the nonlinear aspect of the equation. In a theory of the charge-density-glass mentioned just above, one has to incorporate also the randomness, which makes the problem quite complicated. Numerical experiments, such as those done by Teranishi and Kubo [106], Sokoloff [144] and Littlewood and Rice [116], will be helpful in this context. Recently Pietronero and Strässler [212] have extended such numerical analysis and examined the characteristic behavior of the excess current at $E \gtrsim E_T$ as well as that of the narrow band noise.

It is a formidable task to explain all the phenomena associated with CDW dynamics by a microscopic theory, i.e. directly from the fundamental Hamiltonians such as Equation (3.1) and (4.1). It is only possible at the moment to evaluate the various parameters, which appear in the phenomenological theories or in analysis of experiments. Some of them, such as u_p, v' and $\omega_F(= \omega_M$, or $\omega_{\text{imp}})$, are already explained in the previous sections. For example the phason velocity u_p is weakly dependent on temperature in the PF state, since a_T/Δ^2 in Equation (3.21) is so (within the mean field approximation). Concerned with n_s in Equation (5.5), the density of electrons which participate a sliding motion of CDW, one needs some cautions as discussed in detail by Lee and Rice [112].

The microscopic derivation of the effective phase Hamiltonian described in Section 3.1 is quite analogous to that of the Ginzburg–Landau equation in superconductivity [188, 213]. From this analogy one may simply expect that n_s is proportional to the square of the order parameter, or $n_s \propto \Delta^2$, just below the transition temperature, T_c [7]. However this is not the case experimentally [142, 184]. In the ground state ($T = 0$) all electrons except those in other parts of the Fermi surface are in the CDW condensate so that n_s equals to the total electron density in the above sense. The gap Δ is simply related with the amplitude of the charge density modulation. As the temperature is raised the number of quasiparticles excited across the PF gap increases. Thus an external electric field is fed to accelerate partly such quasiparticles and partly electrons still in the CDW condensate. The microscopic evaluation of the former process leads to the result $n_s \propto \Delta$ near T_c [112], which is also obtained by the argument on the proper Galilean transformation properties of an energy [136]. From this result combined with the expectation that the amplitude of impurity pinning potential, $\omega_F{}^2$, is proportional to Δ^α $(0 < \alpha < 1)$ [112], the threshold field E_T for the non-Ohmic conductivity is expected to diverge at T_c from below (see Equation (5.10) and note $\tilde{\omega}_F{}^2 = \omega_F{}^2/n_s$), which is actually observed in $NbSe_3$ [171] and TaS_3 [214] but not in $K_{0.30}MoO_3$ [178]. The increase of E_T at lower temperatures can be understood qualitatively by the increase of effective pinning potential due to less thermal fluctuations as was discussed by Okabe and Fukuyama [120] in different context. Up to now microscopic arguments on the relaxation mechanism of a CDW motion are very limited as pointed out in Section 5.1.

7. Summary

One of the interesting characteristics of quasi-one-dimensional conductors is the transport phenomenon due to such collective modes as CDW. These collective modes are considered essentially as bosons as far as the low lying excitations are concerned. In CDW the phase of the lattice distortions can describe these excitations and the phase Hamiltonian has been a convenient theoretical tool. Moreover even the electronic excitations can also be described by boson operators in the case of purely one-dimensional systems, since the possible linearization of the Bloch energy band for the description of excitations near the Fermi energy results in the excitation spectrum similar to familiar phonons. This bosonization has naturally introduced the quantum phase variables to represent electrons. Thus the phase Hamiltonian, different from the previous one, has been employed in various discussions. These two phase Hamiltonians can be combined to describe the whole dynamical processes of the complex systems of mutually interacting electrons coupled to the lattice distortions. We have tried to give a comprehensive theoretical review from this particular point of view, and to discuss the results in the context of very fascinating experimental findings observed in the Peierls–Fröhlich state.

Acknowledgements

One of authors (HF) thanks Y. Suzumura, S. Inagaki, J. Hara, T. Nakano and K. Takano for useful discussions, while the other (HT) thanks T. Sambongi and M. Ido for informative discussions about experiments on MX_3.

References

1. H. J. Keller (Ed.) *Chemistry and Physics of One-Dimensional Metals* (1976), Plenum Press, New York.
2. J. T. Devreese. R. P. Evrard and V. E. van Doren (Eds.) *Highly Conducting One-Dimensional Solids* (1979), Plenum Press, New York.
3. J. Bernasconi and T. Schneider (Eds.) *Physics in One Dimension* (1981), Springer Verlag, Berlin, Heidelberg, New York.
4. *J. de Physique*, Proceedings of the International CNRS Colloquim on the Physics and Chemistry of Synthetic and Organic Metals (Les-Arcs, Dec. 1982). **44**, C3 (1983).
5. R. E. Peierls, *Quantum Theory of Solids* (1955), Oxford University Press, Oxford.
6. H. Frölich, *Proc. Roy. Soc.* **A223**, 296 (1954).
7. D. Allender, J. W. Bray and J. Bardeen, *Phys. Rev.* **B9**, 119 (1974).
8. P. A. Lee, T. M. Rice and P. W. Anderson, *Solid State Commun.* **14**, 703 (1974).
9. L. Pietronero, S. Strässler and G. A. Toombs, *Phys. Rev.* **B12**, 5213 (1975).
10. H. Fukuyama, *J. Phys. Soc. Japan* **41**, 513 (1976).
11. H. Fukuyama and P. A. Lee, *Phys. Rev.* **B17**, 535 (1978).
12. M. J. Rice, A. R. Bishop, J. A. Krumhansl and S. E. Trullinger, *Phys. Rev. Lett.* **36**, 432 (1976).
13. N. P. Ong, *Can. J. Phys.* **60**, 59 (1982).
14. G. Grüner, *Comments on Solid State Physics* **10**, 183 (1983).
15. J. Solyom, *Adv. in Phys.* **28**, 201 (1979).
16. S. Tomonaga, *Prog. Theor. Phys.* **5**, 349 (1950).
17. A. Luther and I. Peschel, *Phys. Rev.* **B9**, 2911 (1974).
18. A. Luther and V. J. Emery, *Phys. Rev. Lett.* **33**, 589 (1974).
19. D. C. Mattis, *J. Math. Phys.* **15**, 609 (1974).
20. F. D. M. Haldane, *J. Phys. C.* **12**, 4791 (1979).
21. F. D. M. Haldane, *Phys. Rev. Lett.* **45**, 1358 (1980).
22. R. Heidenreich, B. Schroer, R. Seiler and D. Uhlenbrock, *Phys. Lett.* **54A**, 119 (1975).
23. K. B. Efetov and A. I. Larkin, *Soviet Physics-JETP* **42**, 390 (1976).
24. A. Luther, *Phys. Rev.* **B15**, 403 (1977).
25. Y. Suzumura, *Prog. Theor. Phys.* **61**, 1 (1979).
26. J. Hara, T. Nakano and H. Fukuyama, *J. Phys. Soc. Japan* **51**, 341 (1982).
27. K. Takano, T. Nakano and H. Fukuyama, *J. Phys. Soc. Japan* **51**, 366 (1982).
28. J. Hara and H. Fukuyama, in ref. 4; *J. Phys. Soc. Japan* **52**, 2128 (1983).
29. T. Nakano and H. Fukuyama, *J. Phys. Soc. Japan* **49**, 1679 (1980); **50**, 2489 (1981).
30. S. Inagaki and H. Fukuyama, *J. Phys. Soc. Japan* **52**, 2504 (1983); **52**, 3620 (1983).
31. J. W. Bray, L. V. Interrante, I. S. Jacobs and J. C. Bonner in *Extended Linear Chain Compounds* (1982) vol. 3, Plenum Press, New York.
32. E. Abrahams, P. W. Anderson, D. C. Licciardello and T. V. Ramakrishnan, *Phys. Rev. Lett.* **42**, 673 (1979).
33. P. W. Anderson, *Physica* **117** and **118 B + C**, 30 (1983).
34. Y. Nagaoka and H. Fukuyama (Eds.) *Anderson Localization* (1982) Springer Verlag, Berlin, Heidelberg, New York.
35. S. Takada and S. Misawa, *Prog. Theor. Phys.* **66**, 101 (1981).
36. J. M. Luttinger, *J. Math. Phys.* **4**, 1154 (1963).
37. Yu. A. Bychkov, L. P. Gorkov and I. E. Dzyaloshinskii, *Soviet Physics-JETP* **23**, 489 (1966).
38. T. M. Rice, *Phys. Rev.* **104A**, 1889 (1965).
39. R. Dashen, B. Hasslacher and A. Neveu, *Phys. Rev.* **D11**, 3424 (1975).
40. S. Coleman, *Phys. Rev.* **D11**, 2088 (1975).
41. N. Menyhard and J. Solyom, *J. Low Temp. Phys.* **12**, 529 (1973).
42. J. Solyom, *J. Low Temp. Phys.* **12**, 547 (1973).
43. P. Bak, *Rept. on Prog. in Phys.* **45**, 587 (1982).
44. L. A. Turkevich and S. Doniach, *Ann. of Phys.* **139**, 343 (1982).
45. A. L. Fetter and M. J. Stephen, *Phys. Rev.* **168**, 475 (1968).

46. B. Sutherland, *Phys. Rev.* **A8**, 2514 (1973).
47. W. L. McMillan, *Phys. Rev.* **14**, 1496 (1976).
48. S. Iwabuchi and H. Fukuyama, in ref. 4, p. 1001.
49. J. C. Scott, H. J. Pedersen and K. Bechgaard, *Phys. Rev. Lett.* **45**, 2125. (1980).
50. K. Mortensen, Y. Tomkiewicz, T. D. Schultz and E. M. Engler, *Phys. Rev. Lett.* **46**. 1234 (1981).
51. D. J. Scalapino, Y. Imry and P. Pincus, *Phys. Rev.* **B11**, 2042 (1975).
52. K. B. Efetov and A. I. Larkin, *Soviet Physics-JETP* **39**, 1129 (1974).
53. P. Bak and V. J. Emery, *Phys. Rev. Lett.* **36**, 978 (1976).
54. K. Saub, S. Barisic and J. Friedel, *Phys. Lett.* **56A**, 302 (1976).
55. T. D. Schultz and S. Etemad, *Phys. Rev.* **B13**, 4928 (1976).
56. S. Nakajima and Y. Okabe, *J. Phys. Soc. Japan* **42**, 1115 (1977).
57. J. Yamauchi, K. Sato, S. Iwabuchi and Y. Nagaoka, *J. Phys. Soc. Japan* **44**, 460 (1978).
58. K. Sato, S. Iwabuchi, J. Yamauchi and Y. Nagaoka, *J. Phys. Soc. Japan* **45**, 515 (1978).
59. P. Bak and S. A. Brazovskii, *Phys. Rev.* **B17** 3154 (1978).
60. H. Fukuyama, *J. Phys. Soc. Japan* **45**, 1266 (1978).
61. I. Batistic, G. Theodorou and S. Barisic, *Solid State Commun.* **34**, 499 (1980).
62. L. N. Bulaevskii, A. I. Buzdin and D. I. Khomskii, *Solid State Commun.* **35**, 101 (1980).
63. Y. Suzumura and H. Fukuyama, *J. Phys. Soc. Japan* **49**, 915 (1980); **49**, 2081 (1980).
64. S. A. Brazovskii and I. E. Dzyaloshinskii, *Soviet Physics-JETP* **44**, 1233 (1976).
65. P. A. Lee, T. M. Rice and P. W. Anderson, *Phys. Rev. Lett.* **31**, 462 (1973).
66. A. A. Abrikosov, L. P. Gorkov and I. E. Dzyaloshinskii, *Method of Quantum Field Theory in Statistical Physics* (1963), Prentice-Hall, Englewood Cliffs.
67. S. Kurihara, *J. Phys. Soc. Japan* **44**, 2011 (1978), and *ibid.* 48, 1821 (1980).
68. K. Takano, *Prog. Theor. Phys.* **68**, 1 (1982).
69. E. J. Mele and M. J. Rice, *Phys. Rev.* **B23**, 5397 (1981).
70. H. J. Schulz, *Phys. Rev.* **B18**, 5756 (1978).
71. M. Nakahara and K. Maki, *Phys. Rev.* **B25**, 7789 (1982).
72. J. T. Gammel and J. A. Krumhansl, *Phys. Rev. Lett.* **49**, 899 (1982).
73. A. C. Scott, F. Y. F. Chu and D. W. McLaughlin, *Proc. IEEE* **61**, 1443 (1973).
74. A. R. Bishop and T. Schneider (Ed.) *Solitons and Condensed Matter Physics* (1978), Springer-Verlag, Berlin.
75. T. L. Ho, *Phys. Rev. Lett.* **48**, 946 (1982).
76. B. Horovitz, *Phys. Rev. Lett.* **48**, 1416 (1982); and in ref. 4, p. 1519.
77. J. A. Krumhansl, B. Horovitz and A. J. Heeger, *Solid State Commun.* **34**, 945 (1980).
78. W. P. Su and J. R. Schrieffer, *Phys. Rev. Lett.* **46**, 738 (1981).
79. W. P. Su, J. R. Schrieffer and A. J. Heeger, *Phys. Rev. Lett* **42**, 1698 (1979), and *Phys. Rev.* **B22**, 2099 (1980).
80. M. J. Rice, *Phys. Lett.* **71A**, 152 (1979).
81. H. Takayama, Y. R. Lin-Liu and K. Maki, *Phys. Rev.* **B21**, 2388 (1980).
82. B. Horovitz, *Phys. Rev.* **B22**, 1101 (1980).
83. W. P. Su and J. R. Schrieffer, *Proc. Natl. Acad. Sci. USA* 77, 5626 (1980).
84. D. K. Campbell and A. R. Bishop, *Phys. Rev.* **B24**, 4859 (1981).
85. S. A. Brazovskii, *Soviet Physics-JETP Lett.* **28**, 606 (1978).
86. S. A. Brazovskii and N. N. Kirova, *Soviet Physics-JETP Lett.* **33** 4 (1981).
87. W. P. Su, *Phys. Rev.* **B27**, 370 (1983).
88. J. T. Gammel and J. A. Krumhansl, *Phys. Rev.* **B27**, 7659 (1983).
89. S. A. Brazovskii, I. E. Dzyaloshinskii and S. P. Obukhov, *Soviet Physics-JETP* **45**, 814 (1977).
90. E. Pytte, *Phys. Rev.* **B10**, 4637 (1974).
91. F. D. M. Haldane, *Phys. Rev. Lett.* **45**, 1358 (1980).
92. M. C. Cross and D. S. Fisher, *Phys. Rev.* **B19**, 402 (1979).
93. P. W. Anderson, *Mater. Res. Bull.* **8**, 153 (1973).
94. W. Duffy, Jr., and K. P. Barr, *Phys. Rev.* **165**, 647 (1968).
95. J. C. Bonner and H. W. J. Blöte, *Phys. Rev.* **B25**, 6959 (1982).

96. H. P. van de Braak, W. J. Caspers, P. W. Wiegel and M. W. M. Willeme, *J. Stat. Phys.* **18**, 579 (1978).
97. J. N. Fields, H. W. J. Blöte and J. C. Bonner, *J. Appl. Phys.* **50**, 1807 (1979).
98. J. L. Black and V. J. Emery, *Phys. Rev.* **B23**, 429 (1981).
99. Y. Imry and S. K. Ma, *Phys. Rev. Lett.* **35**, 1399 (1975).
100. L. J. Sham and B. R. Patton, *Phys. Rev.* **B13**, 3151 (1976).
101. D. J. Bergmann, T. M. Rice and P. A. Lee. *Phys. Rev.* **B15**, 1706 (1977).
102. K. B. Efetov and A. I. Larkin, *Soviet Physics-JETP* **45**, 1236 (1977).
103. L. P. Gorkov, *Soviet Physics-JETP Lett.* **25**, 358 (1977).
104. H. Fukuyama, *J. Phys. Soc. Japan* **45**, 1474 (1978).
105. M. V. Feigelman, *Soviet Physics-JETP* **52**, 555 (1980).
106. N. Teranishi and R. Kubo, *J. Phys. Soc. Japan* **47**, 720 (1979).
107. T. R. Koehler and P. A. Lee, *Phys. Rev.* **B16**, 5263 (1977).
108. J. F. Weisz. J. B. Sokoloff and J. E. Sacco, *Phys. Rev.* **B20**, 4713 (1979).
109. P. A. Lee and H. Fukuyama, *Phys. Rev.* **B17**, 542 (1978).
110. J. B. Sokoloff, *Phys. Rev.* **B16**, 3367 (1977).
111. M. L. Boriack and A. W. Overhauser, *Phys. Rev.* **B16**, 5206 (1977).
112. P. A. Lee and T. M. Rice, *Phys. Rev.* **B19**, 3970 (1979).
113. R. A. Klemm and J. R. Schrieffer, *Phys. Rev. Lett.* **51**, 47 (1983).
114. J. B. Sokoloff, *Phys. Rev.* **B23**, 1992 (1981) and in ref. [4], p. 1667.
115. K. Nakanishi, *J. Phys Soc. Japan* **46**, 1434 (1979).
116. P. B. Littlewood and T. M. Rice, *Phys. Rev. Lett.* **48**, 44 (1982).
117. D. Baeriswyl and A. R. Bishop, *J. Phys. C.* **13**, 1403 (1980).
118. T. M. Rice, S. Whitehouse and P. Littlewood, *Phys. Rev.* **B24**, 2751 (1981).
119. M. B. Fogel, S. E. Trullinger, A. R. Bishop and J. A. Krumhansl, *Phys. Rev.* **B15**, 1578 (1977).
120. Y. Okabe and H. Fukuyama, *Solid State Commun.* **20**, 345 (1976).
121. S. T. Chui, *Phys. Rev.* **B19**, 4333 (1979).
122. A. Houghton, R. D. Kenway and S. C. Ying, *Phys. Rev.* **B23**, 298 (1981).
123. H. Fukuyama, T. M. Rice and C. M. Varma, *Phys. Rev. Lett.* **33**, 305 (1974).
124. B. P. Patton and L. J. Sham, *Phys. Rev. Lett.* **31**, 631 (1973), *ibid.* **33**, 638 (1974).
125. S. Kurihara, *J. Phys. Soc. Japan* **44**, 2011 (1978).
125a.. S. Takada and E. Sakai, *Prog. Theor. Phys.* **59**, 1802 (1978).
126. S. Kurihara, *J. Phys. Soc. Japan* **41**, 1488 (1976).
127. K. Okamoto, *Prog. Theor. Phys.* **68**, 1443 (1982).
128. B. L. Altshuler and A. G. Aronov, *Solid State Commun.* **30**, 115 (1979).
129. B. L. Altshuler, A. G. Aronov and P. A. Lee, *Phys Rev. Lett.* **44**, 1288 (1980).
130. H. Fukuyama, *J. Phys. Soc. Japan* **48**, 2169 (1980).
131. S. T. Chui and J. W. Bray, *Phys. Rev.* **B16**, 1329 (1977); **B19**, 4020 (1982).
132. W. Apel, *J. Phys. C* **15**, 1973 (1982).
133. W. Apel and T. M. Rice, *Phys. Rev.* **B26**, 7063 (1982).
134. V. L. Berezinsky, *Soviet Physics-JEPT* **38**, 620 (1974).
135. Y. Suzumura and H. Fukuyama, *J. Phys. Soc. Japan* **52**, 2870 (1983).
136. M. L. Boriack and A. W. Overhauser, *Phys. Rev.* **B17**, 2395 (1978).
137. T. M. Rice, P. A. Lee and M. C. Cross, *Phys. Rev.* **B20**, 1345 (1979).
137a. J. E. Sacco, J. B. Sokoloff and A. Widom, *Phys. Rev.* **B20**, 5071 (1979).
138. G. X. Tessema and N. P. Ong, *Phys. Rev.* **B23**, 5607 (1981).
139. K. Kawabata, M. Ido and T. Sambongi, *J. Phys. Soc. Japan* **50**, 739, 1992 (1981).
140. T. Takoshima, M. Ido, K. Tsutsumi, T. Sambongi, S. Homma, K. Yamaya and Y. Abe, *Solid State Commun.* **35**, 911 (1980).
141. P. Monceau, J. Richard and M. Renard, *Phys. Rev.* **B25**, 931 (1982).
142. J. Bardeen, E. Ben-Jacob, A. Zettl and G. Grüner, *Phys. Rev. Lett.* **49**, 493 (1982).
143. V. Ambegaokar and B. I. Halperin, *Phys. Rev. Lett.* **22**, 1364 (1969).
144. P. E. Lindelof, *Rep. Prog. Phys.* **44**, 949 (1981).
145. P. Fulde, L. Pietronero, W. R. Schneider and S. Strässler, *Phys. Rev. Lett.* **35**, 1776 (1975).

146. W. Dieterich, I. Peschel and W. R. Schneider, *Z. Phys.* **B27**, 177 (1977).
147. B. D. Josephson, *Adv. Phys.* **14**, 419 (1965).
148. G. Wilemski, *J. Statis. Phys.* **14**, 153 (1976).
149. J. Bardeen, *Phys. Rev. Lett.* **45**, 1978 (1980).
150. G. Grüner, A. Zawadowski and P. M. Chaikin, *Phys. Rev. Lett.* **46**, 511 (1981).
151. J. Richard, P. Monceau and M. Renard, *Phys. Rev.* **B25**, 948 (1982).
152. R. M. Fleming and C. C. Grimes, *Phys. Rev. Lett.* **42**, 1923 (1979).
153. G. Grüner, W. G. Clark and A. M. Portis, *Phys. Rev.* **B24**, 3641 (1981).
154. J. F. Currie, J. A. Krumhansl, A. R. Bishop and S. E. Trullinger, *Phys. Rev.* **B22**, 477 (1980).
155. J. A. Krumhansl and J. R. Schrieffer, *Phys. Rev.* **B11**, 3535 (1975).
156. H. Takayama and K. Maki, *Phys. Rev.* **B21**, 4558 (1980).
157. S. E. Trullinger, M. D. Miller, R. A. Guyer, A. R. Bishop, F. Palmer and J. A. Krumhansl, *Phys. Rev. Lett.* **40**, 206 (1978).
158. M. Imada, *J. Phys. Soc. Japan* **50**, 401 (1981).
159. T. Schneider and E. Stoll, *Phys. Rev.* **B23**, 4631 (1981).
160. W. C. Kerr, D. Baeriswyl and A. R. Bishop, *Phys. Rev.* **B24**, 6566 (1981).
161. B. Büttiker and R. Landauer, *Phys. Rev. Lett.* **43**, 1453 (1979).
162. K. Maki, *Phys. Rev. Lett.* **39**, 46 (1977), *Phys. Rev.* **B18**, 1641 (1978).
163. G. Mihaly and L. Mihaly, *Solid State Commun.* **29**, 645 (1979).
164. P. M. Marcus and Y. Imry, *Solid State Commun.* **33**, 345 (1980).
165. M. Weger and B. Horovitz, *Solid State Commun.* **43**, 583 (1982).
166. P. Bak, *Phys. Rev. Lett.* **48**, 692 (1982).
167. P. Monceau, N. P. Ong, A. M. Portis, A. Meerschaut and J. Rouxel, *Phys. Rev. Lett.* **37**, 602 (1976).
168. N. P. Ong and P. Monceau, *Phys. Rev.* **B16**, 3443 (1977).
169. K. Tsutsumi, T. Takagaki, M. Yamamoto, Y. Shiozaki, K. Ido, T. Sambongi, K. Yamaya and Y. Abe, *Phys. Rev. Lett.* **39**, 1675 (1977).
170. R. M. Fleming, *Phys. Rev.* **B22**, 5606 (1980).
171. N. P. Ong and P. Monceau, *Phys. Rev.* **B16**, 3443 (1977).
172. J. C. Gill, *Solid State Commun.* **39**, 1203 (1981).
173. R. M. Fleming, *Solid State Commun.* **43**, 167 (1982).
174. G. X. Tessema and N. P. Ong, *Phys. Rev.* **B27**, 1417 (1983).
175. G. Grüner and A. Zettl, *Phys. Rev.* **B26**, 2298 (1982).
176. G. Grüner, *Mol. Cryst. Liq. Cryst.* **81** 17 (1982).
176a. K. Hasegawa, A. Maeda, S. Uchida and S. Tanaka, *Physica* **117B** & **118B**, 599 (1983).
177. C. Roucau, R. Ayroles, P. Monceau, L. Guemas, A. Meerschaut and J. Rouxel, *Phys. Stat. Sol.* **62a**, 483 (1980).
178. J. Dumas, C. Schlenker, J. Marcus and R. Buder, *Phys. Rev. Lett.* **50**, 757 (1983).
179. M. Maki, M. Kaiser, A. Zettl and G. Grüner, *Solid State Commun.* **46**, 497 (1983).
180. R. Allegyer, B. H. Suits and F. C. Brown, *Solid State Commun.* **43**, 207 (1982).
181. G. Grüner and A. Zettl, in ref. [4], p. 1631.
182. J. Bardeen, *Phys. Rev. Lett.* **42**, 1498 (1979).
183. J. Richard and P. Monceau, *Solid State Commun.* **33**, 635 (1980).
184. M. Oda and M. Ido, *Solid State Commun.* **44**, 1535 (1982).
185. A. P. Portis, *Mol. Cryst. Liq. Cryst.* **81**, 59 (1982).
186. J. L. Tholence and R. Tournier, *J. de Phys.* **35**, C4–229 (1974).
187. P. Monod, J. J. Prejean and B. Tissier, *J. Appl. Phys.* **50**, 1324 (1979).
188. J. Ferre, J. Rajchenbach and H. Maletta, *J. Appl. Phys.* **52**, 1697 (1981).
189. H. Takayama and H. Matsukawa, unpublished.
190. L. Sneddon, M. C. Cross and D. S. Fisher, *Phys. Rev. Lett.* **49**, 292 (1982).
190a. D. S. Fisher, *Phys. Rev. Lett.* **50**, 1486 (1983).
191. M. Oda and M. Ido, personal communications.
192. G. Grüner, L. C. Tippie, J. Sanny, W. G. Clark and N. P. Ong, *Phys. Rev. Lett.* **45**, 935 (1980).
193. R. J. Tucker, *IEEE J. Quant. Electron.* **15**, 1234 (1979).

194. G. Grüner, A. Zettl, W. G. Clark and J. Bardeen, *Phys. Rev.* **B24**, 7247 (1981).
195. B. Horovitz and S. E. Trullinger, *Solid State Commun.* **49**, 195 (1984).
196. M. Weger, G. Grüner and W. G. Clark, *Solid State Commun.* **44**, 1179 (1982).
197. B. A. Huberman and J. P. Crutchfield, *Phys. Rev. Lett.* **43**, 1743 (1979).
198. J. P. Crutchfield, J. D. Farmer and B. A. Huberman, *Phys. Rep.* (*Phys. Lett. Sec. C*) **92**, 47 (1982).
199. S. Chatterjee and S. V. Subramanyam, *Solid State Commun.* **41**, 541 (1982).
200. D. Bennett, A. R. Bishop, S. E. Trullinger, *Z. Phys.* **B47**, 265 (1982).
201. N. P. Ong and P. Monceau, *Solid State Commun.* **26**, 487 (1978).
202. N. P. Ong and J. W. Brill, *Phys. Rev.* **B18**, 5265 (1978).
203. R. J. Wagner and N. P. Ong, *Solid State Commun.* **46**, 491 (1983).
204. S. E. Barnes and A. Zawadowski, *Phys. Rev. Lett.* **51**, 1003 (1983).
205. N. P. Ong and G. Verma, *Phys. Rev.* **B27**, 4495 (1983).
206. N. P. Ong, G. Verma and K. Maki, *Phys. Rev. Lett.* **52**, 663 (1984).
207. J. Richard, P. Monceau, M. Papoular and M. Renard, *J. Phys. C* **15**, 7157 (1982).
208. A. Zettl and G. Grüner, *Solid State Commun.* **46**, 29 (1983).
209. M. Papoular, *Phys. Rev.* **B25**, 7856 (1982).
210 D. W. Ruesink, J. M. Perz and I. M. Templeton, *Phys. Rev. Lett.* **45**, 734 (1980).
211. E. W. Fenton, *Phys. Rev. Lett.* **45**, 736 (1980).
212. L. Pietronero and S. Strässler, *Phys. Rev.* **B28**, 5863 (1983).
213. H. Takayama, *Prog. Theor. Phys.* **48**, 382 (1972).
214. A. H. Thompson, A. Zettl and G. Grüner, *Phys. Rev. Lett.* **47**, 64 (1981).
215. H. Matsukawa and H. Takayama, *Solid State Commun.* **50**, 283 (1984).
216. L. Sneddon, *Phys. Rev. Lett.* **52**, 65 (1984).

[Note added in proof]. In a recent computer work [215] based on the Hamiltonian of Equations (4.3) and (4.9), the total impurity force acting on the CDW phase in the non-Ohmic regime is shown to have a constant part, i.e. the effective potential $V(\theta)$ involves partly a saw-tooth type, although its magnitude is not large enough to reproduce experimental results of $NbSe_3$ (Equation (6.1)). In this context it is worth pointing out that if $\partial\theta/\partial t = \pi J/n_s e = \pi\sigma(E)E/n_s e$ with $\sigma(E)$ of Equation (6.1) substituted into Equation (5.11), then it turns out that the pinning force $-\tilde{\omega}_F{}^2\,\partial V/\partial\theta$ at $E \gg E_T$ is twice as large as that just at $E = E_T$. Such increase of the pinning force after the CDW is de-pinned has not yet explored in the classical medium model for the CDW condensate.

DEPINNING OF CHARGE-DENSITY WAVES BY QUANTUM TUNNELING

JOHN BARDEEN

Department of Physics,
University of Illinois at Urbana-Champaign,
1110 W. Green Street,
Urbana, IL 61801, U.S.A.

1. Introduction

A review of the basic theories that have been given for charge transport by the Fröhlich mechanism of moving charge-density waves is given in this book in the chapter by Fukuyama and Takayama. This chapter will review the theory of and evidence for depinning of incommensurate CDW's that are pinned by weak impurities by quantum tunneling. By weak is meant that a single impurity cannot pin the phase of the wave but pinning results from fluctuations in impurity concentration as suggested by the theories of Fukuyama and Lee [1] and Lee and Rice [2].

Quantum tunneling was suggested in the initial papers on conduction in $NbSe_3$ by Monceau *et al.* [3] who found that the nonlinear dc current, $I(E)$, as a function of the electric field, E, could be expressed approximately in the form:

$$I(E) = \sigma_a E + \sigma_b E \exp[-E_0/E]. \tag{1.1}$$

The exponential suggests Zener tunneling across a small pinning gap. However the model was rejected because the gap required for the small values of the depinning fields E_0 ($\sim$ 1–2 V/cm below the upper transition, T_1, and $\sim$ 0.1–0.2 V/cm below T_2) was found to be much smaller than thermal energy, $k_B T$, and so would be wiped out by thermal effects. In 1979, the author [4a] pointed out that the model could be revived if applied only to the coherent motion of electrons in the condensate of the CDW. In the absence of pinning, an incommensurate CDW should be accelerated freely by an electric field. The Peierls gaps move with the Fermi sea and do not affect the motion of the condensate, but they do inhibit excitation of individual particles. If the number of electrons involved in the coherent motion is sufficiently large ($\sim 10^9$ or 10^{10}), the total energy involved is much larger than $k_B T$.

The discovery that there is a threshold field for nonlinear conduction by Fleming and Grimes, [5] since confirmed by many others, suggested [4b] that electrons can be accelerated and pick up energy from the field only over a distance, L, which is interpreted as the distance over which there is phase coherence of the CDW. This model also suggested that the voltage across L could be treated in analogy with the voltage across a superconductor-insulator-normal (SIN) tunnel junction or an SIS junction. In an important paper, J. R. Tucker [6] extended the Tien–Gordon theory of photon-assisted tunneling in tunnel junctions to discuss the quantum limits of detection and mixing in such

P. Monceau (ed.), Electronic Properties of Inorganic Quasi-One-Dimensional Materials, I, 105–123.

junctions. A complete quantum theory was given for response to combined ac and dc fields, treating $\hbar\omega$ as a quantum of energy.

When taken over bodily and applied to the CDW problem, Tucker's theory made a number of predictions that could be tested experimentally. One of these was a scaling relation between the frequency and field dependence of the conductivity. This relation has been confirmed in a number of experiments on $NbSe_3$ and TaS_3, mainly by Grüner and Zettl [7, 8]. However, other predictions of the theory in its original form failed. In particular photon-assisted tunneling in the sense of observing a rectified signal by biasing below threshold with the energy quantum $\hbar\omega$ to go above threshold was not observed.

In 1982, J. R. Tucker instigated an experimental program at Illinois to study detection and mixing in CDW systems to try to better understand effects of ac and combined ac and dc voltages on CDW systems. Experiments with Miller, Richard and others showed that dc biases below threshold had little or no effect on any ac response. This suggested a revised form of scaling in which scaling is between the frequency and the voltage above threshold rather than to the total applied voltage. This revised form of scaling has been remarkably successful in accounting for a wide range of detection and mixing experiments and gives strong evidence in favor of the tunneling model.

This short history shows that the tunneling model has developed from a close interaction between theory and experiments. It has come in large part intuitively from experiment rather than derived from basic microscopic theory. If the theory is correct, quantum effects of frequencies in the range of $10^6 - 10^{10}$ Hz have been observed in transport phenomena at temperatures as high as 200 K. The CDW system should be a fertile field for the study of macroscopic quantum phenomena.

In Section 2 is described the Fukuyama–Lee–Rice theory of impurity pinning. The classical equations of motion for the phase of the CDW are given in Section 3. Solutions of these equations near threshold and for currents well above threshold are given in Section 4. The tunneling model for depinning is introduced in Section 5. All of the energy and momentum transferred to the electron–phonon system that makes up the CDW must go through the electrons, and it is here that the quantum step is required. Section 6 gives a brief discussion of the application of Tucker's theory and its application to detection and mixing phenomena.

No attempt is made to discuss hysteresis and memory effects or alternative theories that have been derived to account for various CDW phenomena with different models. Other theories of impurity pinning, such as that of Barnes and Zawadowski [9] may be consistent with the tunneling model.

This chapter is based in large part on an invited talk given in March 1983, at a meeting of the American Physical Society and on lectures [10] given at the International School of Physics 'Enrico Fermi', Varenna, Italy, in July 1983 and at an International Symposium at Hokkaido University, Sapporo, Japan, in October 1983.

2. Weak Impurity Pinning

According to the theory of Fukuyama and Lee [1] and Lee and Rice, [2] when the pinning is weak so that a single impurity cannot pin the phase of the CDW, pinning is determined by fluctuations in numbers of impurities in regions whose volume is determined by a balance between the increase in kinetic energy from changes in electron density required to change the phase of the wave and the energy gain from the phase-dependent impurity pinning energy. The Lee–Rice phase-coherent regions are of length L and volume $\Omega = (\xi_\perp/\xi_\parallel)^2 L^3$, where $\xi_\perp$ and $\xi_\parallel$ are coherence distances transverse and parallel to the chain direction. The increase in kinetic energy (K.E.) per electron is of order $\hbar^2/(2mL^2)$, the energy required to confine an electron to the length L, and the negative pinning energy, proportional to the fluctuation, $N_i^{1/2}$, in number impurities is the region, varies as $L^{3/2}$. Here, N_i is the total number of impurities in Ω.

It is found that L varies inversely with the impurity concentration, c_i. In reasonably pure crystals ($c_i \sim 10^{-6}$), L is quite large, of the order of 3×10^{-3} cm for $NbSe_3$, and the phase coherent volume, Ω, is of order 10^{-12} cm^3. The pinning energy is of order $0.1\ \hbar^2/mL^2$ or $\sim 10^{-23}$ ergs/elec. With an electron density of 1.5×10^{21}/cm^3 there are of the order of 10^{10} electrons in a coherent volume. The pinning energy in Ω is of the order of 10^{-13} ergs, larger than the thermal energy at 100 K. Note that the total pinning energy in Ω is proportional to L so that it is the larger Lee–Rice regions that are stable against thermal fluctuations.

The phase can be adjusted to any value between 0 and 2π (mod 2π) at a point in the center of Ω (say $x = 0$) by adding to the phase of the uniform CDW slowly varying functions of the form

$$\phi_A = \frac{\pi}{2} \sin\left(\frac{\pi x}{L} + \phi_0\right) \tag{2.1a}$$

$$\phi_B = \pi + \frac{\pi}{2} \sin\left(\frac{\pi x}{L} + \phi_0\right) \tag{2.1b}$$

where ϕ_0 in the range $-\pi/2 < \phi_0 < \pi/2$. The change δn in the electron density in Ω is positive for this choice of ϕ_0. Equally well, one can adjust the phase to any value at $x = 0$ by taking ϕ_0 in the range $\pi/2 < \phi_0 < 3\pi/2$, in which case the δn is negative.

The phase can be adjusted with either sign of charge, so that there are many metastable configurations. This may be related to the observed hysteresis and memory effects. It should be noted that the one electron per chain in a distance L is sufficient to adjust the phase.

3. Equations of Motion for Phase

The possibility that CDW's formed below a critical temperature may become depinned and move through the lattice was suggested by Fröhlich [11] in 1954 as a model for a one-dimensional superconductor. The moving CDW's transport electrons with the drift velocity v_d and if there were no dissipation the current would persist in time. It is now known that a moving CDW does dissipate energy, so it is not a superconductor. Nevertheless, Fröhlich's model, based on pairing of electrons and holes with wave vectors differing by $2k_F$, does have features in common with the BCS theory of superconductivity in which there is pairing of electrons with a common total momentum. Both give an energy gap at the Fermi surface and may be described by a complex order parameter with amplitude and phase. Both yield a Ginzburg–Landau type theory for the order parameter near T_c.

Fröhlich used an elastic continuum model in which below T_c a phonon with wave vector $Q = 2k_F$ is macroscopically occupied. If the CDW moves with a velocity v_d, the charge density becomes

$$\rho = \rho_0 + \rho_1 \cos 2k_F(x - v_d t). \tag{3.1}$$

Peierls gaps appear at the boundaries of the moving 1D Fermi surface (FS) defined by the wave vectors $-k_F + q$, $k_F + q$, where $\hbar q = mv_d$. In a more general band model, [12] m is the band mass at the FS. At low temperatures, in the absence of quasiparticle excitations, the current is the same as it would be in the absence of the Peierls gaps:

$$j = nev_d, \tag{3.2}$$

a feature in common with a superconductor with a gap.

The moving CDW may be regarded as a macroscopic occupation of a phonon of wave vector $2k_F$ and frequency $\omega_d = 2k_F v_d$. The energy difference between the two sides of the Fermi surface is $2\hbar k_F v_d$, just equal to the energy quantum $\hbar\omega_d$ of the phonon. The CDW is the equilibrium resulting from emission and absorption of $2k_F$ phonons by the electrons. For a uniform v_d the only degree of freedom is a coherent motion of all of the electrons in the volume so as to change v_d. Excitation of individual electrons is inhibited by the Peierls gap.

In a moving CDW the ions oscillate about their equilibrium positions so that the total kinetic energy is much larger than that of the electrons alone. The total K.E. per electron associated with the motion may be written [11]

$$\text{K.E./elec} = \tfrac{1}{2}(m + M_F)v_d^2, \tag{3.3}$$

where $M_F \sim 10^3$ m is the Fröhlich mass representing the K.E. of the ion motion.

A density and drift velocity varying slowly in space and time may be described by the phase, $\phi(x, t)$, of the CDW relative to the lattice. The charge density is defined by

$$\rho = \rho_0 + \rho_1 \cos \phi(x, t). \tag{3.4}$$

The electron density, $n(x, t)$, per chain is given by the local value of $2k_F/\pi$, or [12]

$$n(x, t) = (1/\pi)\partial\phi/\partial x. \tag{3.5}$$

The particle current per chain, nev_d, is given by [12]

$$j(x, t) = -(1/\pi)\partial\phi/\partial t. \tag{3.6}$$

Equality of cross-derivatives gives the equation of continuity

$$\frac{\partial j}{\partial x} + \frac{\partial n}{\partial t} = 0. \tag{3.7}$$

Equations (3.5) and (3.6) imply that each node of the CDW carries two electrons. These equations can be generalized to the three-dimensional case.

To describe slow variations in space and time, one may write

$$\phi(x, t) = 2k_F x + \theta_0(x) + \theta(x, t) \tag{3.8}$$

where θ is a slowly varying function of space and time. The phase $\theta_0(x)$ is that required to minimize the energy from fluctuations in impurity concentration with weak impurity pinning. The pinning energy is then proportional to a periodic function, $V(\theta)$, of the change, θ, in phase from that which gives the energy minimum, $\theta_0(x)$. The equation of motion for θ is generally written in the form (see Chapter 2, Equation (5.1))

$$\frac{\partial^2\theta}{\partial t^2} - u_P^2 \frac{\partial^2\theta}{\partial x^2} + \omega_P^2 \frac{dV}{d\theta} = \frac{2k_F eE}{M_F} = \frac{2k_F e^* E}{m} \tag{3.9}$$

where $u_P = (m/M_F)^{1/2} v_F$ is the phason velocity, ω_P is the pinning frequency, $e^* = (m/M_F)e$ and $V(\theta) \to \theta$ when θ is small. In the semi-classical approach a phenomenological damping term is added to (3.9).

Phase is also an important variable to describe current flow in a superconductor, but the relations are different. The space derivative gives the canonical momentum of the superfluid, $p_s = mv_s = -i\hbar\nabla\theta$ (in the absence of a magnetic field) and the time derivative gives the chemical potential, $\mu = i\hbar\partial\theta/\partial t$. Equality of the cross-derivatives give the equation of motion:

$$\frac{\partial p_s}{\partial t} = -\nabla\mu. \tag{3.10}$$

Zero current flow implies that the phase is constant everywhere so that there is long-range coherence in phase.

In a CDW system there is long-range coherence in phase only if the current-density is the same everywhere, a requirement much more difficult to achieve. Important correlation functions are the phase-phase correlation function, $\langle\theta(x)\theta(x')\rangle$, the density–density correlation function,

$$\langle n(x)n(x') = \pi^{-2}\left\langle\frac{\partial\theta}{\partial x}\ \frac{\partial'\theta}{\partial x'}\right\rangle \tag{3.11}$$

and the current–current correlation function:

$$\langle j(x, t)j(x', t)\rangle = \pi^{-2}\left\langle\frac{\partial\theta(x, t)}{\partial t}\ \frac{\partial\theta(x', t)}{\partial t}\right\rangle. \tag{3.12}$$

The length L defined by (3.11) for the phase or density correlation function, determined by minimizing the impurity pinning energy, is typically of the order of 10^{-3} cm. The length, L_j, determined by the current–current correlation function may be much larger. In a few exceptionally pure and uniform specimens of $NbSe_3$, L_j is of the order of the sample size, [13] 10^{-1} cm.

In the latter, $\theta(x, t)$ is a function only of the time coordinate required to give a uniform current density:

$$\theta(x, t) = \theta(t). \tag{3.13}$$

The equation for the time variation is then

$$\frac{d^2\theta}{dt^2} + \omega_P^2 \frac{dV(\theta)}{d\theta} = \frac{2k_F eE(t)}{M_F}. \tag{3.14}$$

Even when L_j is less than sample size, Equation (3.14) may be used to calculate the acceleration of the current by the field over regions of size L_j. To take dissipation into account in steady state one may balance the acceleration by the field with the loss due to dissipation. In a simple relaxation time approximation,

$$\left(\frac{\partial^2\theta}{\partial t}\right)_{\text{field}} + \frac{1}{\tau^*}\frac{\partial\theta}{\partial t} = 0. \tag{3.15}$$

As we shall see in Section 5, quantum tunneling effects are required to determine the acceleration of the electrons by the field from a quantum version of the semiclassical equation (3.14).

It should be recognized that although in this section the equations are written for a single chain, many parallel chains are involved in the coherent motion of a CDW. When applied to the coupled system of electrons and macroscopically occupied phonons, they are essentially classical. It is believed that quantum tunneling is required only to discuss the acceleration through forces applied to the electrons. In the following section which shall give solutions of these classical equations and in Section 5 discuss the basis for the tunneling model.

4. Solutions of the Equations of Motion

We shall first give the solutions of the equation of motion (3.14) in the absence of applied fields and dissipation. There is an energy integral of

$$\frac{d^2\theta}{dt^2} + \omega_P \frac{dV(\theta)}{d\theta} = 0 \tag{4.1}$$

which is

$$\tfrac{1}{2}(d\theta/dt)^2 + \omega_P^2 V(\theta) = W = \text{const.} \tag{4.2}$$

The first term represents the kinetic energy and the second the potential energy of the phase-dependent pinning energy.

The threshold solution is that corresponding to $W = 0$. Consider the solutions

$$\theta_A = -\tfrac{1}{2}\pi \sin \omega_p t, \qquad \theta_B = -\pi + \tfrac{1}{2}\pi \sin \omega_p t. \tag{4.3}$$

For both, we have from (4.2) with $W = 0$:

$$\omega_p^2 V(\theta) = -\tfrac{1}{2}(\mathrm{d}\theta/\mathrm{d}t)^2 = -(\pi^2 \omega_p^2/8) \cos^2 \omega_p t, \tag{4.4}$$

which gives, with use of (4.3),

$$V(\theta) = -\tfrac{1}{2}((\tfrac{1}{2}\pi)^2 - \theta^2) \quad (-\tfrac{1}{2}\pi < \theta < \tfrac{1}{2}\pi). \tag{4.5}$$

The current is proportional to

$$-\frac{\mathrm{d}\theta_{A,B}}{\mathrm{d}t} = \pm\frac{1}{2}\pi\omega_p \cos \omega_p t. \tag{4.6}$$

where the + sign corresponds to A and the − sign to B. To get a unidirectional current one must alternate the solutions θ_A and θ_B:

$$\theta_A \quad \text{for} \quad -\tfrac{1}{2}\pi < \omega_p t < \tfrac{1}{2}\pi \quad \bmod 2\pi \tag{4.7a}$$

$$\theta_B \quad \text{for} \quad \tfrac{1}{2}\pi < \omega_p t < \tfrac{3}{2}\pi \quad \bmod 2\pi. \tag{4.7b}$$

Note that at the cross-over, $\theta_A = \theta_B$ and the current is zero. The current is proportional to:

$$-(\mathrm{d}\theta/\mathrm{d}t) = \tfrac{1}{2}\pi\omega_p\, |\cos \omega_p t|. \tag{4.8}$$

The pinning energy for this solution is always negative

$$V(\theta) = -(\pi^2/8)\cos^2 \omega_p t = -\tfrac{1}{2}(\pi^2/4 - \theta^2), \quad -\pi/2 < \theta < \pi/2. \tag{4.9}$$

This function, plotted in Figure 1, may be expanded into a series of harmonics:

$$V(\theta) = -\tfrac{1}{2}(\pi^2/6 + \cos 2\theta - \tfrac{1}{4}\cos 4\theta + \tfrac{1}{9}\cos 6\theta - \text{etc.}) \tag{4.10}$$

The amplitudes of the harmonics vary inversely with the square of the order, n.

The zero energy solution ($W = 0$) is that for which the system just has sufficient energy to surmount the potential barriers and represents the dividing line between oscillatory solutions (θ_A or θ_B) for $W < 0$ and progressive motion in either the positive or negative direction for $W > 0$, with the system oscillating between A and B type solutions to keep the pinning energy always negative. The period for unidirectional motion is π in θ or 2π in the argument 2θ.

The solution for large current flow, $\omega_d \gg \omega_p$ is

$$\theta = -\omega_d t - \frac{\omega_p^2}{4\omega_d^2} \sin 2\omega_d t + \text{harmonics}. \tag{4.11}$$

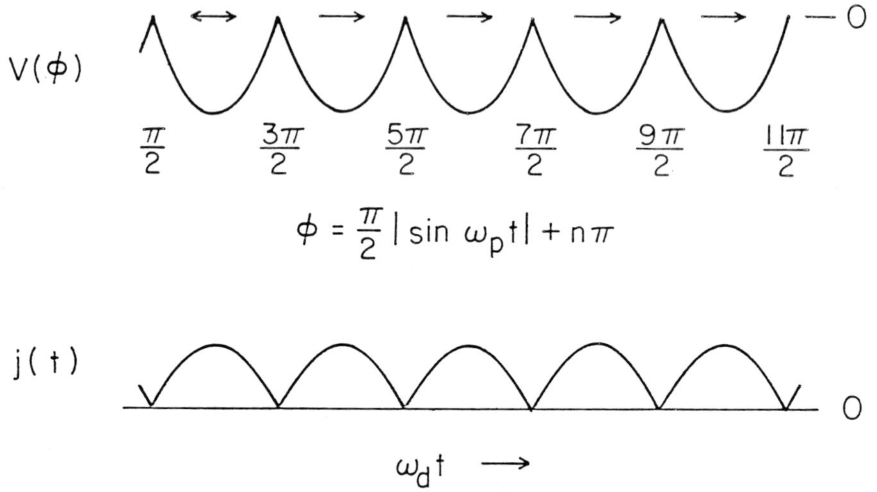

Fig. 1. Pinning potential $V(\phi)$ and current at threshold corresponding to energy just sufficient to surmount peaks at energy $W = 0$. (See Equation (4.9)).)

The current is then proportional to

$$-\frac{\partial\theta}{\partial t} = \omega_d + \frac{\omega_P^2}{2\omega_d}\cos 2\omega_d t + \text{harmonics} \tag{4.12}$$

and the potential proportional to

$$V(\omega_d t) = -\frac{1}{2}\,\omega_P^2\left(\frac{\pi^2}{6} + \cos 2\omega_d t + \text{harmonics}\right) \tag{4.13}$$

as in (4.10) with $\omega_d t = \theta$. Solutions may be given that bridge between (4.8) and (4.12).

It is believed that the oscillating component of (4.12) is a source of 'narrow-band noise', although there may be other contributions. In samples in which the current is coherent in phase throughout the volume, the oscillating component varies inversely with the dc current as predicted by (4.12). The harmonics are found to decrease approximately with the square of the order n, also in accordance with (4.12).

The harmonics are neglected in the sine-Gordon approximation, giving

$$V(\theta) = -\tfrac{1}{4}\,(1 + \cos 2\theta). \tag{4.14}$$

If ω_P^2 is redefined by a factor of two, in the limit $\omega_d \gg \omega_P$,

$$\frac{\partial^2\theta}{\partial t^2} + \tfrac{1}{2}\,\omega_P^2 \sin 2\theta = \frac{2k_F eE}{M_F}\,. \tag{4.15}$$

This equation, with the addition of damping terms, is analogous to that for a resistively shunted Josephson junction. It has been used with success [14] to account for resistivity peaks observed when the dc characteristic is measured in the presence of an applied ac voltage when the ac frequency coincides with the frequency of the oscillatory current or one of its harmonics.

As we shall see in Section 5, Equations (4.1) and (4.15) for the acceleration of the CDW velocity require modification for the quantum-tunneling model. For example, the equations do not give the observed gradual increase in dc conductivity when the voltage increases beyond threshold, but predict an abrupt increase with a current varying as $(E^2 - E_{th}^2)^{1/2}$. The threshold field, E_{th} is that for which the right-hand side of (4.15) is equal to the maximum of the force from the potential $\omega_P^2 V(\theta)$ on the left, or

$$E_{th} = \frac{M_F \omega_P^2}{2k_F e} = \frac{m\omega_P^2}{2k_F e^*}. \tag{4.16}$$

Although the electrons contain only a tiny fraction of the momentum and energy of the CDW system, all supplied to the system from applied electric fields must be transferred through the electrons.

5. The Tunneling Model

As discussed in Section 1, the tunneling model was first suggested by the observation of a nonlinear current increasing with field as $E \exp[-E_0/E]$. It was proposed that in order to displace the Fermi sea of electrons so as to give a current flow, a small pinning gap must be overcome. Zener tunneling of electrons through a small gap would give an expression of this form, but the model was rejected because the required gap is orders of magnitude less than $k_B T$. In 1979, the author [4] pointed out that such a model could be used if applied only to the coherent motion of electrons in the condensate, not to excitations of individual electrons. The latter are inhibited by the much larger Peierls gap. If the Peierls gaps move with the Fermi sea, as in the Fröhlich model, they do not affect the motion of the condensate as a whole.

By equating the wave vectors involved in zero-point motion of the electrons at the pinning frequency, ω_P, and those involved in creating a gap at the Fermi surface, the pinning gap was taken to be equal to $\hbar\omega_P$. The number of electrons involved in the coherent motion, N_e, must be large enough so that $N_e \hbar\omega_P > k_B T$. The phase-coherent region is that of Lee–Rice and involves many parallel chains.

The expression for E_0 in the Zener tunneling probability, $P(E) = \exp[-E_0/E]$, is, for a gap $\hbar\omega_P$,

$$E_0 = \frac{\pi\hbar\omega_P^2}{4v_F e^*} = \frac{\hbar\omega_P}{2\xi_0 e^*}, \tag{5.1}$$

where $2\xi_0$ is twice the Pippard coherence distance

$$2\xi_0 = \frac{4v_F}{\pi\omega_P}. \tag{5.2}$$

In this formalism, the only way the CDW is taken into account is the replacement of e by e^*. Nearly all of the crystal momentum picked up by the electrons from the electric field is transferred to the macroscopically occupied phonons. The effective charge $e* = (m/M_F)e$ reflects that the equation of motion for the displacement of the Fermi sea is

$$\hbar \frac{dq}{dt} = e*E. \tag{5.3}$$

It should be noted that the depinning field, E_0, in (5.1) is close to the classical threshold field. E_{th}, from (4.6) which may be written

$$E_{th} = \frac{\hbar\omega_P^2}{2v_F e*} = \frac{2}{\pi} E_0, \tag{5.4}$$

indicating that the semiconductor model with a pinning gap equal to $\hbar\omega_P$ includes essentially the same physics as the equations of motion for the phase described in Section 4.

In 1979, Fleming and Grimes [5] showed that there is a threshold field for the onset of nonlinear conduction, and this has subsequently been verified in many other studies. The tunneling theory was modified [4] (1980) to give a threshold field by including what corresponds in ordinary semiconductors to a free path effect. Electrons can gain energy and momentum from the field at most over a free path. In our case, this corresponds to the distance L for phase coherence in the Lee–Rice theory. The threshold field is then expected to be given approximately by

$$e*E_{th}L = \hbar\omega_P. \tag{5.5}$$

The value of E_0 also modified by replacing $2\xi_0$ by L. While the L's that appear in E_{th} and E_0 should be of the same order of magnitude, they need not be exactly the same. In $NbSe_3$, $E_0 \cong E_{th}$ below T_2 and $E_0 \cong 2E_{th}$ below T_1. In TaS_3, E_0 is ~ 2–3 times E_{th}.

In the 1980 paper, it was shown that data of Richard and Monceau [15] on the dc current as a function of E in $NbSe_3$ at several different temperatures below T_2 could be fitted well for $E > E_0$ by the expression:

$$I = \sigma_a E + \sigma_b (E - E_{th}) \exp(-E_0/E) \qquad (E > E_0). \tag{5.6}$$

The fit to the data with $E_{th} = E_0$ is shown in Figure 2.

Except for sharing of crystal momentum with the phonons, included by the replacement of e by $e*$, electrons are freely accelerated over phase-coherent distances represented by the Lee–Rice distance, L. This is illustrated in the Zener diagram of Figure 2 which shows the energy bands tilted by the electric field as a function of a space coordinate in the chain-direction. In order to have tunneling, one must go from an occupied state in the lower band to an unoccupied state in the next higher band. The reduction in states available for tunneling accounts for the factor $E - E_{th}$ preceding the exponential tunneling probability in (5.6).

The effective height of the tunneling barrier in real space may be estimated from the

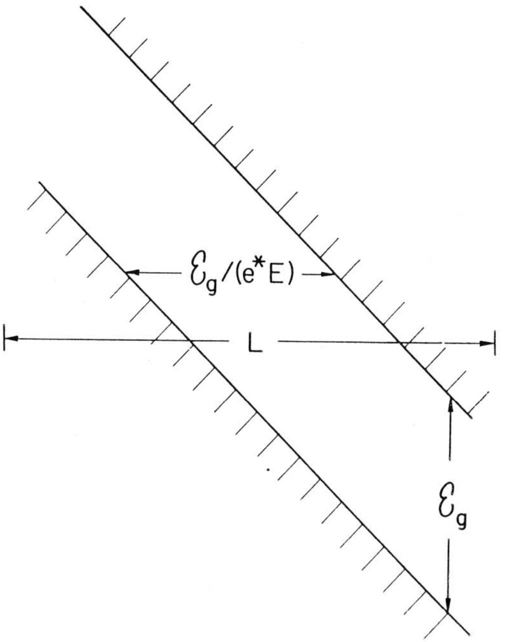

Fig. 2. Zener diagram showing energy bands tilted in space by an electric field, E. The field must be applied within a correlation length L to be effective in accelerating the electrons. (From ref. [4b].)

exponential decay of the wave function in the tunneling region. [7a] The decay is given by the imaginary part, κ of the wave vector, $k_F + i\kappa$

$$\kappa^2 = \frac{m(\Delta^2 - \epsilon^2)}{2\hbar^2 E_F}, \tag{5.7}$$

where the energy gap is $\epsilon_q = \hbar\omega_P = 2\Delta$ and ϵ is measured from midgap. The barrier corresponding to an energy ϵ is:

$$V_B = \frac{(\hbar\omega_P)^2}{16E_F}\left(1 - \frac{\epsilon^2}{\Delta^2}\right)^{1/2} \tag{5.8}$$

where E_F is the Fermi energy. Tunneling occurs across a distance ξ_T determined by

$$e^*E\xi_T = \hbar\omega_P. \tag{5.9}$$

Essentially the same barrier may be obtained from the potential $V(\theta)$ of (4.10) illustrated in Figure 1. The argument is that there is no current flow in the lowest state corresponding to an energy $\frac{1}{2}\hbar\omega_P$. To obtain current flow, one must mix in the wave function from the next quantum state, which is higher by $\hbar\omega_P$. In a field E, this requires that tunneling occurs across the same distance, ξ_T, as in the semiconductor model.

The equation of motion of an electron in the condensate of the CDW may be obtained from (3.14) by replacing $\partial^2\theta/\partial t^2$ by $2k_F\ \partial v_d/\partial t$ and $\mathrm{d}V/\mathrm{d}\theta$ by $(2k_F)^{-1}\mathrm{d}V/\mathrm{d}x$. The pinning energy is then $m\omega_P^2(V(\theta) - V(0))/(4k_F^2)$ and the average over x or θ is

$$\langle V_B \rangle = \frac{(\hbar\omega_P)^2}{16E_F}\langle\theta^2\rangle = \frac{(\hbar\omega_P)^2}{16E_F}\cdot\frac{\pi^2}{12}, \tag{5.10}$$

which is not far from (5.8). The average of (5.8) over ϵ introduces a factor $\pi/4$ in place of $\pi^2/12$. Presumably the threshold field enters in the same way as in the semiconductor model.

The elementary tunneling step is to add a wave vector $2k_F$ to the combined electron-phonon system. A pair of electrons transferred a distance L corresponds to displacing a segment of length L of the CDW by one wavelength. If an electron were transferred from one side the Fermi surface to the opposite, with an increase in wave vector by $2k_F$, it would have to share its momentum with the phonons, giving a net increase of $(e^*/e)2k_F$. Thus the contribution to the current is if the charge were e^* rather than e.

However, this is not the only effect of dissipation of wave vector from the electrons to the macroscopically occupied phonons. It was pointed out in a footnote to the 1980 paper [4] that Maki's theory [16] for the creation of a pair of solitons by tunneling also has an exponential factor $\exp(-E_0/E)$, but that the expression for E_0 differed from that in the depinning problem in that the Fermi velocity, v_F, is replaced by the phason velocity, $u_P = (m/M_F)^{1/2}v_F$. It was suggested that there might be a difficulty in Maki's derivation, but it was shown to be exact at least for one value of a parameter involved. Further, the experimental data indicated that in the Zener theory the Pippard distance $2\xi_0$, roughly v_F/ω_P, should be replaced by the much smaller Lee–Rice distance, L, of order u_P/ω_P. These difficulties were resolved by Wonneberger [17] who extended the theory of Caldeira and Leggett [18] to take into account the effect of dissipation of wave vector by the electrons on the tunneling probability. In a corrected version of the theory, E_0 is similar to the Zener expression, (5.1), but with v_F replaced by u_P.

Only a simplified version of Wonneberger's theory will be given here. Caldeira and Leggett [18] showed that for tunneling through a barrier extending over a distance ξ, the correction to the tunneling probability $\exp(-B_0)$ introduced by damping is to replace B_0 by $B = B_0 + \Delta B$, where

$$\Delta B = \frac{\pi}{4}\eta^*\xi^2. \tag{5.11}$$

Here η^* is a viscosity coefficient such that the viscous force is $\eta^* v$ where v is the velocity of the particle.

We are here concerned with the damping introduced by electrons giving up most of the energy and momentum gained from the electric field to the macroscopically occupied phonons. The energy gained by the electron–phonon system when an electron tunnels across the pinning gap in $(\alpha + 1)\hbar\omega_P$, where $\alpha = M_F/m$ is the Fröhlich mass ratio. If the tunneling distance is $\xi(E)$, the average froce is

$$\langle\eta^* v_F\rangle = eE = \frac{(\alpha + 1)\hbar\omega_P}{\xi(E)}, \tag{5.12}$$

giving $\xi(E) = \hbar\omega_P/e^*E$. The average viscosity coefficient is

$$\eta^* = \frac{(\alpha+1)\hbar\omega_P}{v_F\xi(E)} \tag{5.13}$$

so that

$$\Delta B_{\text{eff}} = \frac{\pi(\alpha+1)\hbar\omega_P\xi(E)}{4v_F} = (\alpha+1)B_0, \tag{5.14}$$

where B_0 is the Zener tunneling probability

$$B_0 = \frac{\pi\hbar\omega_P^2}{4v_Fe^*E}. \tag{5.15}$$

Thus ΔB_{eff} is not a small correction but is much larger than B_0. Caldeira and Leggett used a formalism in which $t \to it$ and τ_B is the 'bounce' time in the inverted potential well. Wonneberger used τ_b as a variational parameter to obtain the tunneling probability. The optimum value is that which makes $B_0 + \Delta B$ a minimum, or to choose τ_b to minimize

$$\frac{\tau_b}{2} + \frac{\tau_{b0}^2}{2\tau_b}(1 + 2(1+\alpha)). \tag{5.16}$$

This gives:

$$\tau_b = \tau_{b0}(3+2\alpha)^{1/2} \tag{5.17}$$

and, neglecting 3 compared with 2α,

$$B = B_0(2\alpha)^{1/2} = B_0(2M_F/m)^{1/2}. \tag{5.18}$$

This correction gives the Zener expression with v_F replaced by $u_P/\sqrt{2}$ and an expression for E_0 similar to that derived by Maki.

Tunneling of an electron adds $2k_F$ to the wave vector of the electron–phonon system in a Lee–Rice region of volume Ω. This is divided between the parallel chains in the region and between the electrons and macroscopically occupied phonons. The various states corresponding to the one degree of freedom in the region may be defined by the total wave vector which is an integral multiple of $2k_F$.

6. Response to AC and Combined AC and DC Fields

The expression (5.6) for the dc current as a function of field has been found to give a good fit to a large amount of data on $NbSe_3$ and TaS_3. Other expressions have been suggested on empirical grounds but (5.6) appears to fit the data as well as any. The threshold field, E_{th}, is determined by a phase coherent length such that $e^*E_{th}L = \hbar\omega_P$.

It is assumed that electrons in the CDW condensate are not scattered except by sharing their momentum by the phonons, indicated by the replacement of e by $e*$. This suggests an analogy with Tucker's [6] theory of photon-assisted tunneling in superconductor-insulator-superconductor (SIS) tunnel junctions, with the voltage $V = EL$ corresponding to the voltage across the SIS junction. The threshold voltage $V_T = E_{th}L$. Tucker showed that the response to ac and combined ac and dc voltages may be derived from the non-linear dc response.

For small signal ac, with $V = V_0 + V_1 \cos \omega t$, Tucker finds for the ac response, $I_{ac} \cos \omega t$,

$$I_{ac} = \frac{eV_1}{2\hbar\omega}\left[I_{dc}\left(V_0 + \frac{\hbar\omega}{e}\right) - I_{dc}\left(V_0 - \frac{\hbar\omega}{e}\right)\right], \tag{6.1}$$

where $I_{dc}(V)$ is the dc current for an applied voltage V. In the classical limit ($\hbar \to 0$) this gives just the usual classical result

$$I_{ac} = V_1\left[\frac{\mathrm{d}I_{dc}(V)}{\mathrm{d}V}\right]_{V=V_0}. \tag{6.2}$$

Thus quantum theory replaces a derivative by a finite difference.

It was suggested initially that to apply to the CDW problem, one replace V_0 by EL and e by $e*$. For $V_0 = 0$, this gives for the response to $V_1 \cos \omega t$,

$$I_{ac}(\omega) = \frac{e*V_1}{\hbar\omega}\, I_{dc}\left(\frac{\hbar\omega}{e*}\right), \tag{6.3}$$

or expressed in terms of conductivities

$$\sigma(\omega) = \sigma_{dc}(\hbar\omega/e*L), \tag{6.4}$$

with $\sigma_{dc}(E)$ defined to be I/E rather than $\mathrm{d}I/\mathrm{d}E$. This gives a scaling relation between ac and dc response.

The only data available at the time this suggestion was made were a few measurements of Grüner *et al.* [19] at different frequencies at one temperature on $NbSe_3$ below T_2. These are shown in the plot of Figure 3. There is good agreement except near threshold. There is no threshold frequency for changes in $\sigma(\omega)$, but $\sigma(\omega)$ increases gradually as the frequency is increased from zero. Many subsequent measurements by Grüner, Zettl and others [7, 8] have shown that the scaling relation holds accurately for $\omega > 2\omega_T$, where ω_T is the frequency corresponding to E_{th} for both $NbSe_3$ and TaS_3 at many different temperatures. The excess in $\sigma(\omega)$ for $\omega \sim \omega_T$ was attributed to absorption by the pinned mode with a peak for $\omega = \omega_T = \omega_P$.

Other predictions of Tucker's photon-assisted tunneling theory failed when applied to the CDW problem. Great success has been achieved in using SIN and SIS junctions as detectors in the submillimeter wave length range by biasing below threshold, $V < V_T$, (in this case from the superconductor gaps) but with a frequency large enough so that

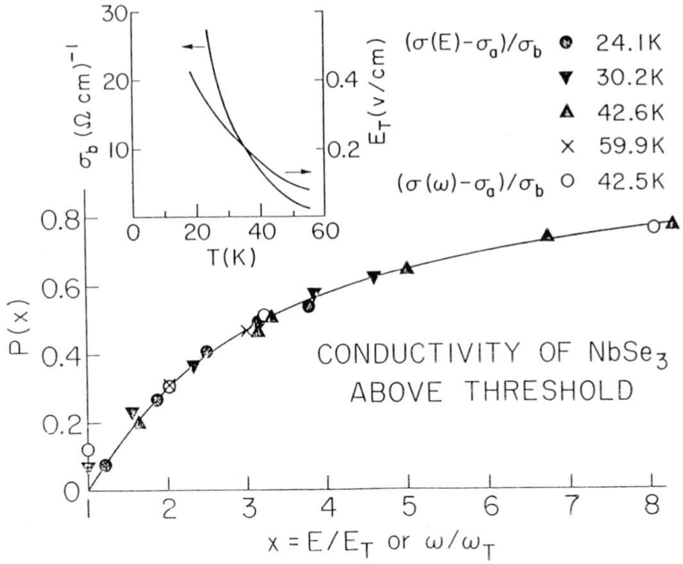

Fig. 3. Field, $\sigma(E)$, and frequency, $\sigma(\omega)$ dependent conductivities of $NbSe_3$ plotted on reduced scales. The solid line is a plot of $P(x) = (1 - 1/x)\exp[-1/x]$ based on tunneling theory with $E_{th} = E_0$. The data points for field dependent conductivity are from J. Richard and P. Monceau, ref. 15, and on ac conductivity from Grüner *et al.* (from ref. 4b).

$V + (\hbar\omega/e)$ is above threshold. It was initially hoped to use the nonlinear characteristics of CDW conductors as detectors in a similar way, to bias below threshold, $V < V_{th}$, and to use the effective voltage from the frequency, $\hbar\omega/e^*$, to bring the total above threshold. Experiments by Grüner and Zettl [7] showed that there is no detected signal unless the total voltage, $V_0 + V_1 \cos \omega t$, is above threshold, V_T.

In 1982, Tucker initiated a program at Illinois to study detection and mixing in $NbSe_3$ and TaS_3. In a series of experiments by Miller, Richard and Tucker, it was found that dc biases below threshold had little or no effect on any ac response. Miller proposed a revised form of the theory in which scaling is to the voltage beyond threshold, $V' = V - V_T$ rather than to V. Negative values of V' correspond to $V + V_T$, with $V < -V_T$. Values of V in the range $-V_T < V < V_T$ correspond to $V' = 0$. This revised form of the theory has been remarkably successful in accounting for a wide range of detection and mixing experiments.

When applied to small-signal ac conductivity, the result is

$$I_{ac} = \frac{eV_1}{2\hbar\omega}\left[I'\left(V' + \frac{\hbar\omega}{e^*}\right) - I'\left(V' - \frac{\hbar\omega}{e^*}\right)\right] \tag{6.5}$$

where

$$I'(V') = I_{\mathrm{CDW}}(V) = \sigma_b V' \exp\left[-E_0/(E_{th} + E')\right] \tag{6.6}$$

and E' is the field corresponding to V'. In the limit $V' = 0$, this gives for the ac response

$$\sigma(\omega) = \sigma_b \exp[-\omega_0/(\omega_{th} + \omega)] \tag{6.7}$$

where ω_0 is the frequency corresponding to E_0 and ω_{th} that corresponding to E_{th}.

The revised-scaling gives a correction to the conductivity from CDW motion at $\omega = 0$, which is physically unrealistic, so that one expects that there is a lower frequency limit for which scaling applies. Below this limit, the ac should be considered a slowly varying dc and merge at very low frequencies to the dc response. The lower limit is probably the order of a few MHz for $NbSe_3$ and TaS_3.

A qualitative justification for the revised scaling may be given in terms of the threshold solution illustrated in Figure 1. At threshold, the system may go equally well to right or the left. If biased with a V' giving a small current to the right and an ac is superimposed such that $V' - (\hbar\omega/e^*)$ is negative, the latter would give a contribution corresponding to a current to the left. Near threshold, very little is required to swing the current from moving right to moving left.

The revised scaling gives not only improved agreement with ac conductivity near threshold, but also removes previous discrepancies with rectification and mixing experiments. In the small signal limit, the rectified current for an applied voltage $V \equiv V_0 + V_1 \cos \omega t$ is:

$$\Delta I = \frac{V_1^2}{2} \left\{ \frac{I'(V_0' + \alpha\omega) - 2I'(V_0) + I'(V_0' - \alpha\omega)}{(\alpha\omega)^2} \right\}. \tag{6.8}$$

The classical second derivative is replaced by a quantum finite difference, with $\alpha = \hbar/e^*$ if V is the voltage across L. If biased below threshold, $V_0' = 0$ and so rectified signal is to be expected.

From (6.8), one can calculate ΔI from the dc characteristic, $I'(V')$ and the scaling parameter, α, with no other adjustable parameters. Good quantitative agreement has been found for measurements over a range of bias voltages, frequencies and temperatures for both TaS_3 [20] and $NbSe_3$ [21]. Frequencies are in the range 10^6–10^9 Hz. What discrepancies were found are largely accounted for by the fact that ac voltages (of the order of 10–20% V_T) were not quite in the small signal limit. A finite amplitude ac depresses the threshold voltage, V_T, and also introduces some broadening of the curves.

Further evidence comes from experiments on harmonic mixing, which classically depends on the third derivative, $d^3 I_{dc}/dV^3$ of the dc current. An applied voltage, V_2, at a frequency of the second harmonic beats against the second harmonic of an ac applied voltage, V_1, arising from the nonlinear characteristic. If the applied voltage is

$$V(t) = V_0 + V_1 \cos(\omega_1 t + \phi) + V_2 \cos \omega_2 t \tag{6.9}$$

there will be a rectified signal proportional to $V_1^2 V^2$ and varying with the phase as $\cos(2\phi + \psi)$. The signal expected is the finite difference third derivative

$$\Delta I = \frac{V_1^2 V^2}{16} \left\{ \frac{I'(V_0' + 2\alpha\omega) - 2I'(V_0' + \alpha\omega) + 2I'(V_0' - \alpha\omega) - I'(V_0' - 2\alpha\omega)}{(\alpha\omega)^3} \right\} \tag{6.10}$$

with the phase shift $\psi = 0$. This differs from the classical overdamped oscillator model where ψ is expected to increase from zero at frequencies small compared with the cross-over frequency $\omega_c = \omega_p^2 \tau$ to $\pi/2$ when $\omega \gg \omega_c$. In experiments of Miller *et al.* [20] on TaS_3, no phase shifts were found for frequencies as high as $\nu_1 = 500$ MHz, $\nu_2 = 1$ GHz, much larger than the cross-over frequency of ~ 80 MHz. When effects of finite amplitude ac are taken into account, there is good quantitative agreement with measurements taken at several different frequencies as a function of the dc bias.

An observed discrepancy is that the theory predicts a zero bias signal while none was observed. However, it was found that the expected signal is recovered if there is a finite frequency difference $\omega_0 = \omega_2 - 2\omega_1$ between the applied frequency, ω_2, and that generated from the fundamental at a frequency ω_1. A more detailed understanding of the response of the CDW when biased below threshold will be required to account for this frequency dependence. Frequencies of the order of 5–10 MHz are required for the harmonic mixing signal to reach saturation in TaS_3 and $NbSe_3$, and these are approximately those expected from theory.

Harmonic mixing was suggested and first studied in $NbSe_3$ in the microwave range by Seeger, Mayr and Phillip [22]. They found signals of the magnitude expected by the tunneling model and in a recent study found that the singal varies as $V_1^2 V_2$ down to values such that $V_1 + V_2 < V_T$ and that there is no phase shift as a function of amplitude of the applied microwave voltages. However, they were unable to measure the absolute value of the phase shift and thus verify that it is zero rather than $\pi/2$ (the latter expected from the overdamped oscillator model).

While Tucker's theory for superconducting junctions applies to ac fields of arbitrary amplitude, there are additional problems when applied to the CDW problem. It is known that large amplitude, high frequency ac fields decrease the threshold voltage for dc currents from motion of CDW's. The threshold becomes even sharper with increasing amplitude of the ac, with the second derivative, d^2I/dV^2 having a narrower and higher peak. Further, Shapiro-like steps or resistivity peaks occur when the frequency of the applied ac coincides with that of the oscillatory current associated with CDW motion or one of the harmonics. These effects have been studied by Zettl and Gruner [14] in experiments on $NbSe_3$ below T_2. They have been interpreted in terms of a sine-Gordon equation similar to (4.15) with a damping term added. To include quantum tunneling with these other effects requires that appropriate corrections be made to the semiclassical (3.14).

It is suggested in (3.15) that one calculate, using quantum tunneling, the acceleration of the current by the field, $\partial^2\theta/\partial t^2$, and to balance this against the decay of the current by damping terms. To get the response to dc, one could simply replace the electric field on the right hand side of (3.14) by

$$EP(E) = (E - E_{th})\exp[-E_0/E]. \tag{6.11}$$

This would give the correct dc current and 'narrow band' noise associated with it. To derive effects of dc combined with large amplitude ac would require some way of calculating the effect of the ac on the dc response as well as interaction with the oscillatory currents of the 'narrow band' noise.

Such a theory may require a more basic derivation of the equations of motion of the CDW from the microscopic theory of the system of interacting electrons, phonons and impurities. Hopefully, this would allow one to treat the coupling of ac and dc fields of arbitrary amplitude.

7. Conclusions and Acknowledgements

There is strong evidence that quasi-one-dimensional metals in which charge-density waves are formed are macroscopic quantum systems that exhibit quantum effects of frequencies in the megaHertz range at temperatures as high as 200 K. The only reason that the evidence is not completely conclusive is that no direct evidence for quantum effects such as the photon steps observed in SIS junctions has been found.

As the materials and experiments become more refined and the theory is further developed, the study of CDW systems should be an excellent field for the study of macroscopic quantum phenomena in a new range of variables.

The author is indebted to many people who have aided in the initiation and development of the quantum tunneling model of CDW depinning. The work started at the University of Karlsruhe in the fall of 2978 where the author was visiting under an award from the Alexander Humboldt Foundation. During this time, he benefited from discussions with A. Schmid. He is particularly grateful to G. Grüner, J. H. Miller, Jr. and J. R. Tucker for helping to provide the close interaction between theory and experiment during which the theory has developed to the present stage. Many others have been very helpful and provided important ideas or experimental information, including R. M. Fleming, A. J. Leggett, W. G. Lyons, K. Maki, P. Monceau, N. P. Ong, J. Richard, R. E. Thorne, W. Wonneberger and A. Zettl.

This chapter was written in part while the author was a visitor at the Institute for Theoretical Physics, University of California, Santa Barbara and he is grateful for the support received.

References

1. H. Fukuyama and P. A. Lee, *Phys. Rev.* **B17**, 535 (1978).
2. P. A. Lee and M. Rice, *Phys. Rev.* **B19**, 3970 (1979).
3. P. Monceau, N. P. Ong, A. M. Portis, A. Meerschaut and J. Rouxel, *Phys. Rev. Lett.* 37, 602 (1976); N. P. Ong and P. Monceau, *Phys. Rev.* **B16**, 3443 (1977).
4. John Bardeen, (a) *Phys. Rev. Lett.* 42, 1498 (1979); (b) *ibid* 45, 1978 (1980); (c) *Mol. Cryst. Liq. Cryst.* 81, 1 (1981).
5. R. M. Fleming and C. C. Grimes, *Phys. Rev. Lett.* 42, 1423 (1979).
6. J. R. Tucker, *J. Quantum Electron.* 15, 1234 (1979).
7. (a) G. Grüner, A. Zettl, W. G. Clark and J. Bardeen, *Phys. Rev.* **B24**, 7247 (1981); (b) A. Zettl, G. Grüner and A. H. Thompson, *Solid State Comm.* 39, 899 (1981).
8. For a review, see G. Grüner, *Physica* **8D**, 1 (1983).
9. S. E. Barnes and A. Zawadowski, *Phys. Rev. Lett.* 51, 1003 (1983).
10. Proceedings of the International School of Physics, 'Enrico Fermi', Varenna, Italy, 1983 (to be published). Proceedings of the International Symposium on Nonlinear Transport and Related Phenomena in Inorganic Quasi-One-Dimensional Metals, Hokkaido Univ., Sapporo, Japan, October 20–22, 1983.
11. H. Fröhlich, *Proc. Roy. Soc. London*, Ser. A **223**, 296 (1954).
12. D. Allender, J. W. Bray and John Bardeen, *Phys. Rev.* **B9**, 119 (1974).
13. John Bardeen, E. Ben-Jacob, A. Zettl and G. Grüner, *Phys. Rev. Lett.* 49, 493 (1982).

14. A. Zettl and G. Grüner, *Phys. Rev.* **B29**, 755 (1984).
15. J. Richard and P. Monceau, *Solid State Comm.* **33**, 635 (1980).
16. K. Maki, *Phys. Rev. Lett.* 39, 46 (1977), *Phys. Rev.* **B18**, 1641 (1978).
17. W. Wonneberger, *Z. Phys.* **B50**, 23 (1983).
18. A. O. Caldeira and A. J. Leggett, *Phys. Rev. Lett.* **46**, 211 (1981).
19. G. Grüner, L. C. Tippie, J. Sanny, W. G. Clark and N. P. Ong, *Phys. Rev. Lett.* **45**, 935 (1980).
20. J. H. Miller, Jr., J. Richard, J. R. Tucker and John Bardeen, *Phys. Rev. Lett.* **51**, 1592 (1983).
21. J. H. Miller, Jr., J. Richard, R. E. Thorne, W. G. Lyons, J. R. Tucker and John Bardeen, *Phys. Rev.* **B29**, 2328 (1984).
22. K. Seeger, W. Mayr and A. Phillip, *Solid State Commun.* 43, 113 (1982).

SOLITONS IN ONE-DIMENSIONAL SYSTEMS

KAZUMI MAKI

Department of Physics,
University of Southern California,
Los Angeles, CA 90089–0484, U.S.A.

1. Introduction

Although a soliton related phenomenon has been discovered by Scott-Russel [1] in 1834, it took a long time for solitons to acquire their present popularity. The object of the present chapter is two-fold: First we would like to familiarize the readers with the notion of solitons. This will be done through a brief historical introduction and with the help of some examples. Then some salient features of the sine-Gordon soliton are summarized in Section 2. Second we shall describe applications of the soliton notion to real systems. Section 3 is devoted to the magnetic systems, while Sections 4, 5 and 6 are devoted to the conducting systems. The general background for the present review may be found in [2–7]. In the following we shall sketch how the notion of soliton is developed and finds application in many branches of physics.

How Scott-Russel [1] discovered the marvellous phenomenon of solitary wave in a water channel belongs already to a folklore. Indeed the stability of a water bump gliding smoothly over the water channel should be quite impressive. However, it took some time until this wonderful phenomenon was formulated as a special solution of the nonlinear equation discovered by Korteweg and de Vries [8]. In their work the importance of the nonlinearity in the solitary wave is delineated for the first time. Since then the subject appeared to be almost forgotten by the physics community, although meantime the sine-Gordon equation (another canonical equation) was discovered by Kochendörfer and Seeger [9] as the continuum version of a model of dislocation introduced by Frenkel and Kontorova [10].

The real renaissance of solitons is due largely in the development of computer technology after World War II. In a well-known computer experiment Fermi *et al.* [11] discovered to their surprise that most of one-dimensional anharmonic lattices are non-ergodic; the systems appeared to return to their initial configuration after a finite lapse of time. Later Toda [12] demonstrated that this recurrence phenomenon is due to solitary waves in anharmonic lattices. In 1965, while studying the collision between two solitary waves of the Korteweg–de Vries (KdV) equation numerically, Zabusky and Kruskal [13] discovered the intriguing stability of these solitary waves; the solitary waves reappear after colliding without change in their shapes and velocities. The only effect of the collision is a finite displacement from their original trajectories. This feature prompted Zabusky and Kruskal to call these stable solitary waves as 'solitons' in analogy to elementary particles. Then in 1967 Gardner *et al.* [14] succeeded in showing that the initial value problem of the KdV equation is solved completely by a few steps of linear operations (inverse scattering method). Therefore the remarkable stability of solitons

P. Monceau (ed.), Electronic Properties of Inorganic Quasi-One-Dimensional Materials, I, 125–193.

is partially due to the special character of the KdV equation; the KdV equation is completely integrable. Then in rapid succession it was discovered that there is a large class of nonlinear equations in 1 + 1 dimensions (one space and one time dimension), which are integrable [15–17]. Interested readers can consult an excellent review on this subject by Scott *et al.* [18]. So far we have described the soliton physics as a branch of mathematical physics. However, there is another strand of development on solitons, which focus on their topological aspect.

It was first realized by Finkelstein [19] that some of the nonlinear solutions of field equations can be classified in terms of homotopy classes; the solitons are classified according to the mapping between the real n-dimensional physical space surrounding the soliton and the manifold formed by the ground state field configuration. Although Finkelstein called these solitons 'kinks', it is now common practice among field theorists to call these solitons 'topological' solitons or simply solitons. This approach has been recently generalized to quantum field theory [20], liquid crystal and superfluid ^{3}He [21, 22].

The topological solitons are of particular importance in condensed matter physics. A condensed phase is characterized by an order parameter [23]. Furthermore, the ground states of a condensed phase are in general highly degenerate; there is a finite subspace in the order parameter space corresponding to the ground states. In this circumstance the topological solitons play a crucial role in the physics of the condensed states. Typical condensed states are: the superconducting state in metals, superfluid ^{4}He, superfluid ^{3}He, ferromagnet, antiferromagnet, charge density wave (CDW) state and spin density wave (SDW) state; although, in some of the condensates, two order parameters can coexist. The first three condensates are superfluid. Among other solitons the superfluid state allows linear solitons; the Onsager–Feynman vortices in superfluid ^{4}He and the Abrikosov vortices in superconductors [24] are the most well-known examples. In quasi-two-dimensional systems, the linear solitons become point-like objects and play a crucial role in the destruction of the two-dimensional order [25, 26]; the two-dimensional crystal melts by unbinding of dislocation pairs, while the two-dimensional superfluid is destroyed by unbinding of vortex pairs.

Far more common topological solitons are the domain walls, which exist whenever the ground states of the condensate allow only discrete symmetry. Domain walls in ferromagnets [27], composite solitons in superfluid ^{3}He-A [28], **n**-solitons in superfluid ^{3}He-B [29, 30], discommensurations [31, 32] and ϕ-solitons in CDW states [33], and solitons in *trans*-polyacetylene [34, 35] are well-known examples. In the quasi-one-dimensional systems the domain wall is a point-like object with microscopic energy. Then, as first realized by Krumhansl and Schrieffer [36], solitons play a fundamental role in the thermodynamics of the quasi-one-dimensional systems; the crossover between the topologically disordered high temperature phase to the quasi-ordered low temperature phase is controlled by thermally activated solitons. Indeed, recent excitement in soliton physics is, in part, due to the experimental observation of magnetic solitons in quasi-one-dimensional magnetic systems [37–39].

In the case of quasi-one-dimensional conductors, on the other hand, evidence for solitons appears rather scarce, except in *trans*-polyacetylene. In spite of lively controversy about the existence of solitons both in pristine-polyacetylene and doped polyacetylene, the present author believes that the soliton is the central notion which can correlate a large amount of experimental data on polyacetylene.

Furthermore the phase vortices in the CDW states in quasi-one-dimensional conductors like $NbSe_3$ [40, 41] may be of importance in the interpretation of anomalous transport properties of these systems. We shall describe later in detail these solitons in CDW states.

Soliton physics is further enriched, recently, by field theorists, who consider solitons as models of particles in quantum field theory [19, 20]. An impressive advance has been achieved recently in the quantum field theory of solitons in 1 + 1 dimensions [42–44]. There is now an exciting possibility that some of the predictions in quantum field theory in 1 + 1 dimensions may be confronted with experiment in quasi-one-dimensional condensed matter.

2. Sine-Gordon Solitons

2.1. Some Examples

A sine-Gordon (SG) equation is one of the canonical nonlinear equations, which not only have soliton solutions but also is completely integrable [3]. This implies that the initial value problem of the classical sine-Gordon equation is completely solvable [17]. More recently the quantum sine-Gordon system in 1 + 1 dimensions has been shown to be solvable by the Bethe Ansatz [45]. Therefore the sine-Gordon equation in soliton physics plays a role similar to the hydrogen atom in quantum mechanics. Furthermore, as we shall see shortly, the sine-Gordon equation is ubiquitous in condensed matter physics [2, 6]. A number of equations describing magnetic systems [46–49] and charge density wave (CDW) condensates [31, 32] reduce to the sine-Gordon equation in certain limits.

A. *Mechanical Model* [2]

First let us consider a series of pendula connected by elastic strings as shown in Figure 1, which is readily made by attaching pins to a rubber band. The angle ϕ_i, the ith pendulum, from the vertical direction, obeys the following equation of motion

$$I\frac{d^2\phi_i}{dt^2} = -mg\sin\phi_i + K(\phi_{i+1} - 2\phi_i + \phi_{i+1}) \tag{2.1}$$

where I, m and K are the moment of inertia, the mass of the pendulum, and the elastic constant, respectively. When the angle ϕ_i depends slowly on i, we can take the continuum limit of Equation (2.1) and we obtain;

$$\phi_{tt} = C_0^2\phi_{xx} - \omega_0^2\sin\phi \tag{2.2}$$

where

$$C_0 = (K/I)^{1/2}a, \qquad \omega_0 = (mg/I)^{1/2} \tag{2.3}$$

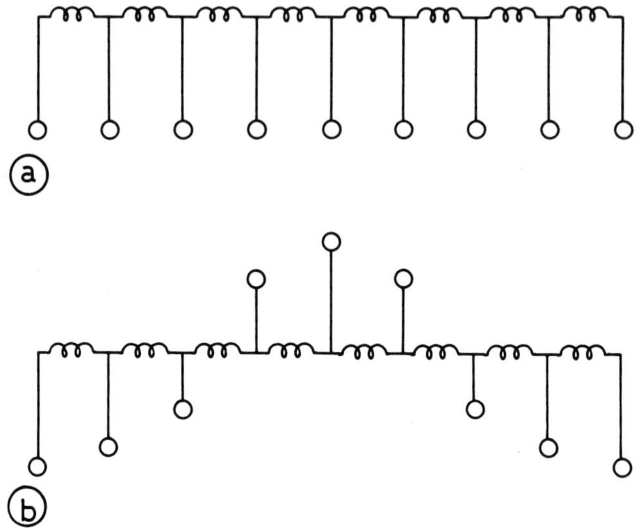

Fig. 1. A series of pendula connected by springs are shown. (a) the equilibrium configuration and (b) a configuration with one soliton.

and a is the distance between two adjacent pendula. Here suffices t and x on ϕ mean partial derivative in t and x (= ia) respectively. A soliton in the present model is a 2π-twist in a series of pendula, as shown in Figure 1b.

B. *One Dimensional Model of Dislocation*

Frenkel and Kontrova [10] studied a one dimensional model, shown in Figure 2, to describe the motion of a slip in a crystal. It consists of two chains of atoms; the lower chain is assumed to be fixed to the regular lattice position while the atoms in the upper chain can move. The lower chain of atoms create a periodic potential of the form

$$V(\phi_i) = A\,[1 - \cos(2\pi\phi_i/a)] \tag{2.4}$$

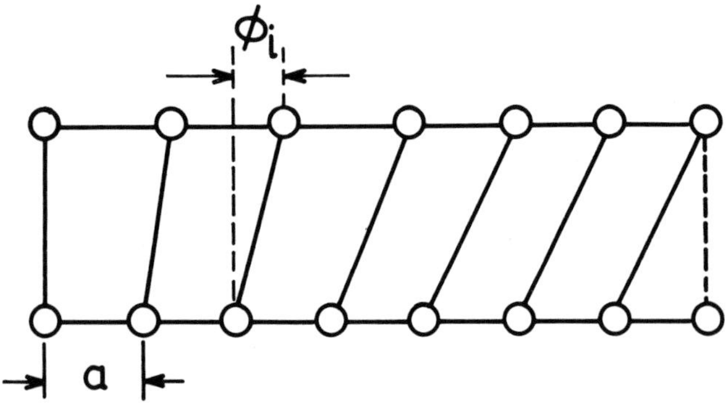

Fig. 2. The upper chain of atoms contains one slip (dislocation), while the lower chain of atoms are fixed at the regular positions.

for the ith atom in the upper chain with displacement ϕ_i from the equilibrium position. The equation of motion for ϕ_i is then given by

$$m\phi_{i,\,tt} = -\frac{2\pi A}{a}\sin(2\pi\phi_i/a) + K(\phi_{i+1} - 2\phi_i + \phi_{i-1}) \tag{2.5}$$

where m is the mass of the ith atom and K is the elastic constant between two atoms in the upper chain. Again the continumm limit [3] of Equation (2.5) reduces to Equation (2.2) where now C_0 and ω_0 are given by

$$C_0 = (K/m)^{1/2}a, \qquad \omega_0 = \frac{2\pi}{a}(A/m)^{1/2}$$

and

$$\phi = 2\pi\phi_i/a \tag{2.6}$$

Here the soliton corresponds to a propagating slip (or dislocation) with rest energy:

$$E_s = \frac{4a}{\pi}(KA)^{1/2} \tag{2.7}$$

The present model describes lossless motion of dislocation; the slip moves uniformaly without change in its velocity. This continuum version has been extensively studied by Seeger and his collaborators [9, 50].

C. *Magnetic Flux in a Josephson Junction*

Let us first consider two superconductors separated by a thin insulating barrier (say an oxide film of thickness 10–20Å). As shown by Josephson [51] the pair of electrons can tunnel from one superconductor to another leading to weak coupling between the wave functions Ψ_1 and Ψ_2 describing the condensates of the two superconductors across the junction. When ϕ $(= \phi_1 - \phi_2) \neq 0$, where ϕ_1 and ϕ_2 are the phases of Ψ_1 and Ψ_2, there is a flow of supercurrent J per unit area across the junction

$$J = J_0 \sin\phi \tag{2.8}$$

where J_0 is a constant depending on the characteristic of the junction and temperature. Furthermore if a voltage V is applied across the junction the phase difference ϕ increases with time as

$$\phi_t = \frac{2e}{\hbar}V \tag{2.9}$$

These are the celebrated dc and ac Josephson effect, respectively. Now let us consider a Josephson junction with a large area, which extends in the x–y plane. A large area

Josephson junction is characterized in terms of a transverse (scalar) voltage V and a transverse current $\mathbf{J}$ (the surface current on the junction) which obey [52, 53]

$$\begin{aligned}
&\boldsymbol{\nabla} V = L\mathbf{J}_t \\
&\boldsymbol{\nabla} \cdot \mathbf{J} = -CV_t - J_0 \sin\phi \\
&\boldsymbol{\nabla}\phi = -\frac{2e}{\hbar} L\mathbf{J} \\
&\phi_t = \frac{2e}{\hbar} V
\end{aligned} \tag{2.10}$$

where $\mathbf{J}$ and $\boldsymbol{\nabla}$ are two dimensional vectors lying in the surface. The capacitance C and the reactance L of the junction are given by:

$$C = \epsilon/d, \qquad L = (2\lambda_L + d) \tag{2.11}$$

where ϵ is the dielectric constant, d the gap between the two superconductors and λ_L is the London penetration depth. Eliminating V and $\mathbf{J}$ from Equation (2.9), we obtain

$$LC\phi_{tt} - \nabla^2\phi = -\frac{2e}{h} LJ_0 \sin\phi \tag{2.12}$$

the two-dimensional SG equation. The state of the junction is completely determined by ϕ. Of technical importance is the Josephons transmission line, which consists of two thin tapes of superconductors separated by thin oxide layer. In this limit Equation (2.11) reduces to the (one-dimensional) sine-Gordon equation. The soliton here describes a moving magnetic flux line along the junction. The flux or the fluxon carries a single flux quantum

$$\Phi = \int_{-\infty}^{\infty} V\mathrm{d}t = \frac{h}{2e}\,[\phi(\infty) - \phi(-\infty)] = \frac{h}{2e} \equiv \Phi_0 \tag{2.13}$$

with $\Phi_0 = h/2e \cong 2 \times 10^{-7}$ gauss cm^2

These propagating fluxons may be used as a signal unit in a future high speed computer. So far, in deriving Equation (2.11), the voltage loss across the junction has been completely neglected. If the loss mechanism is included, the moving fluxon will decelerate or sometimes a fluxon and an antifluxon will annihilate each other upon collision.

Besides these semi-macroscopic examples we have described, it is known that some of quasi-one-dimensional planar ferromagnets and planar antiferromagnets in a magnetic fields are described to a good approximation by the sine-Gordon equation, as we shall see in Section 2. Furthermore, a sine-Gordon equation will provide the simplest model to describe a commensurate-incommensurate CDW transition, which will be analysed in Section 4.

2.2. Classical Solutions

We shall consider again Equation (2.2), which can be derived from the Lagrangian;

$$\mathscr{L} = \frac{A}{2} \int dx \, \{\phi_t^2 - C_0^2 \, \phi_x^2 - 2\,\omega_0^2 \,(1 - \cos\phi)\} \tag{2.14}$$

where coefficient A is an arbitrary scaling factor in the classical theory. In the quantum field theory A is of prime importance, since the quantum action is measured by $\hbar$ and A provides the scale of the soliton mass. One of the fundamental features of the sine-Gordon system is that its equilibrium configuration (the vacuum in the quantum field theory) is infinitely degenerate;

$$\phi = 0, \pm 2\pi, \pm 4\pi, \ldots . \tag{2.15}$$

Corresponding to this degeneracy the SG system has solutions;

$$\phi_s(x, t) = 4 \tan^{-1}(e^{ms}) \tag{2.16}$$

$$\phi_{\bar{s}}(x, t) = 4 \tan^{-1}(e^{-ms}) \tag{2.17}$$

where $m = \omega_0/C_0$, $s = \gamma(x - vt)$ and

$$\gamma = (1 - v^2/c_0^2)^{-1/2} \tag{2.18}$$

They are mobile domain walls separating two equilibrium configurations; ϕ_s and $\phi_{\bar{s}}$ are called soliton and antisoliton. Besides isolated solitons, the soliton–soliton scattering and the soliton-antisoliton scattering solutions are given by [50, 54]

$$\tan(\phi/4) = v \frac{\sinh(m\gamma x)}{\cosh(m\gamma t)} \tag{2.19}$$

and

$$\tan(\phi/4) = v^{-1} \frac{\sinh(m\gamma v t)}{\cosh(m\gamma x)} \tag{2.20}$$

respectively where $\gamma = (1 - v^2/c_0^2)^{-1/2}$. Equation (2.19), for example, describes two solitons with velocity v and $-v$ in the limit $|t| \to \infty$;

$$\tan(\phi/4) \cong \begin{cases} v[\exp(m\gamma(x + vt)) - \exp(-m\gamma(x - vt))] & \text{for} \quad t \to -\infty \\ v[\exp(m\gamma(x - vt)) - \exp(-m\gamma(x + vt))] & \text{for} \quad t \to +\infty \end{cases} \tag{2.21}$$

Therefore Equation (2.19) implies that solitons with velocity v and $-v$ are displaced in their position by

$$\Delta = -2(m\gamma)^{-1} \ln v \tag{2.22}$$

after the collision. More generally it is shown that the only effect of collision between two solitons or between a soliton and an antisoliton is a spatial displacement (i.e. a phase shift) compared to their unperturbed trajectories in the absence of collision.

Moreover the bound states of soliton and antisoliton are given by

$$\tan(\phi/4) = u^{-1} \frac{\sin(m\tilde{\gamma}ut)}{\cosh(m\tilde{\gamma}x)} \tag{2.23}$$

where $\tilde{\gamma} = (1 + u^2/c_0{}^2)^{-1/2}$ and u is a parameter. It is easy to see that the solution (2.23) can be obtained from Equation (2.20) by continuing analytically v to iu. The quantum mechanical analog of the solution corresponding to a purely imaginary velocity (or momentum) is the bound state. Since Equation (2.23) describes a localized oscillation of ϕ_t or ϕ_x, these solutions are called breathers or bions. Furthermore making use of the inverse scattering method [17] any localized initial perturbation results in solitons, antisolitons and breathers after a lapse of time. In this sense we may think of solitons, antisolitons and breathers as basic constituents of the sine-Gordon system.

In the case of the sine-Gordon system an exact N-soliton, M-antisoliton solution can be constructed [18].

2.3. Statistical Mechanics

In the one-dimensional system, the soliton given by Equation (2.16) has energy

$$E_s(v) = E_s\gamma, \qquad E_s = 8A\omega_0C_0 \tag{2.24}$$

When the sine-Gordon system is in thermal equilibrium, solitons can be treated as particles [36], and indeed the classical statistical mechanics of the sine-Gordon system can be formulated in terms of solitons [55, 56]. However, we shall describe here an alternative method, the transfer integral method (TIM) developed by Scalapino *et al.* [57]. This method reduces calculation of the free energy and the correlation lengths to solving an eigenvalue equation as long as we are interested in the classical limit. Of course the results thus derived can be interpreted in terms of solitons. First we shall introduce the Hamiltonian corresponding to Equation (2.14)

$$\mathcal{H} = \frac{A}{2} \int \mathrm{d}x \{\phi_t{}^2 + c_0{}^2\phi_x{}^2 + 2\omega_0{}^2 (1 - \cos\phi)\} \tag{2.25}$$

where π the conjugate operator to ϕ is given by

$$\pi = A\phi_t \tag{2.26}$$

The thermodynamics of the sine-Gordon system is determined from the partition function defined by

$$Z(=e^{-\beta F}) = \int \mathrm{D}\pi(x) \int \mathrm{D}\phi(x) e^{-\beta H} \tag{2.27}$$

where $D\pi(x)$ and $D\phi(x)$ are functional integrals over the static configuration of $\pi(x)$ and $\phi(x)$, respectively. In the classical limit the $D\pi(x)$ and $D\phi(x)$ integrations are performed separately, and the free energy F is given as [57, 58]

$$F = \frac{T}{l} \ln \left(\frac{C_0}{lT} \right) + A\omega_0{}^2 \epsilon_0 \tag{2.28}$$

where l is the lattice constant (the $D\pi(x)$ integral is carried out after discretizing $\pi(x)$ as π_i, and l^{-1} refers to the degrees of freedom associated with the system) and ϵ_0 is the lowest eigenvalue of the eigenequation;

$$\left(-\frac{1}{2m*} \frac{d^2}{d\phi^2} + (1 - \cos \phi) \right) \psi_n(\phi) = \epsilon_n \psi_n(\phi) \tag{2.29}$$

and

$$m* = (A\omega_0 C_0 \beta)^2 = (\beta E_s/8)^2. \tag{2.31}$$

Of particular interest is ξ_1, which is the correlation length associated with the correlation function sensitive to the topological order [59]. Indeed ξ_1 is expressed in terms of the soliton density as [59]

$$\xi_1 = 4n_s \tag{2.32}$$

where n_s is the total soliton density, to be defined later.

Recently Equation (2.29) has been studied both numerically [60] and perturbationally [61–63]. In the classical limit F_ϕ can be expanded in a double series as [61–63]

$$\begin{aligned} F_\phi = A\omega_0{}^2\epsilon_0 = E_s m \Bigg\{ & \tfrac{1}{2}t - \tfrac{1}{4}t^2 - \tfrac{1}{8}t^3 - 3t^4/16 - 53t^5/2^7 - \\ & - \frac{297t^6}{2^8} - \frac{3971t^7}{2^{10}} - (t/2\pi)^{1/2} e^{-1/t} \left(1 - \frac{7t}{8} - \frac{59t^2}{2^7} - \frac{897t^3}{2^{10}} - \right. \\ & \left. - \frac{75005t^4}{2^{15}} - \frac{1916273t^5}{2^{18}} - \frac{113580575t^6}{2^{22}} \right) + \\ & + \pi^{-1} e^{-2/t} \left(\ln \frac{4\gamma}{t} - \frac{5t}{4} \left(1 + \ln \frac{4\gamma}{t} \right) - \right. \\ & \left. - \frac{t^2}{4} \left(23 + 2 \ln \frac{4\gamma}{t} \right) + \cdots \right) \Bigg\} \end{aligned} \tag{2.33}$$

where $m = \omega_0/C_0$, $t = (\beta E_s)^{-1}$ and $\gamma = 1.781$, the Euler constant. The first term in the bracket of Equation (2.33) is interpreted as being due to anharmonic phonons, while the second term is due to thermal solitons [note that $n_s \propto e^{-1/t}$]. The last term is due to the soliton–soliton interaction [63]. Indeed the total soliton density is given within the same approximation by [61, 62]

$$n_s = m(2/\pi t)^{1/2} e^{-t^{-1}} \left(1 - \frac{7t}{8} - \frac{59t^2}{2^7} - \frac{897t^3}{2^{10}} \cdots\right) \tag{2.34}$$

The earlier version of the ideal soliton gas phenomenology [55, 56] was able to account for only the lowest order term in t in the coefficient of Equation (2.34). However, it is recently shown that the higher order corrections in t can be found [64], by making use of the method of collective coordinates developed by Gervais and Sakita [65].

Equation (2.32) together with Equation (2.34) indicates that the inverse of the correlation length ξ_1^{-1} increases exponentially as the temperature decreases below E_s. Furthermore, the specific heat of the sine-Gordon system is obtained from Equation (2.33) as [63]

$$\begin{aligned} C_s &= -T\frac{\partial^2 F}{\partial T^2} \\ &= l^{-1} + m\left\{\frac{t}{2} + \frac{3t^2}{4} + \frac{9t^3}{4} + \frac{265t^4}{2^5} + \frac{4455t^5}{2^7} + \cdots + \right. \\ &\quad + 4(2\pi)^{-1/2} t^{-5/2} e^{-1/t}\left(1 - \frac{15t}{8} - \frac{203t^2}{2^7} - \frac{2985t^3}{2^{10}} - \right. \\ &\quad \left. - \frac{275165t^4}{2^{15}} \cdots\right) - 32\pi^{-1}t^{-3}e^{-2/t}\left(\ln\frac{4\gamma}{t} - \frac{9t}{4}\left(1 \ln\frac{4\gamma}{t}\right) + \right. \\ &\quad \left.\left. + t^2\left(\frac{11}{8} - \frac{13}{32}\ln\frac{4\gamma}{t}\right)\right)\right\} \end{aligned} \tag{2.35}$$

The specific heat is shown as function of t in Figure 3, which is taken from [63].

The specific heat has a broad peak around $T = 0.22E_s$ indicating the crossover from the topologically disordered high temperature phase to the topologically ordered low temperature phase. It is clear from the Figure that the two-soliton term is essential to obtain the specific heat consistent with the numerical result. In particular, the peak in the specific heat at $t = 0.22$ will be shifted to a much higher value if the two-soliton term is neglected. The topological order is destroyed by the presence of thermally excited solitons. Therefore the existence of the sine-Gordon solitons in the one-dimensional system is most clearly seen in the anomaly in the specific heat and in the temperature dependence of the coherence length ξ_1.

2.4. Topological Conservation

We shall now describe a remarkable conservation law associated with the sine-Gordon system [20]. Without loss of generality we can assume that $\phi(\infty, t)$ and $\phi(-\infty, t)$ are at one of the ground states. Then we have

$$Q = \frac{1}{2\pi}\int_{-\infty}^{\infty} dx\phi_x(x, t) = n - m = N \tag{2.36}$$

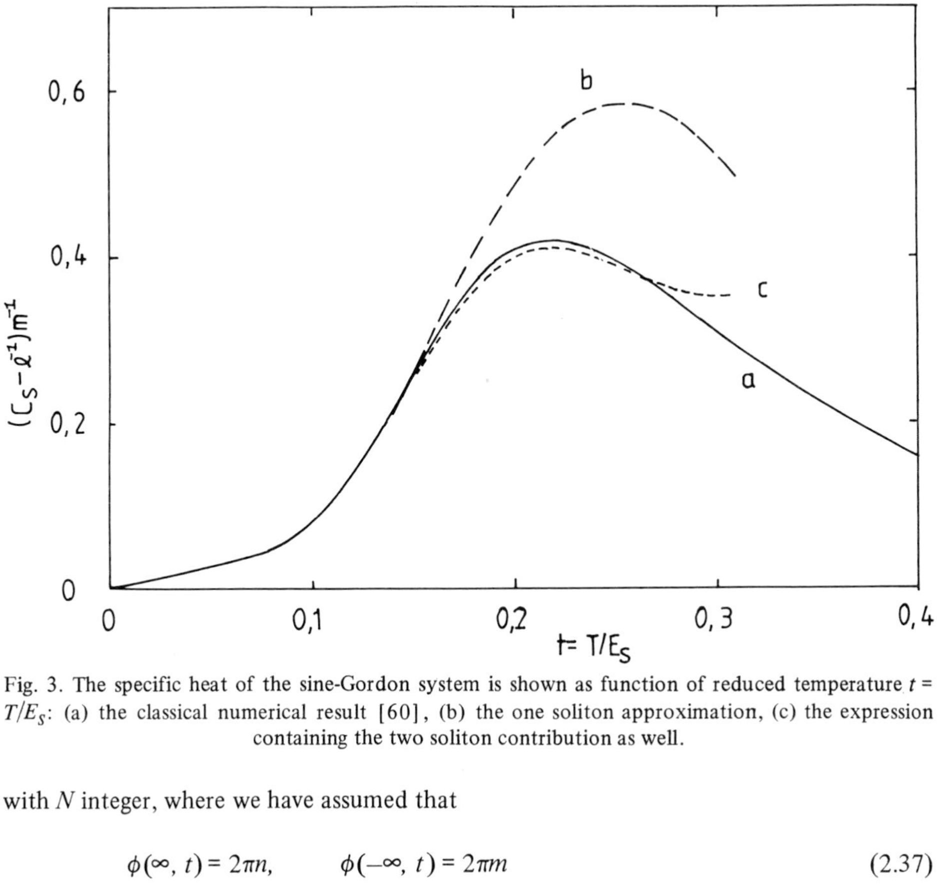

Fig. 3. The specific heat of the sine-Gordon system is shown as function of reduced temperature $t = T/E_S$: (a) the classical numerical result [60], (b) the one soliton approximation, (c) the expression containing the two soliton contribution as well.

with N integer, where we have assumed that

$$\phi(\infty, t) = 2\pi n, \qquad \phi(-\infty, t) = 2\pi m \tag{2.37}$$

The quantity Q is the topological charge of the system and the integer N is the difference between the soliton number n_s in the system

$$N = n_s - n_{\bar{s}} \tag{2.38}$$

The topological charge Q is conserved in the presence of any local perturbation on the SG system and therefore plays a very similar role to the electric charge in quantum electrodynamics for example. In terms of homotopy the topological charge indexes the mapping between two physical points $x = \infty$ and $x = -\infty$ to a discrete set of ground states of the sine-Gordon system.

2.5. Quantum Field Theory

The sine-Gordon Lagrangian (2.14) may be considered to describe a quantum system in 1 + 1 dimensions. Then it is more convenient to rewrite Equation (2.14) as

$$\mathscr{L} = \frac{1}{2} \int \mathrm{d}x \left\{ \phi_t{}^2 - \phi_x{}^2 - \frac{2m^2}{g^2} (1 - \cos g\phi) \right\} \tag{2.39}$$

where

$$m = \omega_0, \ g = A^{-1/2} \tag{2.40}$$

and the present ϕ corresponds to $\sqrt{A}\phi$ in Equation (2.14). Furthermore we take $C_0 = 1$ for simplicity. If we assume that Equation (2.39) describes a quantum system, we have to eliminate divergences associated with the quantum fluctuation. In the case of the sine-Gordon system, this is done [43] simply by renormalizing mass m; we can rewrite Equation (2.39) as

$$\mathscr{L} = \frac{1}{2} \int \mathrm{d}x \left\{ \phi_t{}^2 - \phi_x{}^2 - \frac{2m_0^2}{g^2} (1 - N(\cos g\phi)) \right\} \tag{2.41}$$

where

$$m_0{}^2 = m^2 e^{-g^2 D/2}, \ D = \langle \phi^2 \rangle \tag{2.42}$$

and the normal product is defined by

$$V(\phi) = N \left\{ \exp \frac{1}{2} \frac{\partial^2}{\partial \phi^2} V(\phi) \right\} \tag{2.43}$$

Diagrammatically the normal product eliminates all the single loop contractions associated with $V(\phi)$. In general, D diverges logarithmically with the cut-off momentum Λ

$$D = \frac{1}{2\pi} \ln \left(\frac{2\Lambda}{m_0} \right)$$

and therefore in the quantum field theory only m_0 has physical meaning. On the other hand in applying a field theoretic model to the quasi-one-dimensional condensed systems, the systems in general has a natural cut off associated with the underlying crystalline lattice. Furthermore, in this circumstance both m and m_0 are observable [66]. We shall discuss the related examples in Section 3.

Making use of the semi-classical approximation Dashen *et al.* [42] have shown that the soliton energy is now given by

$$E_s = \frac{8m_0}{g^2} - \frac{m_0}{\pi} \tag{2.44}$$

where the second term is the mass correction due to the quantum fluctuation. Furthermore, the breathers are quantized; they are bosons with masses given by

$$E_n = 2E_s \sin(ng'^2/16) \tag{2.45}$$

with $n = 1, 2, \ldots < 8\pi g'^{-2}$ and

$$g'^2 = g^2 \, (1 - g^2/8\pi)^{-1} \tag{2.46}$$

A simplified derivation of Equation (2.45) is following; making use of the Bohr–Sommerfeld quantization condition, we obtain;

$$\int_{-\infty}^{\infty} \mathrm{d}x \int \pi(x)\mathrm{d}\phi(x) = \int_{-\infty}^{\infty} \mathrm{d}x \int_{0}^{T} \mathrm{d}t\pi(x)\phi_t(x) = 2\pi n \tag{2.47}$$

with n integer and T is the period of the breather solution. Now inserting the breather solution (2.23) slightly modified to be consistent with the Lagrangian (2.39)

$$\pi(x, t) = \phi_t(x, t) = \frac{4m\tilde{\gamma}}{g} \cos(m\tilde{\gamma}ut) \left(1 + u^{-2} \frac{\sin^2(m\tilde{\gamma}ut)}{\cosh^2(m\tilde{\gamma}x)}\right)^{-1} \tag{2.48}$$

with

$$\tilde{\gamma} = (1 + u^2)^{-1/2}$$

into Equation (47) and integrating over x and t (note that $T = 2\pi(m\tilde{\gamma}u)^{-1}$), we obtain

$$u^{-1} = \tan\left(\frac{ng^2}{16}\right) \tag{2.49}$$

which yields

$$E_n = 2E_s\tilde{\gamma} = 2E_s \sin\left(\frac{ng^2}{16}\right) \tag{2.50}$$

Finally the fluctuation of the quantum field around the classical solution, Equation (2.48), gives rise to additional corrections, which lead to Equation (2.45). Further, Dashen *et al.* [42] established that Equation (2.45) gives the correct bound state spectrum of the soliton and antisoliton up to the order of g^6 by perturbation analysis. Finally Dashen *et al.* [42] established that there is no other bosonic excitation at low temperatures; the breathers exhaust the bosonic excitations of the system. Later the above mass spectrum was confirmed by Luther [67] who exploited the relation between the bound state spectrum of the *XYZ* chain [68] and that of the sine-Gordon system. Therefore as far as the low temperature behavior is concerned the sine-Gordon system is described completely in terms of breathers, solitons and antisolitons; there is no room for the phonon modes which plays an important role in the thermodynamics of the sine-Gordon system in the classical limit (see Subsection 2.3).

Indeed, following the pioneering work of Bergknoff and Thacker [45], who solved the energy spectrum of the sine-Gordon system alias Massive Thirring model by Bethe Ansatz, it is possible to formulate the thermodynamics of the sine-Gordon system in terms only of solitons and breathers [69–73].

How the breathers are transformed into phonons in the classical limit is an intriguing question, which is discussed in [72]. In the weak-coupling limit the thermodynamics of the sine-Gordon system can be constructed by the functional integral technique [55]. in particular, it is shown that the classical statistical mechanics breaks down when $T \simeq m$ the mass of the phonon (or the mass of the lowest breather mode). Therefore in analysing the observed specific heat of quasi-one-dimensional systems, it is rather important to know the phonon mass to see if the classical theory is applicable or not. Such comparison of the present theory with observed specific heat in the quasi-one dimensional magnetic system will be discussed in Section 3.

3. Solitons in Magnetic Systems

3.1. Introduction

There are many magnetic insulators in which the interspin interaction in one direction (we call this direction the chain direction) is much stronger than the interactions in other directions. Most of these insulators undergo a phase transition into a three-dimensional magnetically-ordered state at low enough temperature T_N. However, in many cases we can describe these magnetic systems as one-dimensional systems when the temperature is somewhat above the Néel temperature T_N. For a general review on such magnetic systems the reader may consult Steiner *et al.* [74]. We shall describe here three typical magnetic systems where, in our opinion, the existence of magnetic solitons is clearly established, namely $CsCoCl_3$, $(CH_3)_4NMnCl_3$(TMMC) and $CsNiF_3$.

3.2. Ising-like Antiferromagnetic Chain

Perhaps the simplest magnetic system, which has soliton like excitations is the Ising-like antiferromagnetic chain described by the following Hamiltonian;

$$\mathcal{H} = 2J \sum_n S_n^z S_{n+1}^z + 2\epsilon J \sum_n (S_n^x S_{n+1}^x + S_n^y S_{n+1}^y) \tag{3.1}$$

where $\mathbf{S}_n$ is the spin operator with $S = \frac{1}{2}$ at site n, and $|\epsilon| \ll 1$ and $J > 0$ (antiferromagnetic coupling). The magnetic properties of quasi-one-dimensional systems like $CsCoCl_3$ and $CsCoBr_3$ [74] are well described by Equation (3.1). The Néel ground state of this system is doubly degenerate. In the limit of small ϵ the ground state configurations are shown Figures 4 (a) and (b). Further, as was first shown by Villain [75], there are low energy excitations which are shown in Figure 4(c) and (d). These excitations are magnetic solitons (or domain walls, which separate two distinct Néel ground states). The second term in Equation (3.1) transfers the soliton from one site to another and within the tight binding approximation the soliton energy is given by

$$E_s(k) = J[1 + 2\epsilon \cos(2ak)] \tag{3.2}$$

where k is the wave vector associated with the moving soliton and a is the lattice constant. The velocity of soliton with momentum k is given by

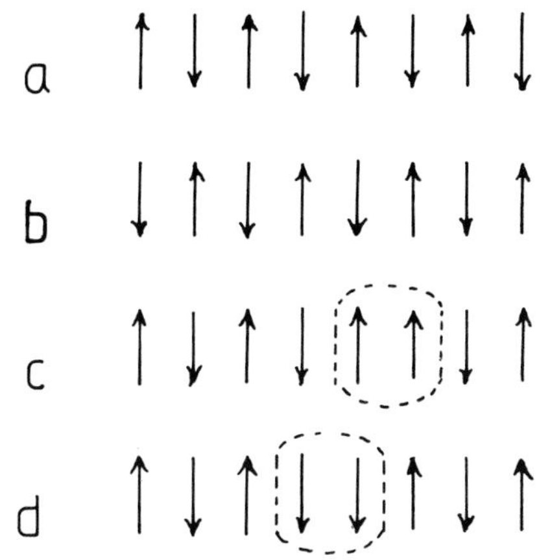

Fig. 4. The ground state spin configurations (a) and (b) of Ising-like antiferromagnetic chain. (c) and (d) are the soliton configurations, where the arrows indicate the spin direction.

$$v = \frac{\mathrm{d}E_s(k)}{\mathrm{d}k} = -\,4\epsilon aJ \sin(2ak) \tag{3.3}$$

and the density of soliton with momentum k is given by the usual Boltzmann distribution when $\beta J \gg 1$

$$n_s(k) = \exp(-\beta E_s(k)) \tag{3.4}$$

The system described by Equation (3.1) has spin wave excitations in addition to solitons. However, the spin waves have larger excitation energies [76].

Villain [75] has predicted that the longitudinal spin correlation function is dominated by the soliton. In particular, for $|q - \pi/a| \gg K$, where q is the transferred momentum and K^{-1} is the longitudinal correlation length in the Ising limit (a more accurate expression including the correction due to ϵ will be given later)

$$e^{-Ka} = \tanh(\tfrac{1}{2}\beta J) \tag{3.5}$$

Villain obtained the dynamical structure factor associated with the single soliton excitation;

$$\begin{aligned} S_{zz}(q, \omega) &= (2\pi Z)^{-1} \frac{2aK}{(Ka)^2 + 4\cos^2(\frac{1}{2}aq)} \times \\ &\quad \times \frac{e^{\beta\omega/2}}{(\Omega_q^2 - \omega^2)^{1/2}} \cosh[\tfrac{1}{2}(\Omega_q^2 - \omega^2)^{1/2} \cot(aq)], \quad \text{for } |\omega| < \Omega_q \\ &= 0, \quad \text{for } |\omega| > \Omega_q \end{aligned} \tag{3.6}$$

where $\Omega_q = 4|\epsilon J \sin(aq)|$

$$Z = N^{-1} \sum_k \exp[-2\beta J\epsilon \cos(2ak)] \tag{3.7}$$

This structure factor has a sharp cut-off at $\omega = \pm\Omega_q$ and has been observed in recent neutron scattering experiments on $CsCoCl_3$ and $CsCoBr_3$ by Yoshizawa *et al.* [77] and by Nagler *et al.* [78], respectively. When $|q - \pi/a| \lesssim K$ on the other hand, Equation (3.6) is no longer applicable, since the structure factor is dominated by the multi-soliton term in this limit [79]. Within the ideal soliton gas approximation for solitons [36], the asymptotic expression for large m limit of the spin correlation function is given by [79];

$$\langle [S^z_{n+m}(t), S^z_n(t)] \rangle = \exp\left\{ i\pi m - 4\int_{-\pi/2a}^{\pi/2a} \frac{dk}{2\pi} n_s(k)|x - v_k t| \right\} \tag{3.8}$$

where $x = ma$, and $n_s(k)$ and v_k have been defined in Equations (3.4) and (3.3), respectively. A factor of 4 in the second term in the exponential of Equation (3.8) arises from the fact that there are two types of solitons as shown in Figures 4(c) and (d), and that a soliton between the n and the $n + m$ site changes the sign of the correlation function. Unfortunately this additional factor 2 was neglected in the expression given in [79]. Equation (3.8) is well approximated as

$$\langle [S^z_{n+m}(t)\, S^z_n(t)] \rangle = \exp\{ i\pi x/a - K(x^2 + v_0{}^2 t^2)^{1/2} \} \tag{3.9}$$

where

$$K = 4\bar{n}_s = 2a^{-1} e^{-\beta J} I_0(2\epsilon\beta J) \tag{3.10}$$

and

$$v_0 = 4a(\pi\beta I_0(2\epsilon\beta J))^{-1} \sinh(2\epsilon\beta J) \tag{3.11}$$

where $I_0(z)$ is the modified Bessel function. Here $\bar{n}_s$ is the soliton density, K^{-1} is the correlation length, which includes the correction due to the ϵ-dependent term in Equation (3.1), and v_0 is the thermal velocity of solitons. In the low temperature limit ($\beta J \gg 1$) and $\epsilon \to 0$, K in Equation (3.10) reduces to that in Equation (3.5). The corresponding dynamical structure factor is then given by

$$S_{zz}(q, \omega) = \frac{K}{2\pi v_0} \left[(q - \pi/a)^2 + (\omega/v_0)^2 + K^2 \right]^{-3/2} \tag{3.12}$$

We note that this dynamical structure factor with $K = 4\bar{n}_s$ is quite general and finds application in a number of quasi-one-dimensional antiferromagnetic systems.

3.3. Planar Antiferromagnetic Chain in a Magnetic Field

Another magnetic system of great current interest is an antiferromagnetic chain of spins described by

$$\mathcal{H} = -2J\sum_n \mathbf{S}_n \cdot \mathbf{S}_{n+1} - 2D\sum_n S^z_n S^z_{n+1} - H\sum_n S^x_n \tag{3.13}$$

where $J < 0$ and the chain direction is taken as the z direction. This Hamiltonian with $S = 5/2$, $J = -15K$ and $D = 2.5K$ quite well describes a quasi-one-dimensional compound $(CH)_3NMnCl_3$(TMMC) [80, 81]. The last term in Equation (3.13) describes the magnetic energy in a magnetic field H in the x direction (we took $\mu_B = 1$ for simplicity, so H has the dimension of energy). Furthermore, according to Walker *et al.* [80], TMMC can be considered as a planar model below $T = 20$ K. We are interested here in this low temperature regime. Then we can parameterize $\mathbf{S}_n$ as [47–49];

$$\mathbf{S}_n = ((S^2 - \Pi_n^2)^{1/2} \cos\phi_n, \qquad (S^2 - \Pi_n^2)^{1/2} \sin\phi_n, \quad \Pi_n) \tag{3.14}$$

with $[\Pi_n, \quad \phi_n{}'] = -i\delta_{nn'}$

Substituting Equation (3.14) in Equation (3.13) and neglecting the higher-order terms in $\Pi_n S^{-1}$, we obtain in the continuum limit;

$$\mathcal{H} = (2a)^{-1} \int dz \{\tfrac{1}{4}(2|J| - D)^{-1} \phi_t^2 + 2a^2 |J| S^2 \phi_z^2 + \tfrac{1}{8} |J|^{-1} H^2 \cos^2\phi\} \tag{3.15}$$

where a is the lattice constant.

The ground state of Equation (3.15) is given by $\phi = \pm\pi/2$. (The spin in the ground state is perpendicular to the external field $\mathbf{H}$.) As is immediately apparent, ϕ obeys a sine-Gordon equation. A soliton is given by

$$\phi(z, t) = \pi/2 + 2 \tan^{-1} \{\exp[m\gamma(z - vt)]\} \tag{3.16}$$

with the soliton energy

$$E_s(v) = SH\gamma \tag{3.17}$$

where

$$m = H(4a|J|S)^{-1}, \qquad C = 2aS[2(2|J| - D)|J|]^{1/2} \quad \text{and}$$

$$\gamma = (1 - v^2/c^2)^{-1/2} \tag{3.18}$$

In the antiferromagnetic chain the soliton describes a π-twist in the spin as shown in Figure 5. As in the example of the preceding subsection, the spin correlation function $\langle [S^y(z, t) S^y(0, 0)] \rangle$ reflects directly the topological order of the system; the sign of this correlation function changes whenever a soliton crosses a line connecting two points (z, t) and $(0, 0)$. This situation is analoguous to the ϕ–ϕ correlation function in the ϕ^4 theory discussed by Krumhansl and Schrieffer [36]. The soliton gives rise to a large central peak in this correlation function [47–49]. Such a central peak has been observed through the inelastic neutron scattering experiment on TMMC in magnetic fields by Boucher *et al.* [38]. The correlation function is given [47, 48] as $\langle [S^y(z, t) S^y(0, 0)] \rangle = \exp(i\pi z/a) S^2 \langle \sin^2\phi \rangle \exp\{-2N(z, t)\}$ where

$$N(z, t) = \frac{1}{\pi} \int_{-\infty}^{\infty} dp |z - vt| n_s(v) \tag{3.19}$$

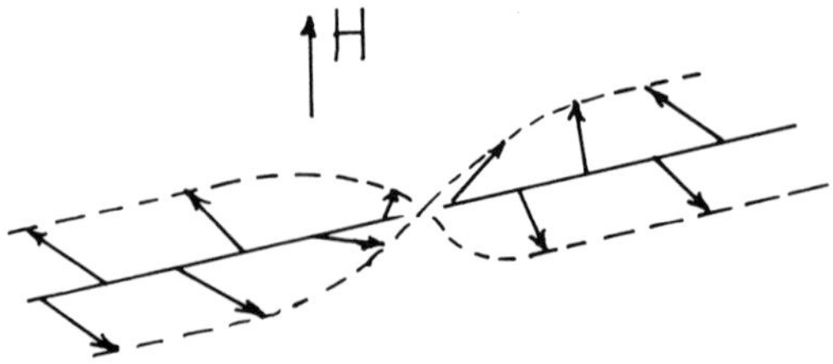

Fig. 5. The spin configuration of a soliton in planar antiferromagnetic chain in a magnetic field.

and $n_s(v)$ is the density of soliton with velocity v given by [55, 56, 64]

$$n_s(v) = 2\beta m \exp(-\beta E_s(v)) \tag{3.20}$$

where $E_s(v)$ and m have been defined in Equations (3.17) and (3.18), respectively. Here Equation (3.20) is the expression appropriate only in the classical limit (see Section 2). Note a similarity between Equation (3.19) and Equation (3.8). $N(z, t)$ may be approximated as [48]

$$N(z, t) = 2\bar{n}_s (z^2 + v_0^2 t^2)^{1/2} \tag{3.21}$$

with

$$\bar{n}_s = \frac{1}{2\pi} \int_{-\infty}^{\infty} \mathrm{d}p \; n_s(v) \cong \frac{2m}{c} \left(\frac{\beta E_s}{2\pi}\right)^{1/2} \left(1 - \frac{7(\beta E_s)^{-1}}{8} - \frac{59(\beta E_s)^{-2}}{2^7}\right) \exp(-\beta E_s) \tag{3.22}$$

and

$$v_0 \cong (\tfrac{1}{2} \pi \beta E_s)^{-1/2} \tag{3.23}$$

where $\bar{n}_s$ is the total soliton density (we include a few higher-order terms in $(\beta E_s)^{-1}$ following Equation (2.34)) and v_0 is the thermal velocity of solitons. As in the case of the preceding example, there are two kinds of solitons (right-handed and left-handed twist). Finally the Fourier transform of Equation (3.19) yields

$$S_{yy}(q, \omega) = S^2 \langle \sin^2 \phi \rangle \, 2\pi K v_0^{-1} \, [(q - \pi/a)^2 + (\omega/v_0)^2 + K^2]^{-3/2} \tag{3.24}$$

where

$$K = 4\bar{n}_s \tag{3.25}$$

which is formally the same as Equation (3.12).

Both the ω and q dependence of the central peak of TMMC in magnetic fields have been studied in great detail by Regnault *et al.* [82], which confirms the theoretical expression (3.24). In particular, both the width and the shape of the central peak is well described by Equation (3.24) with $\bar{n}_s$ and v_0 given by Equations (3.22) and (3.23), respectively, except that the classical soliton energy given by Equation (3.17) has to be replaced by the renormalized soliton energy $E_s^{\text{ren}} \cong 0.77\ E_s^{\text{cl}}(E_s^{\text{cl}} = SH)$. The latter discrepancy may be interpreted by the quantum correction [66], although the estimated quantum correction appears to be somewhat too large ($E_s^{\text{ren}} = 0.60 E_s^{\text{cl}}$). Another possible source of discrepancy is that, in a high magnetic field, the off-plane spin fluctuation may no longer be neglected and this fluctuation will give rise to an additional renormalization in $\bar{n}_s$ for example [83]. Experiments in a higher magnetic field can, in principle, clarify these different contributions.

The above neutron scattering experiment not only demonstrates the existence of solitons, which are thermally activated in TMMC in a magnetic field for $2\ \text{K} < T < 5\ \text{K}$ and $H/T \gtrsim 10\text{kOe/k}$, but also establishes that the magnetic solitons behave like free particles. Therefore the ideal soliton gas model provides an excellent description of magnetic properties of TMMC. More recently Boucher and Renard [39] have measured the nuclear spin relaxation rate T_1^{-1} of ^{15}N atoms in TMMC in a magnetic field. Since T_1^{-1} is proportional to the local spin correlation function, we have

$$T_1^{-1} \propto \int dq\, S_{yy}(q, \omega) \propto v_0^{-2} K[(\omega/v_0)^2 + K^2] \sim (v_0{}^2 K)^{-1} \tag{3.26}$$

which is actually observed experimentally.

The above free-particle-like behavior of solitons may be destroyed by introducing impurities in TMMC, since the impurities will scatter solitons. Then when the soliton mean free path l becomes $l\bar{n}_s \ll 1$, soliton behaves diffusively [79]. Such diffusive regimes have been studied in [84–86]. In particular, $S_{yy}(q, \omega)$ and T_1^{-1} in this regime are given by [79, 86];

$$S_{yy}(q_1, \omega) \propto 8KD^{-1}[(K^2 + (q - \pi/a)^2)^2 + 24(\omega/D)^2]^{-1} \tag{3.27}$$

$$T_1^{-1} \propto 2K[K^4 + 20(\omega/D)^2]^{-3/4} \tag{3.28}$$

where $D = lv_0$ and v_0 have been defined in Equation (3.23). Recently both neutron scattering and nuclear spin relaxation experiments on TMMC doped by 0.3% to 1% Cu atoms in a magnetic field have been performed by Boucher *et al.* [87]. They observed T_1^{-1} given in Equation (3.28), when the soliton mean free path l is about 100 × the atomic constant, which indicates clearly that, in this system, the soliton behavior is diffusive.

Recently Borsa [88] reported the specific heat of TMMC in the presence of magnetic fields. However, we shall defer the discussion of the specific heat to Subsection 3.5.

3.4. Planar Ferromagnetic Chain in a Magnetic Field

The magnetic properties of quasi-one-dimensional magnets like $CsNiF_3$ are well described by the Hamiltonian;

$$\mathcal{H} = -2J \sum_n \mathbf{S}_n \cdot \mathbf{S}_{n+1} + A \sum_n (S_n^z)^2 - H \sum_n S_n^x \tag{3.29}$$

where $\mathbf{S}_n$ is the spin operator at site n and the chain direction is taken along the z direction. Here a magnetic field $H/g\mu_B$ is applied perpendicular to the chain direction. For $CsNiF_3$ we have $S = 1$, $J = 11.8K$, $A = 9.5K$ and $g = 2.4$ [37]. This is the first magnetic system in which observation of magnetic solitons was reported [37]. At low temperatures where the spin fluctuation is mainly restricted to the $x - y$ plane (i.e. $T \lesssim (AJ)^{1/2}$), Equation (3.29) is recast as [46];

$$\mathcal{H} = -2JS^2 \sum_n \cos(\phi_{n+1} - \phi_n) + A \sum_n \Pi_n^2 - HS \sum_n \cos\phi_n \tag{3.30}$$

where we have parametrized

$$\mathbf{S}_n = [(S^2 - \Pi_n^2)^{1/2} \cos\phi_n, \qquad (S^2 - \Pi_n^2)^{1/2} \sin\phi_n, \quad \Pi_n]$$

with

$$[\Pi_n, \quad \phi_{n'}] = -i\delta_{nn'} \tag{3.31}$$

and higher-order terms in Π_n/S are neglected. Finally in the continuum limit Equation (3.30) reduces to the sine-Gordon Hamiltonian;

$$\mathcal{H} = a^{-1} \int dz \, [(4A)^{-1} \phi_t^2 + (aSJ)^2 \phi_z^2 - SH \cos\phi] \tag{3.32}$$

where a is the lattice constant. Here the planar symmetry of the ground state is broken by an external field H. The sine-Gordon soliton of Equation (3.32) is given by

$$\phi(z, t) = 4 \tan^{-1} [\exp(m\gamma(z - vt))] \tag{3.33}$$

with

$$m = a^{-1}(H/2JS)^{1/2} \tag{3.34}$$

The soliton energy is given by

$$E(v) = 8S^{3/2}(2JH)^{1/2}\gamma \tag{3.35}$$

The soliton describes a moving 2π-twist in magnetization as shown in Figure 6 and carries local magnetization given by

$$\delta\mathbf{M}(z, t) = S[-\mathrm{sech}^2 s, 2 \,\mathrm{sech}\, s \tanh s, (2aSA)^{-1} m\gamma v \,\mathrm{sech}\, s] \tag{3.36}$$

As shown by Mikeska [46], the magnetic soliton gives rise to an additional contribution to the spin correlation functions;

$$S_{xx}(q, \omega) \propto 2n_s(v) [(2\pi cq/m) \,\mathrm{cosech}(\pi cq/2m)]^2 \tag{3.37}$$

$$S_{yy}(q, \omega) \propto 2n_s(v) [(2\pi cq/m) \,\mathrm{sech}(\pi cq/2m)]^2 \tag{3.38}$$

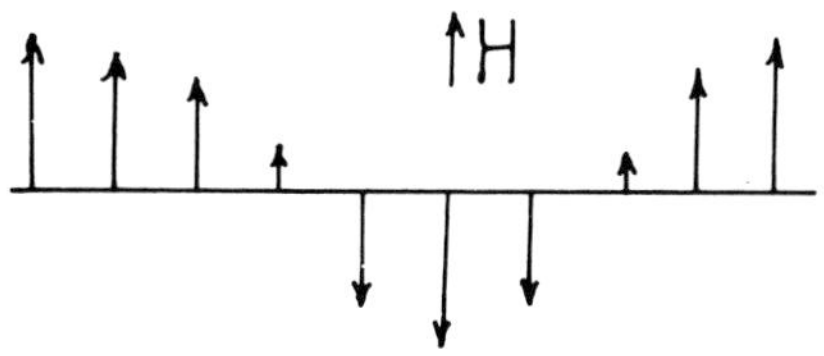

Fig. 6. The soliton in a planar ferromagnetic chain in a magnetic field.

where $n_s(v)$ is the density of solitons with velocity v,

$$N_s(v) \cong 2\beta m \exp(-\beta E_s(v)) \tag{3.39}$$

in the classical limit and v in Equations (3.37) and (3.38) should be taken to be $v = \omega/q$. The central peak given by Equation (3.37) has been observed in $CsNiF_3$ in a magnetic field by inelastic neutron scattering by Kjems and Steiner [37]. Except the fact that the soliton energy required to describe the experimental data is somewhat smaller (by about 20–30%) than that expected from Equation (3.35), Equation (3.37) quite well describes the observed neutron scattering cross-section. The discrepancy in the soliton energy can be understood as the effect of the mass renormalization due to the quantum fluctuation [66]. Although there is still controversy as to whether the observed central peak is due to the thermal magnons or the thermal solitons [89], the author believes that the thermal magnon contribution is too small to account for the observed central peak.

More recently the specific heat measurement of $CsNiF_3$ has been reported by Ramirez and Wolf [90]. We shall discuss the specific heat of the quasi-one-dimensional planar magnets in the following subsection.

3.5. Specific Heat of the Planar Magnets in a Magnetic Field

As already mentioned the specific heats of TMMC and $CsNiF_3$ in magnetic fields were reported [88, 90]. Initial attempts to describe the observed peaks in the specific heat in terms of the single soliton term (the second term in Equation (2.35)) lead to confusion. The observed peaks appeared to imply too small a soliton energy. However, the inclusion of the two soliton term (see Equation (2.35)) improved the agreement greatly. On the other hand, the classical expression (2.35) cannot describe the observed rapid decreases in the specific heat for $t \leqslant 0.2$, which are actually due to the quantum freezing of the magnons in these systems. A more satisfactory comparison is made with the specific heat of the sine-Gordon system within the Bethe Ansatz [91]. Some of these fits are shown in Figure 7, which is taken from [91]. As may easily be seen the theory appears to describe quite well, not only the peak positions, but also the general temperature-dependence of the specific heat. In drawing the figure the following renormalized soliton energies $E_s^r = 0.7E_s$ is used for $CsNiF_3$. For TMMC the classical theory with $E_s^r = 0.84E_s$ appears to describe the observed specific heat, although the predicted peak in the specific heat appears to be somewhat lower than that observed. These renormalized soliton energies are not inconsistent with the ones due to the quantum fluctuation [66].

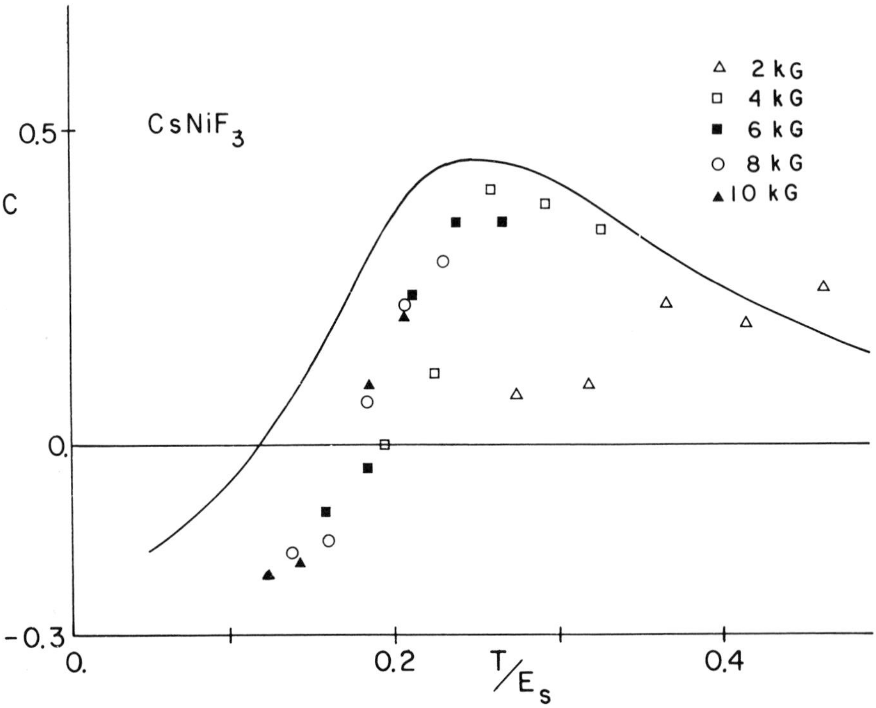

Fig. 7. The specific heat of $CsNiF_3$ (planar ferromagnetic chain) in magnetic fields [90] are compared with the Bethe Ansatz calculation [91].

4. Solitons in Conducting Systems

4.1. Peierls – Fröhlich Instability

In the following we shall describe solitons in conducting systems. We are interested mostly in quasi-one-dimensional systems, where the system undergoes a Peierls–Fröhlich transition [92, 93] at nonzero temperature; the resulting condensed state is called the Peierls–Fröhlich state or the charge density wave (CDW) state. A recent work on low dimensional systems is found in [7]. We shall start with a one-dimensional system described by the Fröhlich Hamiltonian [93]

$$\mathcal{H} = \sum_{ks} \epsilon_k C_{ks}^+ C_{ks} + \sum \omega_q^0 b_q^* b_q + ig \sum_q (\omega_q^0/2)^{1/2} C_{k+qs}^+ C_{ks}(b_q + b_q^*) \quad (4.1)$$

where C_{ks}^+ and b_q^* are the creation operators for a Bloch electron with momentum k and spin s and a longitudinal phonon with momentum q; and ϵ_k and ω_q^0 are bare energies of the electron and phonon, respectively. Furthermore the electron–phonon coupling constant g is taken to be constant for simplicity. Due to the electron–phonon interaction the phonon dispersion is modified from ω_q^0 [94, 95] as

$$\omega_q = \omega_q^0 (1 - \pi(q))^{1/2} \quad (4.2)$$

where

$$\pi(q) = \frac{2q^2}{\pi} \int \mathrm{d}p \, \frac{f_p(1 - f_{p+q})}{\epsilon_{p+q} - \epsilon_p}$$

$$= \frac{1}{\pi v_F} g^2 \frac{1}{x} \ln \left| \frac{1+x}{1-x} \right| \quad \text{at} \quad T = 0 \text{ K} \tag{4.3}$$

and $x = q/2P_F$.

Here f_p is the Fermi distribution function and

$$\epsilon_p = \frac{1}{2m} p^2 - \mu$$

is used in evaluating Equation (4.3). At $T = 0$ K, $\pi(q)$ diverges logarithmically for $q = 2P_F$ and P_F is the Fermi momentum. This singularity in the phonon dispersion indicates that the one-dimensional normal electron state becomes unstable at low temperatures. This instability is known as Peierls–Fröhlich instability [92, 93]. At high temperatures, the logarithmic divergence is completely smeared out by the temperature. As the temperature is decreased, the phonon dispersion starts to exhibit a sharp dip around $q = 2P_F$. This is the Kohn anomaly [95], and has been observed in a number of quasi-one-dimensional systems. See [96] for a review. With lowering temperature, this dip becomes sharper, and ultimately the system becomes unstable against formation of the charge density wave (CDW) state when $\omega_Q = 0$ with $Q = 2P_F$, within the mean field approximation.

Strictly speaking both the Midgal approximation and the mean field approximation are not valid in the one-dimensional system due to the fluctuations neglected in the mean field theory. However, no real physical systems are strictly one-dimensional. Furthermore, even in the one-dimensional system, the short-range correlation with the local order parameter close to the mean field value develops at low temperature, say below $T_c/4$ [97], where T_c is the transition temperature determined within the mean field theory. Therefore we shall continue to analyse the charge density wave (CDW) condensate in terms of the mean field theory in spite of its limitations.

4.2. THE CHARGE DENSITY WAVE (CDW) STATE

The mean field theory of the charge density wave state is worked out in a classical paper by Lee *et al.* [98]. The order parameter describing the CDW condensate is introduced by

$$\Delta = g(2\omega_Q)^{1/2} \langle b_Q \rangle \tag{4.5}$$

Substituting this into Equation (4.2) where we replace b_Q by $\langle b_Q \rangle$, we obtain;

$$\mathcal{H}_{\mathrm{MF}} = g^{-2} |\Delta|^2 + \sum_s \int \mathrm{d}x \, \Psi_s^\dagger(x) (-i\sigma_3 v_F \partial_x + \sigma_1 \Delta_1 + \sigma_2 \Delta_2) \Psi_s(x) \tag{4.6}$$

where we have introduced a spinor representation for electrons [99] as

$$\Psi_{ks} = \begin{Bmatrix} C_{k+\frac{1}{2}Qs} \\ C_{k-\frac{1}{2}Qs} \end{Bmatrix} \tag{4.7}$$

and $\Psi_s(x)$ is the Fourier transform of Ψ_{ks} and $\Delta = \Delta_1 + i\Delta_2$. Here σ_i are the Pauli matrices and we have neglected $k^2/(2m)$ in comparison with $v_F k$. The thermal Green's function for the electron is now given by

$$G(k, \omega_n) = (i\omega_n - \xi_k \sigma_3 - \Delta_1 \sigma_1 - \Delta_2 \sigma_2)^{-1} \tag{4.8}$$

where $\xi_k = v_F k$ and $\omega_n = (2n+1)\pi T$ is the Matsubara frequency. Then Δ is determined from the self-consistent equation

$$\Delta = g^2 T \sum_n \int \frac{\mathrm{d}k}{2\pi} \operatorname{Tr}\{(\sigma_1 + i\sigma_2) G(k, \omega_n)\}$$

$$= \frac{g^2}{2\pi} \Delta \int \mathrm{d}k \frac{\tanh(\frac{1}{2}\beta E_k)}{E_k} \tag{4.9}$$

where

$$E_k = (\xi_k^2 + |\Delta|^2)^{1/2} \tag{4.10}$$

This is essentially the BCS gap equation [100]. In particular, at $T = 0$ K, we obtain

$$\Delta = 2We^{-1/\lambda} \tag{4.11}$$

with

$$\lambda = g^2(\pi v_F)^{-1} \quad \text{and} \quad W = Q^2/8m \tag{4.12}$$

where W is the half band width. Within the mean field theory the temperature dependence of Δ is the same as the BCS energy gap of a superconductor. Δ vanishes at T_c given by

$$T_c = \frac{2\pi}{\gamma} We^{-1/\lambda} \tag{4.13}$$

On the other hand, in the vicinity of the transition temeprature the prediction of the mean field theory becomes less reliable for the quasi-one-dimensional system.

In the CDW state, the electron charge density develops an oscillatory behavior given by

$$\rho_{\mathrm{CDW}}(x) = N_0 \lambda^{-1} R_e[\Delta \exp(i(Qx + \phi))] = N_0 \lambda^{-1} \Delta \cos(Qx + \phi) \tag{4.14}$$

where ϕ is the phase of Δ.

As in a superconductor, the Pauli susceptibility decreases rapidly in the CDW state [101]:

$$\chi_{\mathrm{CDW}} = \chi_N \left(1 - 2\pi T \sum_{n=0}^{\infty} \frac{\Delta^2}{(\omega_n^2 + \Delta^2)^{3/2}}\right) = \chi_N Y(T) \tag{4.15}$$

where $Y(T)$ is the Yosida function. Furthermore, the electric conductivity of the CDW state is approximately given by [102]

$$\sigma_{\mathrm{CDW}} = 2\sigma_n(1 + e^{\beta\Delta})^{-1} \tag{4.16}$$

in the clean limit $l \gg \xi$, where l is the electron mean free path, $\xi = v_F/\Delta$ is the BCS coherence length and $\beta = (k_B T)^{-1}$. The temperature dependences of Δ, χ_{CDW} and σ_{CDW} are shown in Figure 8, where we made use of the table given by Mühlschlegel [103]

4.3. Phase (ϕ) – Hamiltonian

The dynamics of the CDW is described in terms of the time-dependent order parameter $\Delta(t)$. As has been pointed out by Lee *et al.* [98], the order parameter has two degrees of freedom in the linear regime; the phase mode and the amplitude mode, as in a superconductor. However, unlike the superconductor the former mode is gapless (i.e. the Goldstone mode) unless the CDW is pinned either by the commensuration energy or by the impurities.

The dynamic of $\Delta(x, t)$ can be formulated as follows [99]: Assuming that $\Delta(x, t)$ is given by

$$\Delta(x, t) = (\Delta + \delta)e^{i\phi} \tag{4.17}$$

where Δ is the equilibrium order parameter and δ and ϕ are fluctuations from the equilibrium value and depend on x and t.

In the absence of the pinning potential a constant ϕ is completely eliminated by a phase change in the electron field such as

$$\tilde{\Psi} = U^+ \Psi U \tag{4.18}$$

with $U = \exp(\frac{1}{2}\sigma_3\phi)$, which is the consequence of the global gauge invariance. However, when ϕ depends both on x and t, the above transformation gives rise to the extra terms in the electronic part of the Hamiltonian (4.6); Equation (4.6) becomes

$$\mathcal{H}_e = \mathcal{H}_e^0 + \mathcal{H}_e^{\mathrm{fl}} \tag{4.19}$$

$$\mathcal{H}_e^0 = \sum_s \int \mathrm{d}x\, \Psi_s^+(x)\{-iv_F\partial_x\sigma_3 + \Delta\sigma_1\}\Psi_s(x) \tag{4.20}$$

$$\mathcal{H}_e^{\mathrm{fl}} = \sum_s \int \mathrm{d}x\, \Psi_s^+(x)\{\delta(x, t)\sigma_1 + \tfrac{1}{2}v_F\phi_x + (i/2)\phi_t\sigma_3\}\Psi_s(x) \tag{4.21}$$

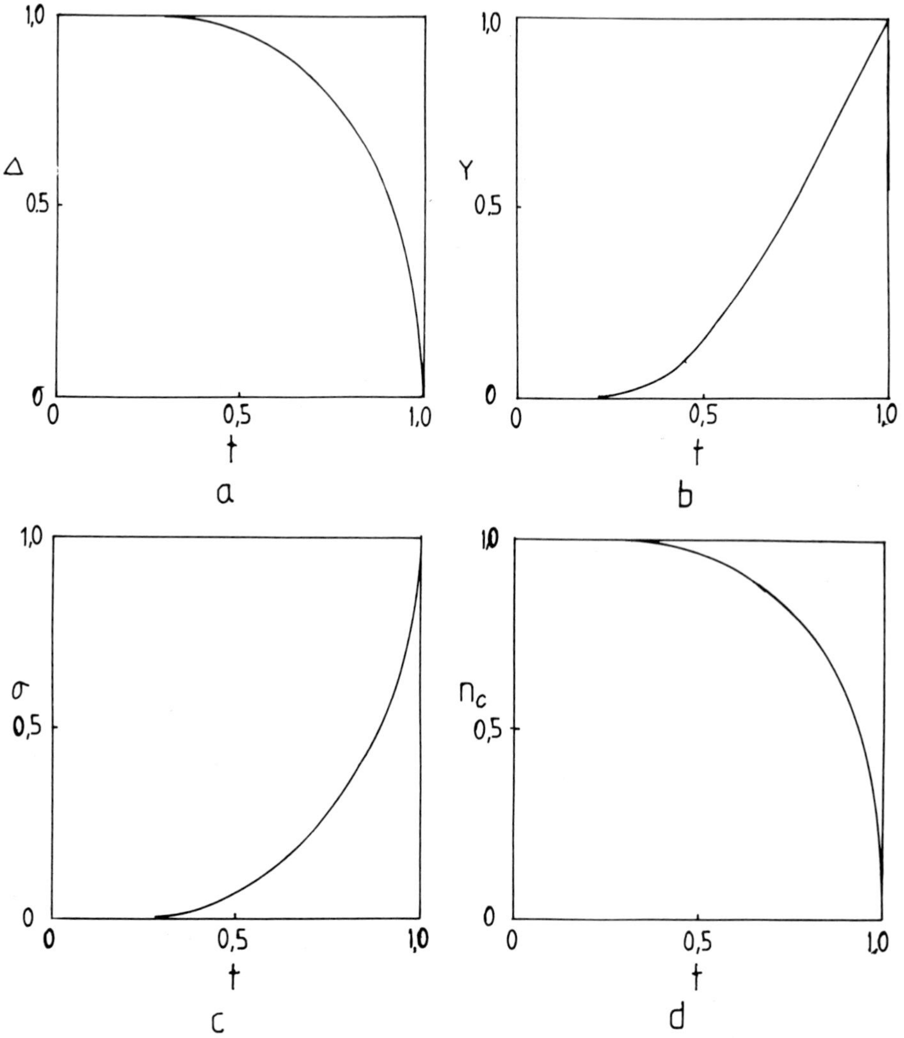

Fig. 8. The mean field theory results for the energy gap $\Delta(t)$, the Yosida function $Y(t) = 1 - n_s(t)$, the Ohmic conductivity $\sigma_a(t)$ and the condensate density $n_c(t)$ are shown as functions of reduced temperature $t = T/T_c$.

Now assuming that δ and ϕ vary slowly both in space and time compared with the characteristic electronic length ξ and time Δ^{-1}, we can treat $\mathcal{H}_e^{\mathrm{fl}}$ perturbationally. The second-order perturbation yields an effective Hamiltonian

$$\mathcal{H}_{ph} = \frac{1}{\pi v_F \lambda \omega_Q{}^2} \int \mathrm{d}x \, (\delta_t^2 + \lambda\omega_Q^2\delta^2 + \tfrac{1}{3}\lambda\omega_Q^2 v_F^2 (2\Delta)^{-2} \left(\frac{\partial\delta}{\partial x}\right)^2 + \Delta^2 \, [\phi_t^2 + c^2\phi_x^2]) \quad (4.22)$$

where

$$c^2 = \lambda\omega_Q^2 v_F^2 (2\Delta)^{-2} \quad (4.23)$$

at $T = 0$ K. More generally the coefficients of δ^2, δ_x^2 and ϕ_x^2 have an additional temperature dependence. The Hamiltonian (4.22) predicts the dispersions of the amplitude mode and the phase mode as

$$\omega_a^2 = \lambda\omega_Q^2 \left[1 + \frac{1}{3}\left(\frac{v_F k}{2\Delta}\right)^2\right] \tag{4.24}$$

$$\omega_{ph}^2 = c^2 k^2$$

These results agree with those obtained by Lee *et al.* [98], while Equation (4.22) differs from that obtained by Brazovskii and Dzyaloshinskii [99], who did not include the spin degrees of freedom. Noting the fact that the electric current and the electric density operators are given by

$$j(x) = ev_F \Psi_s^+(x)\sigma_3 \Psi_s(x) \tag{4.25}$$

and

$$\rho(x) = e\Psi_s^+(x)\Psi_s(x)$$

the extra current and charge associated with ϕ_t and ϕ_x are given by

$$j(x) = \frac{e}{\pi} n_c(T)\phi_t \tag{4.26}$$

$$\rho(x) = -\frac{e}{\pi}(1 - Y(T))\phi_x \tag{4.27}$$

where [40]

$$n_c(T) = 1 - \frac{\beta}{4}\int_{-\infty}^{\infty} \mathrm{d}\xi \, \frac{\xi^2}{\xi^2 + \Delta^2} \, \mathrm{sech}^2\left(\frac{\beta}{2}(\xi^2 + \Delta^2)^{1/2}\right) \tag{4.28}$$

and $Y(T)$ is the Yosida function already defined in Equation (4.15). In particular, in the vicinity of $T = T_c$, $n_c(T)$ is given by $n_c(T) = \pi\Delta(T)/4T_c$ while at $T = 0$ K, $n_c(T) = 1$. The temperature dependence of $n_c(T)$ is shown in Figure 8(d).

4.4. Commensuration Energy

So far we have neglected the pinning potential which will pin the phase of the order parameter. In real quasi-one-dimensional systems, the phase always appears to be pinned. As discussed by Lee, Rice and Anderson [98], the pinning energy comes from the commensurability energy and the impurity potential. We shall consider here the commensurability energy. The effects of the impurity potential will be discussed in Section 6. When the period of the charge density wave is nearly commensurate with the periodicity of the underlying ionic lattice with degree M, there exists an extra energy, the commensurability energy. First let us consider the case of the exact commensurability; the ionic lattice gives rise to a potential energy

$$V(x) = V\cos(Gx) \tag{4.29}$$

where $G = MQ$ and M is an integer. Then the potential energy

$$\mathcal{H}_c = \sum_s \int \mathrm{d}x\, V(x)\psi_s^+(x)\psi_s(x) \tag{4.30}$$

yields a commensuration energy in the CDW state of the form

$$E_{\text{com}} = -\frac{A\omega_0^2}{M^2}\int \mathrm{d}x\, \cos(M\phi(x)) \tag{4.31}$$

where $\quad A = \Delta^2/(\pi v_F \lambda \omega_Q^2) \quad$ and $\quad \omega_0^2 \sim V\Delta^{M-2}/\epsilon_F^{M-1}$ (4.32)

Combining the commensuration energy with the phase Hamiltonian (4.22), we obtain

$$\mathcal{H}_{ph} = \frac{A}{2}\int \mathrm{d}x \left\{\phi_t^2 + c^2\phi_x^2 + \frac{2\omega_0^2}{M^2}(1-\cos M\phi)\right\} \tag{4.33}$$

where we have added a constant so that the ground state energy is zero.

Now in the nearly commensurate case, we replace the relation $G = MQ$ by

$$G = M(Q+q) \quad \text{with} \quad |q| \ll Q \tag{4.34}$$

In this more general situation Equation (4.33) is rewritten as

$$\mathcal{H}_{ph} = \frac{A}{2}\int \mathrm{d}x \left\{\theta_t^2 + c^2(\theta_x - q)^2 + \frac{2\omega_0^2}{M^2}(1-\cos M\theta)\right\} \tag{4.35}$$

where we have introduced a new phase by

$$\theta = \phi + qx \tag{4.36}$$

4.5. ϕ-SOLITONS, DISCOMMENSURATIONS

First consider the commensurate case ($q = 0$) where the phase Hamiltonian is given by Equation (4.33). As is easily seen, the ground state of the system is highly degenerate and is given by

$$\phi = 0, \pm\frac{2\pi}{M}, \pm\frac{4\pi}{M}, \ldots \tag{4.37}$$

Furthermore ϕ obeys a sine-Gordon equation

$$\phi_{tt} - c^2\phi_{xx} = -\frac{\omega_0^2}{M}\sin(M\phi) \tag{4.38}$$

The ϕ soliton is given by [33]

$$\phi(x, t) = \frac{4}{M} \tan^{-1} [\exp m\gamma(x - vt)] \tag{4.39}$$

with the soliton energy

$$E_s(v) = E_s\gamma = 8A\omega_0 cM^{-2}\gamma \tag{4.40}$$

where $\quad m = \omega_0/c \quad$ and $\quad \gamma = (1 - v^2/c^2)^{-1/2}$ (4.41)

The soliton moves with uniform velocity v and carries charge given by

$$Q_s = \frac{e}{\pi} \int_{-\infty}^{\infty} \mathrm{d}x\, \phi_x = 2eM^{-1} \tag{4.42}$$

In general the ϕ soliton carries a fractional charge for $M \geqslant 3$. Indeed the fractional charge for $M = 3$ agrees with the charge of one type of solitons in the trimerized system considered by Su and Schrieffer [104]. Furthermore, as seen from Equation (4.27), the charge Q_s is not even simple fraction at nonzero temperatures. The soliton charge decreases continuously as the temperature is raised and vanishes when the CDW state disappears.

The soliton energy E_s is usually much smaller than the quasi-particle energy Δ. In this circumstance it is expected that the ϕ soliton is the dominant charge carrier at low temperature in the quasi-one dimensional conducting systems. For example it has been suggested in the past that the ϕ soliton would dominate the low temperature electric conductivity of TTF–TCNQ, KCP and TaS_3 [33, 105]. However, the author does not think that the evidence for the ϕ-soliton is very strong.

In the weakly incommensurate case the ground state of the Hamiltonian (4.35) is constructed in terms of a soliton lattice [106, 107].

$$\cos(\tfrac{1}{2}M\theta) = sn(mk^{-1}x\,|\,k) \tag{4.43}$$

where $sn(u\,|\,k)$ is the Jacobian elliptic function and k is a parameter. This describes a soliton lattice with the soliton lattice constant l;

$$l = 2k\,K(k)\,m^{-1} \tag{4.44}$$

and $K(k)$ is the complete elliptic integral. Substituting Equation (4.43) into Equation (4.35), the soliton lattice energy per unit length is calculated as

$$E_{sl} = \frac{AC^2}{2l} \int_0^l \mathrm{d}x \left\{\left(\frac{2m}{Mk} dn - q\right)^2 + \frac{4m^2}{M^2} (cn)^2\right\}$$

$$= \frac{E_s}{l} \left\{\frac{1}{k} (E(k) - \frac{1}{2}(1 - k^2)K(k)) - \frac{\pi}{4} (Mq/m) + \frac{1}{8} (Mq/m)^2\, k\, K(k)\right\} \tag{4.45}$$

where $E(k)$ is the complete elliptic integral.

$$E(k) = \int_0^{\pi/2} d\theta \, (1 - k^2 \sin^2 \theta)^{1/2}$$

$$K(k) = \int_0^{\pi/2} d\theta \, (1 - k^2 \sin^2 \theta)^{-1/2} \tag{4.46}$$

Then the parameter k is chosen so as to minimize Equation (4.45). Although the general soliton can be found only numerically, the critical point where the soliton lattice disappears is studied analytically when q is changed. Indeed, for small l the ground state does not allow solitons; the ground state is commensurate. As q increases to a critical value q_c, the soliton lattice appears suddenly and the soliton density increases rapidly with $q - q_c$. Making use of the asymptotic expressions for the elliptic integrals for k approaching unity;

$$\begin{aligned} E(k) &= 1 + \tfrac{1}{2} k'^2 (\ln(4/k') - \tfrac{1}{2}) + \cdots \\ K(k) &= \ln(4/k') + \tfrac{1}{4} k'^2 (\ln(4/k') - 1) + \cdots \end{aligned} \tag{4.47}$$

where $k' = (1 - k^2)^{1/2}$, we can rewrite Equation (4.45) as

$$E_{sl} \cong E_s \left[\left(\frac{M}{4} q^2 \right) m^{-1} + l^{-1} \left(1 - \frac{\pi}{4} \frac{M}{m} q + 4e^{-ml} \right) + O(e^{-2ml}) \right] \tag{4.48}$$

where we made use of the relation

$$k = 1 - 8e^{-ml} \tag{4.49}$$

Equation (4.48) indicates that in the dilute limit solitons are interacting by a repulsive potential which decays exponentially with their separation l. Minimizing Equation (4.48) with l, we find

$$4(ml + 1)e^{-ml} = \left(\frac{\pi}{4} \frac{M}{m} q - 1 \right) \tag{4.50}$$

which implies that there is no soliton where $q < q_c = 4m/\pi M$. For $q > q_c$, the soliton lattice with the soliton density l^{-1} appears, where l increases as

$$l^{-1} \cong m \left[\ln \left(\frac{\pi}{4} \frac{M}{m} q - 1 \right) \right]^{-1} \tag{4.51}$$

near the threshold.

In a three-dimensional system with weak interaction in the transverse direction, the soliton lattice is a parallel sheet of discommensurations which carry extra charge $Q = 2e/M$

for each chain [31, 32]. A possible charge transport due to moving discommensurations in the CDW state of the quasi-one-dimensional metals has been recently proposed by Bak [108]. However, it appears that discommensurations are not likely to give a periodically oscillating current in the CDW system because of their microscopic size. Another transport mechanism associated with discommensuration is proposed by Weger and Horovitz [109]. However the author believes that the model is unrealistic as they assumed that the discommensurations in TaS_3 are pinning free. The author does not know of any experiment which indicates pinning-free motion of solitons or discommensurations.

Even in polyacetylene, which we shall describe in the next section, solitons appear to be always pinned, as far as the infrared absorption experiment can tell us.

5. Solitons in Polyacetylene

5.1. Dimerization of Polyacetylene

We shall now consider the electronic properties of polyacetylene. Polyacetylene is a chain-like molecule formed by carbon and hydrogen. Recent excitement concerning this system lies, on the one hand, in the ease of controlling the conductivity of polyacetylene by several orders of magnitude by simple doping with cation or anion and, on the other hand, in the theoretical elegance of the model describing it. Also, polyacetylene seems to provide the prototype of the organic conductors whose electronic properties are modified by doping. Also there is a close similarity between the CDW state we have discussed in Section 4 and the dimerization of polyacetylene. Indeed, the dimerization of conjugated chains is in most cases the simplest manifestation of the Peierls instability. For general information on polyacetylene we encourage readers to consult excellent reviews [110–114].

In the following we are primarily interested in *trans*-$(CH)_x$. Usually acetylene is polymerized into *cis*-$(CH)_x$ by making use of a catalytic reaction. Then *cis*-$(CH)_x$ is isomerized into *trans*-$(CH)_x$ either by thermalization or by doping with impurity, although the details of the isomerization are not well understood. Both *cis*-$(CH)_x$ and *trans*-$(CH)_x$ have planar configurations, as shown in Figure 9, where the back bone is formed by the carbon atoms. Furthermore, the σ electrons of the carbon atom form a closed shell while one π electron per carbon atom contributes to the bonding.

Both pristine (i.e. undoped) *trans*-$(CH)_x$ and doped *trans*-$(CH)_x$ exhibit unusual properties. First of all, polyacetylene cannot be dissolved in ordinary liquids like water or alcohol. On the other hand polyacetylene accepts a great variety of cations and anions as dopants. In the pristine state *trans*-polyacetylene is a good insulator with an energy gap $2\Delta \sim 1.5$ eV. However, by doping polyacetylene with ions its conductivity increases rapidly, reaching the conductivity of metals like Cu or Al around a few percent doping of ions at room temperature. The extraordinary adjustability of conductivity by simple doping suggests some technical applications of polyacetylene.

Perhaps more puzzling are the free spins of 10^{-3}–10^{-4} per carbon atom in *trans*-polyacetylene, as seen by electron spin resonance (ESR) experiments [115–118], while in *cis*-polyacetylene the free spin density is one or two orders of magnitude lower. Apparently the free spins are generated in the process of isomerization. Furthermore, these

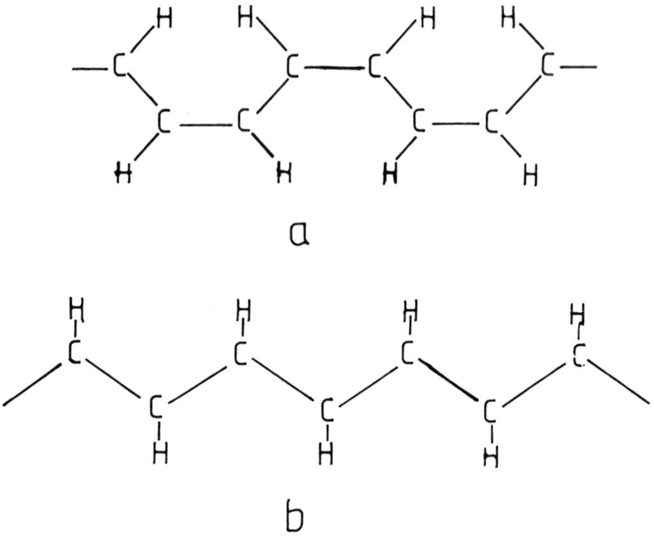

Fig. 9. Two chemical isomers of polyacetylene chain are shown (a) *cis*-$(CH)_x$, (b) *trans*-$(CH)_x$.

spins appear to be quite mobile, at least at room temperature. In order to understand these puzzles a soliton model has been proposed by Su, Schrieffer and Heeger (SSH) [34] and by Rice [35]. In the following we shall describe in greater detail the SSH model.

5.2. SU, SCHRIEFFER, HEEGER MODEL

Although in general polyacetylene has a rather complicated morphology, at the microscopic level it looks like tangled spaghetti with a strand diameter of 200Å. Each strand of spaghetti then consists of bundles of $(CH)_x$-chains aligned parallel to the axis of the strand. Here we shall consider a single chain of *trans*-polyacetylene as shown in Figure 9b. The SSH model is given by the following Hamiltonian;

$$\mathcal{H} = \tfrac{1}{2}M \sum_n \dot{y}_n^2 + \tfrac{1}{2}K \sum_n (y_n - y_{n+1})^2 - \sum_{n,\,s} t_{n,\,n+1}(C_{ns}^+ C_{n+1s} + C_{n+1s}^+ C_{ns}) \qquad (5.1)$$

and

$$t_{n,\,n+1} = t_0 - \alpha(u_{n+1} - u_n)$$

where y_n is the displacement of the nth CH ions from the equilibrium configuration corresponding to the undimerized state projected along the chain direction and C_{ns}^+ is the creation operator of a π-electron at the nth site having spin s. We note that Equation (5.1) describes the simplest one-dimensional electron–phonon system, where the direct electron–electron interaction is neglected. SSH analysed the Hamiltonian (5.1) within the mean field approximation (i.e. the fluctuations in y_n's are neglected) and found:

1. The ground state is dimerized and doubly degenerate. Two ground states which are referred as the A phase and the B phase are shown in Figure 10, where the double bond implies the distance between the adjacent carbon atoms is shorter than that with the single bond.

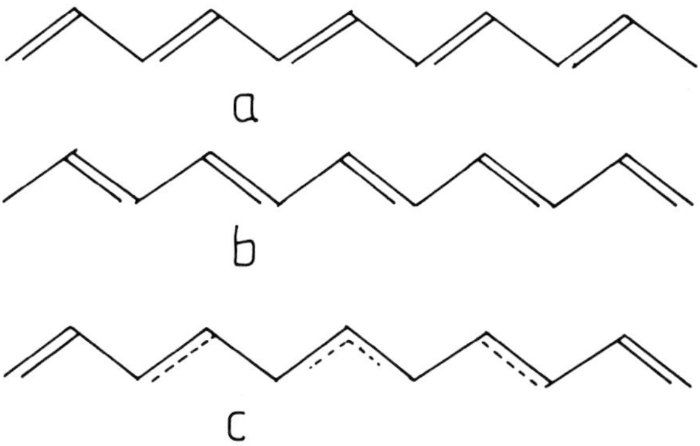

Fig. 10. Two ground state configurations of *trans*-$(CH)_x$ and a soliton configuration are shown in (a) (b) and (c).

2. There are topological solitons (or kinks), which are the domain walls between two distinct ground states. Furthermore, the soliton energy E_s is independent of the charge of the soliton and smaller than the quasi-particle energy Δ. This implies that extra electrons or holes introduced by doping with impurities enter into polyacetylene as charged solitons rather than as quasi-particles (soliton doping).

3. The natural soliton carries spin $\frac{1}{2}$, while the charged soliton is spinless. (Spin charge anomaly.) We believe that this is the most beautiful consequence of the SSH model. Indeed this is the simple realization of the Jackiw and Rebbi mechanism [119] found in the quantum field theory in 1 + 1 dimensions. In the Jackiw–Rebbi model they found that the soliton has a fractional (i.e. $\frac{1}{2}$) fermion charge. In the SSH model the spin degeneracy of the electron masks this anomaly but the same effect manifests as the spin charge anomaly.

This result indicates that the puzzling free spins found experimentally are neutral solitons. I would like to mention that a similar defect, though in a more crude form, has been already considered by Pople and Walmsley [120] long before the SSH model. How neutral solitons are produced in the process of isomerization is speculated by Su [121] as a finite-chain effect.

We shall derive the above results within the continuum version of the SSH model [122–124]. For example, the continuum model allows a number of exact solutions for the dimerization pattern, which are treated analytically.

5.3. Continuum Limit

Let us first introduce the staggered displacement by $\tilde{y}_n = (-)^n y_n$. In the equilibrium dimerized state $\tilde{y}_n$ is either y_0 or $-y_0$ corresponding to the A or the B phase. In more general situations $\tilde{y}_n$ is assumed to vary slowly with n. The SSH Hamiltonian in the continuum limit is then given by [123]

$$\mathcal{H}_c = (2g^2)^{-1} \int dx\, (\dot{\Delta}^2 + \omega_0^2 \Delta^2(x)) + \sum_s \int dx\, \psi_s^+(x) [-iv_F \sigma_3 \partial_x + \sigma_1 \Delta(x)]\, \psi_s(x) \quad (5.2)$$

where

$$\begin{aligned} \Delta(x) &= 4\alpha\tilde{y}(x), & g &= 2\alpha\omega_Q (a/K)^{1/2} \\ \omega_Q &= 2(K/M)^{1/2}, & v_F &= 2t_0 a \end{aligned} \quad (5.3)$$

$\tilde{y}(x)$ is the continuum version of $\tilde{y}_n$ and the spinor $\psi_s(x)$ consists of the right- and left-travelling electrons, as in Equation (4.7)

$$\psi_s(x) = \begin{Bmatrix} u_s(x) \\ v_s(x) \end{Bmatrix}$$

with

$$c_{ns} \to a^{1/2} \{ [\exp(ink_F a)]\, u(x) - i\, [\exp -ink_F a]\, v(x) \} \quad (5.4)$$

Note also the similarity between Equations (5.2) and (4.20). Indeed, within the continuum limit, the SSH model describes a typical one-dimensional electron system with Peierls instability. However, the dimerization order parameter given in Equation (5.3) is real, unlike the general case described in Section 4. This is the particular feature of the half-filled band where $Q = 2k_F = \pi/a$.

The basic approximation introduced in Equation (5.2) is that the spatial variation of $\Delta(x)$ is slow compared with a, the lattice constant; we neglect all higher-order terms in a/ξ, where $\xi = v_F/\Delta$ is the characteristic length controlling the spatial variation of $\Delta(x)$. Then the mean field approximation is equivalent to $\dot{\Delta} = 0$; the ionic displacement is static. Then, minimizing Equation (5.2) with respect to $\Delta(x)$ and $\psi_s(x)$, we obtain a set of self-consistent equations;

$$\Delta(x) = g^2 \omega_Q^{-2} \sum_s \langle \psi_s^+(x) \sigma_1 \psi_s(x) \rangle = g^2 \omega_Q^{-2} \sum_{sn}{}' (u_{ns}^*(x) v_{ns}(x) + \text{c.c}) \quad (5.5)$$

$$\epsilon_n \psi_n(x) = (-iv_F \sigma_3 \partial_x + \sigma_1 \Delta(x)) \psi_n(x) \quad (5.6)$$

where

$$\psi_n(x) = \begin{Bmatrix} u_n(x) \\ v_n(x) \end{Bmatrix} \quad (5.7)$$

and the sum in Equation (5.5) is carried out up to the Fermi level, which we took to be $\epsilon_F = 0$.

The set of Equations (5.5) and (5.6) is very similar to the Bogoliubov–de Gennes (BdG) equation for inhomogeneous superconductors [125]. Furthermore, Equation (5.6) is the same as the equation which appears in the inverse scattering method for the sine-Gordon equation [17]. This allows us to find a number of exact solutions of Equations (5.5) and (5.6). Finally the mean field energy of the particular solution $\Delta(x)$ is given by

$$E_{\mathrm{MF}}[\Delta(x)] = (2g^2)^{-1}\omega_Q^2 \int \mathrm{d}x\, \Delta^2(x) + \sum_{s,\,n}{}' \epsilon_{ns} \tag{5.8}$$

where the sum over energy ϵ_{ns} is carried out up to the Fermi level $\epsilon_F = 0$.

The dimerized ground state is doubly degenerate and is described by $\Delta(x) = \Delta$ (or $-\Delta$) where Δ is a constant. The eigenvalues ϵ_n are then given by

$$\epsilon_k = \pm E_k, \qquad E_k = ((v_F k)^2 + \Delta^2)^{1/2} \tag{5.9}$$

and

$$\psi_k = \begin{Bmatrix} u_k \\ v_k \end{Bmatrix} \frac{1}{L^{1/2}} e^{ikx} \tag{5.10}$$

where

$$\begin{aligned} u_k &= \left[\frac{1}{2}\left(1 + \frac{v_F k}{\epsilon_k}\right)\right]^{1/2} \\ v_k &= \left[\frac{1}{2}\left(1 - \frac{v_F k}{\epsilon_k}\right)\right]^{1/2} \end{aligned} \tag{5.11}$$

and L is the length of the polyacetylene chain and k is related to the total momentum of the quasi-particle P by $P = k_F \pm k$.

Here $\pm$ in Equation (5.9) refers to the electron state in the conduction band and that in the valence band, respectively. Substituting Equation (5.11) into Equation (5.5), we obtain

$$\Delta = 4g^2\omega_Q^{-2} \int_{-k_F}^{k_F} \frac{\mathrm{d}k}{2\pi} \frac{\Delta}{2E_k} = 2\lambda\Delta \ln\left(\frac{2W}{\Delta}\right) \tag{5.12}$$

where $W = v_F k_F$ is the band width and $\lambda = g^2/\pi v_F \omega_Q^2$. From Equation (5.12) we find the BCS expression

$$\Delta = 2We^{-1/2\lambda} \tag{5.13}$$

The ground state energy per unit length is similarly obtained as

$$E_{\mathrm{MF}}[\Delta]L^{-1} = -\frac{1}{\pi v_F} W(W^2 + \Delta^2)^{1/2} \cong -\frac{1}{\pi v_F}\left(W^2 + \frac{1}{2}\Delta^2\right) \tag{5.14}$$

where the second term in Equation (5.14) is the dimerization energy.

The electron density of states is obtained similarly as

$$N(\omega) = \int_{-k_F}^{k_F} \frac{dk}{2\pi} \delta(E - \omega) = \frac{1}{\pi v_F} \frac{|\omega|}{(\omega^2 - \Delta^2)^{1/2}} \theta(|\omega| - \Delta) \tag{5.15}$$

for one spin component. The density of states is the same as in a superconductor with energy gap Δ. The frequency-dependent conductivity is also obtained from

$$\sigma(\omega) = \mathrm{Im}\,[\omega^{-1} \langle [j, j] \rangle (\omega)] \tag{5.16}$$

where $j = e\sigma_3 v_F$ and $\langle [A, B] \rangle (\omega)$ is the Fourier transform of the retarded product. Substituting the expression for j into Equation (5.16) we obtain [129, 130]

$$\begin{aligned} \sigma_0(\omega) &= 2\pi e^2 v_F \omega^{-2} \int dk \frac{\Delta^2}{E_k} \delta(2E - \omega) \\ &= 4\pi e^2 v_F \omega^{-2} [(\omega/2)^2 - \Delta^2]^{-1/2} \theta(|\omega| - 2\Delta) \end{aligned} \tag{5.17}$$

There is no electromagnetic absorption for $|\omega| < 2\Delta$ implying that the dimerized state is an insulator. For $|\omega| > 2\Delta$ the conductivity has a square root singularity at the absorption threshold.

5.4. Solitons in Polyacetylene

Now we shall consider the soliton. A soliton localized at $x = 0$ is given by

$$\Delta(x) = \Delta \tanh(x/\xi) \tag{5.18}$$

which interpolates between the A and B phase smoothly (see Figure 10c). It is shown that Equation (5.18) with $\xi = v_F/\Delta$ satisfies the self-consistent set of Equations (5.5) and (5.6).

There is a bound state at the center of the energy gap with

$$\epsilon_B = 0, \qquad \psi_B(x) = \left\{ {1 \atop -i} \right\} \tfrac{1}{2} \xi^{-1/2} \operatorname{sech}(x/\xi) \tag{5.19}$$

Furthermore, the scattering states are given by

$$\epsilon_k = \pm E_k, \qquad \Psi_k(x) = \tfrac{1}{2}(L - \xi \cos^2 \phi_k)^{-1/2} \begin{Bmatrix} 1 - \dfrac{v_F k + i\Delta(x)}{\epsilon_k} \\ -i\left(1 + \dfrac{v_F k + i\Delta(x)}{\epsilon_k}\right) \end{Bmatrix} e^{ikx} \tag{5.20}$$

where $\phi_k = \tan^{-1}(v_F k/\Delta)$.

Then it is easy to verify that Equation (5.5) is satisfied independently of occupation of the midgap states at $\epsilon = 0$, since $u(x)$ and $v(x)$ are relatively pure imaginary. The soliton energy is calculated from Equation (5.8) as

$$E_s = E_{\mathrm{MF}}[\Delta(x))] - E_{\mathrm{MF}}[\Delta] = \frac{2}{\pi}\Delta \tag{5.21}$$

again independent of occupation of the midgap states, implying that E_s is independent of the charge of the soliton. We note in passing that $E_s < \Delta$ which implies soliton doping. The charge of the soliton is determined as follows. Since the midgap state associated with the soliton is composed of half of the state in the conduction band and half of the state in the valence band, a deficit of a half-electron state is created for both spin components in the valence band. This implies that the neutral soliton is obtained by singly occupying the midgap state, which compensates the one electron deficit of two spin states in the valence band. Since this electron in the midgap state can take the up or the down spin state, the neutral soliton has a spin $\frac{1}{2}$. On the other hand the positively charged soliton is obtained by emptying the midgap state, which corresponds to a spinless object. Similarly, the negatively charged soliton is obtained by doubly occupying the midgap state again resulting in a spinless object. This charge–spin anomaly is shown schematically in Figure 11. The extra spin or charge distribution associated with the neutral soliton or the charged soliton is given by

$$\rho_{\mathrm{spin}}(x) = (4\xi)^{-1}\,\mathrm{sech}^2(x/\xi)$$

or

$$\rho_{\mathrm{charge}}(x) = \pm e(4\xi)^{-1}\,\mathrm{sech}^2(x/\xi) \tag{5.22}$$

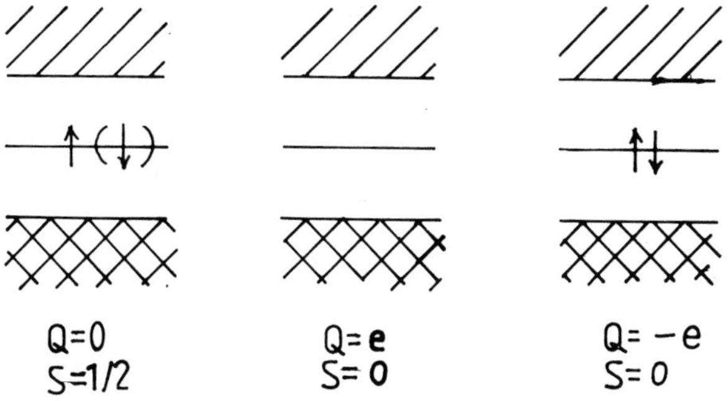

Fig. 11. The occupation of the midgap states for neutral and charged solitons are schematically shown. For neutral soliton a single electron occupies the midgap state resulting in the spin doublet, while the charged solitons have even electrons (0 or 2) in the midgap state resulting in the spin singlet state.

If we go back to the original discrete model, we discover that the spin or the charge resides only on the even or the odd lattice sites depending on the position of the center of the soliton [34], as the on-site density is given by $f_\pm(x)$ where $f_\pm(x) = u(x) \pm iv(x)$. For example for the spin density we have;

$$\rho^{\mathrm{even}}_{\mathrm{spin}}(x) = |f_+(x)|^2, \qquad \rho^{\mathrm{odd}}_{\mathrm{spin}} = |f_-(x)|^2 \tag{5.23}$$

Substituting Equation (5.19) into Equation (5.23), we find

$$\rho^{\mathrm{even}}_{\mathrm{spin}}(x) = \xi^{-1}\,\mathrm{sech}^2(x/\xi), \qquad \rho^{\mathrm{odd}}_{\mathrm{spin}}(x) = 0 \tag{5.24}$$

The midgap state associated with the soliton is most readily seen by optical absorption as appearence of the midgap absorption [126–129]. Although the midgap absorption coefficient is easily obtained in terms of electronic wave functions given by Equations (5.19) and (5.20) [128], in the analysis of the interband absorption the boundary condition for the electronic wave functions plays the crucial role [130]. The boundary condition (BC) for the wave function is formulated as follows. Let us consider a polyacetylene ring which consists of even number N of the (CH)-site. In the absence of solitons we have

$$\Delta(x + L) = \Delta(x) \tag{5.25}$$

with $L = Na$. This leads to the ordinary boundary condition (BC) for electronic wave functions

$$\psi(x + L) = \psi(x) \tag{5.26}$$

In the presence of a single soliton the sign of $\Delta(x)$ is flipped after passing the soliton, resulting in

$$\Delta(x + L) = -\Delta(x) \tag{5.27}$$

Then the BC for electronic wave functions given by Equation (5.26) is no longer consistent with Equation (5.27). The BC consistent with a single soliton is given by

$$\psi(x + L) = \sigma_3 \psi(x) \tag{5.28}$$

which gives rise to the same phase shift of ϕ_k for both $u(x)$ and $v(x)$ components in the presence of the soliton. On the other hand the k selection rules for electrons in the conduction band and those in the valence band are given by [130]

$$\begin{aligned} kL &= 2n\pi + \phi_k \quad \text{for} \quad \epsilon > 0 \\ kL &= (2n+1)\pi + \phi_k \quad \text{for} \quad \epsilon < 0 \end{aligned} \tag{5.29}$$

The above boundary condition is crucial not only for the calculation of the frequency dependent conductivity but also other two point correlation functions. We will not give the derivation of the frequency dependent conductivity here, but the final expression [130];

$$\sigma(\omega) = \sigma^{(1)}(\omega) + \sigma^{(2)}(\omega) \tag{5.30}$$

$$\sigma^{(1)}(\omega) = \frac{e^2 v_F}{4\Delta} \pi^2 \xi n_s \left(z^2 - \frac{1}{4}\right)^{-1/2} \operatorname{sech}^2\left[\pi\left(z^2 - \frac{1}{4}\right)^{1/2}\right] \tag{5.31}$$

$$\sigma^{(2)}(\omega) = \frac{e^2 v_F}{2\Delta} z^{-2}(z^2 - 1)^{-1/2} \{1 - \xi n_s(2 + z^{-2} + f(z))\} \tag{5.32}$$

where

$$f(z) = -2 \lim_{\epsilon\to 0} \left\{\pi \int_{\epsilon}^{(z^2-1)^{1/2}} \mathrm{d}y \cosh^2 \pi y \frac{z^2}{(z^2-y^2)^2+y^2} \left(\frac{z^2-y^2}{z^2-y^2-1}\right)^{1/2} - \right.$$

$$\left. -\frac{1}{\epsilon} z^{-1}(z^2-1)^{-1/2}\right\} \tag{5.33}$$

$z = \omega/(2\Delta)$

and n_s is the soliton density on the $(CH)_x$ chain. The first term in Equation (5.30) involves the midgap state and has the threshold energy at $\omega = \Delta$, while the second term is due to the interband excitation. In the limit $n_s = 0$ Equation (5.30) reduces to Equation (5.17), the conductivity for the perfectly dimerized state. The frequency-dependent conductivity also satisfies the sum rule

$$\int_{-\infty}^{\infty} \mathrm{d}\omega\, \sigma(\omega) = e^2 v_F \tag{5.34}$$

independent of the soliton density.

Within the present model the soliton is considered to be extremely mobile [34]. For this purpose we shall estimate the soliton mass. Assuming that a moving soliton with velocity v is given by

$$\Delta(x, t) = \Delta \tanh\left[(x - vt)/\xi\right] \tag{5.35}$$

the first term of Equation (5.2) yields the kinetic energy

$$E_{\mathrm{kin}} = \tfrac{2}{3}(\Delta/g)^2 \xi^{-1} v^2 \tag{5.36}$$

implying the soliton mass

$$m_s = \frac{4}{3}(\Delta/g)^2 \xi^{-1} = \frac{2}{3\pi}(\Delta^3/\lambda\omega_Q^2 \epsilon_F)m \tag{5.37}$$

Substituting appropriate values for Δ, ω_Q and ϵ_F, the soliton mass is found to be only a few times of the electron mass m. Therefore the soliton in polyacetylene looks like extremely mobile object within the SSH model.

5.5. Other Solutions

We shall consider some other exact solutions of Equations (5.5) and (5.6). The first is the soliton lattice, which is the solution with lowest energy when extra electrons or holes are added to the system. The second is the polaron, which is considered as a bound state of soliton and antisoliton.

A. *Soliton Lattice*

In the absence of external field, solitons of the same charge (positive or negative) on a single $(CH)_x$ chain form a soliton lattice in order to minimize the total energy. The corresponding solution is given in terms of Jacobian elliptic functions [131–133];

$$\Delta(x) = \Delta k sn(u|k) sn(u + K|k) \tag{5.38}$$

with $u = (k\xi)^{-1}x$, $an_s^{-1} = k\xi K(k)$
The soliton lattice energy per unit length is now given by

$$E_{sl}L^{-1} = \frac{2\Delta_0}{\pi k} n_s \left[E(k) - \frac{1}{2}k'^2 K(k)\right] \tag{5.39}$$

where $E(z)$ and $K(z)$ are again complete elliptic integrals and Δ_0 is the dimerization order parameter of the perfectly dimerized state. From Equation (5.39) the chemical potential for the soliton is obtained as

$$\mu_s = \frac{\partial}{\partial n_s} E_{sl} = 2\Delta(\pi k)^{-1} E(k) \tag{5.40}$$

which is always smaller than the energy gap $E_{gap} = \Delta k^{-1}$. Therefore, within the continuum limit, electrons or holes are introduced as solitons independent of the soliton density. However, the continuum limit will be no longer valid for $n_s\xi > 1$. The density of states is found to be rather simple in the soliton lattice and is given by [133]

$$N(\epsilon) = \frac{1}{\pi v_F}\left(\frac{\epsilon^2}{\Delta^2} - \frac{E(k)}{K(k)}\right)\phi(\epsilon) \tag{5.41}$$

$$\phi(\epsilon) = \Delta^2 \operatorname{Re}[(\Delta'^2 - \epsilon^2)(\Delta^2 - \epsilon^2)]^{-1/2} \tag{5.42}$$

where

$$\Delta = k'\Delta_0 \quad \text{and} \quad \Delta' = \Delta_0 k' k^{-1} \tag{5.43}$$

The midgap states form an electron band which extends between $-\Delta' \leqslant \epsilon \leqslant \Delta'$.

B. *Polaron*

The existence of a polaron solution to the SSH model was first discovered by Su and Schrieffer [134] in their numerical analysis of one electron injected into the perfectly dimerized polyacetylene. The electron introduces a local depression of the dimerization order parameter $\Delta(x)$ and moves together with this depression. Subsequently the analytical solution within the continuum model was found by Brazovskii and Kirova and by Campbell and Bishop [135, 136]. Indeed such a solution has been considered much earlier within the Gross–Neveu model by Dashen *et al.* [137]. The semi-classical approximation of the Gross–Neveu model is identical to the mean field approximation of the continuum limit of the SSH model.

The solution is given by

$$\Delta(x) = \Delta_0 - \kappa_0 v_F(t_+ - t_-) \tag{5.44}$$

where

$$t_\pm = \tanh\left[\kappa_0(x \pm x_0)/\xi\right]$$
$$\kappa_0 = 1/\sqrt{2}, \qquad x_0 = \sqrt{2}(\sqrt{2} - 1)\xi \tag{5.45}$$

with

$$E_p = \frac{2\sqrt{2}}{\pi}\Delta_0 \sim 0.900\,\Delta_0 \tag{5.46}$$

Note that $E_p < \Delta$ implying the quasi-particle is unstable against formation of the polaron. The analytic expressions for the electronic wave functions are also known [136]. We only remark here that the polaron introduces two bound states inside the energy gap with energy $\pm\omega_0 = \pm\Delta_0/\sqrt{2}$. Unlike the soliton the polaron has the normal charge-spin relation and is devoid of topological charge. Therefore the polaron may play an important role in the soliton doping, since the solitons are created only by pairs on a single polyacetylene chain due to the topological constraint. On the other hand a single electron can be transformed into a polaron in polyacetylene with emission of phonons. In this circumstance if we limit ourselves to a single polyacetylene chain, the first electron enters the chain as polaron. The second electron then splits the polaron into a pair of negatively charged solitons. Recently an analysis has been reported, which indicates that a pair of polarons are unstable against formation of a pair of solitons [138]. More recently it has been reported that polarons are seen in the optical absorption spectrum of lightly doped polyacetylene around ω = 1.47 and 1.6 eV [139], which is comparable to the optical structure calculated by Fesser *et al.* [140]. Absence of the topological constraint for polarons implies that the polaron does not require the degenerate ground state unlike the soliton. Therefore polarons can exist and will play an important role in many other polymers like *cis*-$(CH)_x$. [135] and polyparaphenylene [141]. This warrants a further study on polarons in polymers.

5.6. Comparison with Experiments

We shall describe briefly the properties of pristine and lightly doped *trans*-polyacetylene.

A. *Magnetic Properties*

As already mentioned the free spins seen by ESR [115–118] are identified with neutral solitons. Furthermore the width of the spin resonance in *trans*-$(CH)_x$ is found to change from 10 Gauss at low temperatures to less than 1 Gauss at room temperature, while that in *cis*-$(CH)_x$ is about 10 Gauss independent of the temperature. This can be interpreted as mobile spins in *trans*-$(CH)_x$ (motional narrowing), whereas the spins are immobile in *cis*-$(CH)_x$. This conclusion is strengthened by dynamic nuclear polarization measurements by Nechtshein *et al.* [142]. The observed enhancement of the NMR amplitude in *trans*-$(CH)_x$ by pumping at and near the electronic Larmor frequency $\nu_e = \gamma_e H$ exhibits the Overhauser effect, indicating that the spin in *trans*-$(CH)_x$ is mobile at room temperature. By contrast, the same experiments in *cis*-$(CH)_x$ indicate the existence of the localized spin. Furthermore measurements of the proton spin relaxation rate T_1^{-1} show that T_1^{-1} is proportional to $\nu_n^{-1/2}$ for pristine *trans*-$(CH)_x$ over the frequency range 10–340 MHz, where ν_n is the proton Larmor frequency. If the proton spin relaxation is mainly due to the coupling to the electron spin, the above result indicates one dimensional electron spin diffusion with diffusion constant $D = 10^{-2}$ cm²/sec. Furthermore, from the deviation from the $\nu_N^{-1/2}$ law, the anisotropy in D is inferred as 10^6. This result is readily interpreted as being due to the soliton diffusion along the $(CH)_x$ chain. However, more recent experiments on T_1 in deuterated polyacetylene [143] revealed that D depended strongly on the deuteron concentration, if the above interpretation is used. Since such a strong concentration dependence of D is quite unlikely, which suggests that the nuclear dipole interaction between proton spins plays a substantial role in T_1 in *trans*-$(CH)_x$. Furthermore, all the electron spins in *trans*-$(CH)_x$ appear to be immobilized below 10 K [143].

Within the SSH model the soliton interacts with both optical and acoustic phonons [144]. However, below the room temperature the scattering with the optical phonon becomes negligible as the thermal optical phonon population decreases exponentially. Then emission and absorption of the acoustic phonon gives rise to the soliton diffusion constant $D \propto T^{1/2}$, where T is the temperature. Therefore this model cannot account for the localization of spins at low temperatures. At room temperature the theory predicts $D \simeq 10^{-1}$ cm²/sec, which may be comparable to the one deduced from the T_1 measurement. However, besides the fact that the deduction suffers a criticism, a quite different value for $D = 10^{-5}$ cm²/sec is obtained from the spin echo experiment [145]. Therefore the spin dynamics in undoped *trans*-$(CH)_x$ may be more complicated than originally thought.

The spin susceptibility of pristine *trans*-$(CH)_x$ follows the Curie law between 1.5 K and 300 K [146, 147]. Upon doping, the number of Curie law spins N_c decreases from 400 ppm for $y = 0$ to about 120 ppm for $y = 0.002$ and continues to fall rapidly to small density for $y = 0.05$, whereas the onset of Pauli susceptibility is not observed before $y =$ 0.07 is reached [148]. Here y is the dopant (AsF_5) concentration. This behavior is of particular interest for the soliton model. The above result indicates that metallic electron band will develop only for $y > 0.07$, whereas the electric conductivity for polyacetylene becomes almost metallic for $y > 0.02$ at least at the room temperature [148]. This suggests strongly existence of a new conduction mechanism due to the charged soliton.

B. *Optical Properies*

The absorption spectrum of pristine $(CH)_x$ of the either isomeric form is typical for a semiconductor with well defined energy gap. The energy gap 2Δ for *cis*-$(CH)_x$ is above 2 eV, whereas that for *trans*-$(CH)_x$ is about 1.5 eV [149]. In contrast to the theoretical prediction Equation (5.17) the observed absorption spectrum does not exhibit the square root singularity at the threshold, most probably due to the interchain coupling. The absorption spectra in the visible and near infrared (IR) show drastic changes upon doping [128, 129]. In general at low doping levels, a new absorption is built up in the region between 0.5 and 0.8 eV while the interband absorption is suppressed. This new absorption is interpreted as due to the midgap states introduced by solitons. Although the earlier result [128] appears to indicate that a dopant molecule introduces more than one soliton [126], the latter result obtained by the electrochemical doping is more consistent with the theoretical calculation [130] as far as relative intensities of the midgap absorption and the interband absorption are concerned. The observed spectrum shape of the midgap absorption is much broader than the SSH model prediction as in the case of the interband absorption.

Upon doping new infrared lines are generated at 1400 and 900 cm^{-1} in *trans*-$(CH)_x$. The position of lines are insensitive to dopants and the intensities increase linearly with dopant concentration [150, 151]. Similar lines are found in $(CD)_x$ at 1120 and 780 cm^{-1} [152]. The phonon modes in the presence of a soliton has been analysed numerically by Mele and Rice [153], which describe most of the features observed experimentally. Now the lower modes are interpreted as being due to the sliding mode, which is pinned by the Coulomb potential of dopant molecule [154]. The author believes that the higher mode is nothing but the optical phonon with small wave vector k, which can couple to the IR wave due to the symmetry breaking introduced by soliton.

C. *Transport Properties*

The pristine *trans*-$(CH)_x$ is a semiconductor with well defined energy gap. Upon doping the electric conductivity increases dramatically and reaches the metallic conductivity for $y \simeq 0.01$ at room temperature [155]. However the nature of the transition from the semiconducting to the highly conducting state is not well understood, although experimentally it is found at $y \simeq 0.01$ [156].

For example as seen from the Pauli susceptibility [148], the density of states at the Fermi surface appears to be zero at the concentration $y = 0.01$. We shall limit ourselves to the dilute concentration limit ($y < 0.01$) of dopants where solitons are expected to play an important role. An attractive model is proposed by Kivelson [157] where the electronic transport is controlled by phonon assisted hopping between the electronic midgap states associated with solitons. This model applies only when both neutral and charged solitons are present and therefore it is valid only for dilute concentration of dopants. With this model Kivelson finds the dc conductivity and the thermoelectric power;

$$\sigma_{\text{hop}} = A\,\frac{e^2\gamma(T)}{k_B TN}\left(\frac{\xi}{R_0^2}\right)\frac{y_n y_{ch}}{(y_n + y_{ch})^2}\exp(-2BR_0/\xi) \tag{5.47}$$

and

$$S = K_B/e\left[\frac{1}{2}\left(3 + T\frac{\partial}{\partial T}\ln\gamma(T)\right. + \ln(y_n/y_{ch})\Big)\right] \tag{5.48}$$

where R_0 is the distance between charged impurities $C_{im} = (\frac{4}{3}\pi R_0^3)^{-1}$, and y_n and y_{ch} are the concentrations of the neutral and the charged solitons per carbon atom and A and

B are constants of the order of unity Finally $\gamma(T)$ describes the temperature-dependent hopping rate. By assuming that $\gamma(T) \simeq 500$ eV $(T/300\text{ K})^{10}$, Kivelson can account for the large and almost temperature-independent thermoelectric power observed experimentally [156]. The ac conductivity within this model gives a characteristic frequency dependence $\sigma_{ac} \sim \omega[\ln(2\omega/T(T))]$,[4] where $\Gamma = y_n y_{ch}(y_n + y_{ch})^{-2}\gamma(T)/N$. This frequency dependence is studied in detail by Epstein *et al.* [158] and they find that the model describes consistently both the dc conductivity and the ac conductivity of dilutely doped polyacetylene for $y_{ch} = 2.35 \times 10^{-4}$ and for $T = 300$–84 K, if $\gamma(T)$ is chosen as $\gamma(T) = 500$ eV $(T/300\text{ K})^{14.7}$.

D. *Photogeneration of Soliton Pairs*

Perhaps the most unambigious evidence for solitons is provided by photogeneration of solitons. In their numerical study of the SSH model, Su and Schrieffer [128] show that an electron–hole pair injected in the perfectly dimerized $(CH)_x$ is converted very rapidly within 10^{-13} sec into charged soliton pairs. This possibility for *cis*-$(CH)_x$ and *trans*-$(CH)_x$ is studied experimentally by Lauchlan *et al.* [159]. They discovered the photocurrent in *trans*-$(CH)_x$, which rises rapidly with the energy of exciting photon for $h\nu \gtrsim 1$ eV, while no photoconductivity is observed in *cis*-$(CH)_x$. On the other hand, a large photoluminescence with the peak value around $h\nu \simeq 1.9$ eV is observed in *cis*-$(CH)_x$, while only a neglible (at most two orders of magnitude smaller) photoluminescence is seen near 1.5 eV in *trans*-$(CH)_x$. The photoconductivity in *trans*-$(CH)_x$ is identified as due to the photogenerated charged solitons. On the other hand the absence of photoconductivity is interpreted as absence of free solitons in *cis*-$(CH)_x$. As the two dimerization configurations in *cis*-$(CH)_x$ are nondegenerate, the soliton pairs created in *cis*-$(CH)_x$ are confined; the pair acquires an additional energy associated with the difference in the energies of the two dimerization configurations per unit cell multiplied by their separation [135]. Therefore all photogenerated soliton pairs are in bound states in *cis*-$(CH)_x$ and they cannot contribute to the conductivity. Ultimately the pair of solitons in the bound state annihilate each other giving rise to the photoluminescence. Although this general picture is very plausible, it is not clear if the charge carriers in *trans*-$(CH)_x$ are indeed solitons, except the fact that the threshold energy for the photoconductivity is roughly 1 eV $(=2\,E_s)$ rather than 1.5 eV $(=2\Delta)$. Within the SSH model the photogeneration probability for solitons is calculated by Sethna and Kivelson [160] and Zhao-bin and Lu [161]. In spite of the difference in their assumption, they predict that the transition rate increases exponentially above the threshold $h\nu = 2\,E_s$, which appears to describe qualitatively the frequency dependence of the photoconductivity.

The more direct evidence of the soliton pair creation is provided by detection of other characteristics associated with solitons; the appearance of the infrared active modes, the midgap absorption, and the reduction in the interband absorption. All of these characteristics have been seen in recent experiments [162–167]. The midgap absorption and the interband bleaching are observed within 10^{-13} sec after charged defects are generated by the photons with $h\nu$ from 1.2 to 1.9 eV [162–164]. Furthermore, from the time dependence of the photodichroism it is concluded that solitons are mobile with linear diffusion constant $D \simeq 10^{-2}$ cm^2/sec at room temperature [163]. Although the higher infrared mode with 1370 cm^{-1} has been observed, the infrared mode with 900 cm^{-1} is shifted down to 500 cm^{-1} [166]. If we assume that this is the pinned sliding mode [154], the above experiment implies that the charged solitons in *trans*-$(CH)_x$ are still pinned in the absence of impurity. We shall discuss later the possible origin of this pinning. Therefore these experimental results appear to provide convincing evidence for solitons in *trans*-$(CH)_x$ polyacetylene. More recently whether these photogenerated charged defects carry spins or not is tested by ESR experiment [168]. Flood and Heeger find that the spin density is at most 10^{-2} of the charged carrier density. Since the conversion rate from the charge soliton pair S^+ and S^- into the neutral soliton pair $S^0{}_\uparrow$ and $S^0{}_\downarrow$ appears to be extremely slow [169], the ESR result will provide the definitive evidence for solitons.

In summary there are bulk of experimental data, which are interpreted in terms of solitons. If we limit ourselves to an individual experiment like the appearence of the midgap states, for example, the electronic bound states introduced by impurities can mimic the absorption spectrum [170]. However, it is then difficult to understand how these bound states are created by photoexcitation.

5.7. Extension of the Model

We have seen that the SSH model can account for a vast variety of experimental results of pristine and dilutely doped *trans*-polyacetylene. However, from the theoretical point of view some drastic assumptions are introduced in constructing and solving the model. Major assumptions made in the SSH theory are;

1. The system is one-dimensional
2. The Coulomb interaction between electrons is neglected except possible renormalization effect on t_0 and K in Equation (5.1)
3. The quantum fluctuation of the lattice is ignored.

In order to remove the assumptions 1 and 2, the SSH model has to be modified, while the assumption 3 can be incorporated within the SSH model. Therefore, we shall start with the assumption 3.

A. *Quantum Fluctuations*

The quantum fluctuations are studied both perturbationally and numerically [171–174]. The perturbation treatment indicates that the dimerization pattern is not destroyed by the quantum fluctuation of the ionic lattice. Furthermore, the soliton and the polaron are stable excitations even in the presence of quantum fluctuations [171]. The most comprehensive treatment on this question is done by Fradkin and Hirsch [174], who make use of the Monte Carlo method on the one hand and the renormalization method on the other hand. They have shown that the dimerized state in the SSH model is the ground state irrespective to the electron phonon coupling strength and the ion mass. For this stability the spin degree of freedom associated with the electron is of crucial importance. For example for the hypothetical spinless electron, the dimerized state is not always the ground state. This result is not hard to understand, since in the extreme quantum limit, the continuum limit of the SSH model is exactly the Gross-Neveu model, of which exact results are known [175]. In this model for $N = 1$, the fermion does not have mass while, for $N \geqq 2$ the fermion becomes massive. Here N is the degrees of freedom associated with the fermion. Therefore, though quantum fluctuation may be important numerically, this does not destroy the general feature of the SSH theory.

B. *Coulomb Correlation*

Until recently, the effect of the Coulomb interaction is a most controversial subject. To begin with we do not know what is the proper potential to use. The long range potential used by quantum chemists to analyse short polyenes appears to lead to unacceptable results [176]. If we limit ourselves to the on-site Coulomb interaction of the Hubbard type, it was believed that the ground state becomes antiferromagnetic for moderate value of U. On the other hand the soliton like state appears to exist independent of the nature of the ground state [177, 178].

Recently it is shown that the on-site Coulomb potential has minor effect on the dimerization order parameter as long as U is less than the band width W [179, 180]. From the experimental point of view, the on-site Coulomb energy of the order of $U = 3$ eV is required to interpret a recent electron nuclear–nuclear triple resonance experiment [181] in terms of the neutral soliton. The above experiment indicates that the spin density at the odd and the even sites are of opposite sign and $\rho_+ = 0.06$ and $\rho_- = -0.02$ where ρ is normalized with the spin density of the single electron. In the original SSH model, the spin density of the nuetral solitons is located either on the odd carbon sites or the even carbon sites. Inclusion of the Coulomb term gives rise to a spin density on the alternate sites with opposite sign [182]. Indeed $U = 3$ eV will give the observed spin density if the soliton is confined in a finite range of the order of fifty unit lattice [183].

Furthermore, the peak in the midgap absorption in the doped polyacetylene is shifted downward by about 0.2 eV. This is again consistent with $U = 3$ eV, as the doubly occupied midgap state corresponding to the negatively charged soliton will be pushed up about $\Delta E = U/2\xi$ in comparison with the singly occupied midgap state [166].

C. *Three Dimensional Order*

A recent X-ray scattering experiment [184] suggests clearly that the dimerization pattern in *trans*-$(CH)_x$ exhibits three dimensional order. If the observed local order extends into the three dimensional space, there should be no free solitons; all solitons are confined. Therefore in order to estimate the confinement potential, it is very important to study the origin of this three dimensional order. The most likely candidate for this is the interchain electron hopping [185]. Within this model, we obtain the confinement energy

$$E_c(x) = W(x/a) \tag{5.49}$$

where

$$W = 4t_\perp^2/\pi t_0 \tag{5.50}$$

and $t_\perp$ is the interchain hopping integral while t_0 is the intrachain hopping integral given in Equation (5.2) and x is the distance between a pair of solitons on the same chain. Making use of $t_\perp$ used in band-structure calculation of polyacetylene [186] it is found that

$$W \simeq 30\text{–}3 \text{ K} \tag{5.51}$$

where $t_\perp/t_0 = 30^{-1}–100^{-1}$. A very similar $t_\perp$ is deduced in a recent analysis of the pressure independent of interband transition energy [187]. It is noteworthy that a similar confinement energy $W \simeq 3.5$ K is obtained by Baughman and Moss [188], when modeling the interchain coupling in terms of Coulomb or dispersion force. Although the above value of W may be modified as in our model two chains of polyacetylene are assumed to lie on the same plane, whereas in reality they are not as seen in the X-ray scattering data [184], we believe that the above value is quite reliable. First of all, due to the confinement potential, all neutral solitons are confined near the edge of the polyacetylene crystal. This is not only consistent with the bond lengths determined by the nuclear spin nutation spectroscopy [189], but also the confinement potential provides the necessary potential well to localize the neutral soliton within 50 atomic units [183]. Indeed when $W = 3$ K, the neutral soliton is localized within 50a from the edge of polyacetylene [190]. Due to the confinement potential all photogenerated charged soliton pairs S^+ and S^- are in the bound state.Then it is quite natural to identify the infrared mode with 500 cm^{-1} associated with the photogenerated solitons [166] to the excitation of the bound pair from the ground state to the first excited level. The potential energy between S^+ and S^- is given by

$$V(x) = W\frac{x}{a} + \frac{8}{\pi}\Delta \exp(-2x/\xi) \tag{5.52}$$

where the second term arises from the repulsive potential between two solitons [191] as modified by Rice [192] who takes into account the proper symmetry of the electronic wave functions in the midgap states. Identifying $E_1 - E_0 = 500$ cm^{-1} yields now $W = 18$ K, where the soliton mass M_s is assumed to be twice the electron mass [190]. Although this W is still within the estimated range, it is six times larger than that deduced for the neutral soliton. We do not believe that this is a serious defect of the model, because it is quite likely that the neutral solitons reside in less crystalline area of polyacetylene and therefore they have a weaker confinement potential.

The presence of the confinement potential suggests that the physical properties of *cis*-$(CH)_x$ and *trans*-$(CH)_x$ are after all quite similar. The only difference is that the former has much stronger confinement potential resulting in no photoconductivity at room temperature. A recent experiment by Blanchet *et al.* [167] indeed shows that the photoconductivity in *trans*-$(CH)_x$ disappears quite suddenly below 150 K. Therefore we expect that below 150 K all photogenerated solitons are in the bound states and disappear by recombination. This implies that the photoluminescence should be observable for *trans*-$(CH)_x$ below 150 K with $\omega \simeq 200–300$ cm^{-1}. The fact that the infrared mode with $\omega = 500$ cm^{-1} disappears above 150 K [167] further confirms the general picture. The solitons in *trans*-$(CH)_x$ are confined below $T = 150$ K and are liberated and free above $T = 150$ K. This picture is also consistent with the disappearance of the solid state effect at 150 K in ESR signal of pristine *trans*-$(CH)_x$ [143]. The liberation of solitons at $T = 150$ K appears to be a collective effect, since it is very difficult to describe such a sudden transition in terms say of a single pair. Clearly a further work on the liberation of solitons is desirable.

5.8. Future Development

Some of the properties of *trans*-$(CH)_x$ described above are consequence of the half-filled band. When the Fermi momentum is 1/3 of π/a as studied by Su and Schrieffer [104], the solitons have fractional charges. The continuum model for the trimerized chain is considered recently by Hara *et al.* [193] and Gammel and Krumhansl [194]. Whether these fractionally charged solitons are observable by transport measurements is not known. Furthermore, the stability of the fractionally charged solitons in the presence of the Coulomb interaction appears to require further study.

A more rigorous analysis of the SSH model (Equation (5.1)) is done by Sutherland [195] for arbitrary electron filling factor. In particular, he finds that the solitons discussed above exist not only for the half-filled band but also for k_F close to the half-filled band. Furthermore, when the band is close to the third-filled he finds in general the solitons with fractional charge as in the Su–Schrieffer model.

A somewhat different model is considered by Brazovskii *et al.* [196]. The model is same as the SSH model (Equation (5.1)) except that $t_{n,\, n+1}$ is replaced by

$$t_{n,\, n+1} = t_0 \exp\left[-\alpha t_0^{-1}\left(u_{n+1} - u_n\right)\right] \tag{5.53}$$

Expanding the exponent in powers of α, the linear term gives the SSH model. They found that this model (the BDK model) is exactly solvable within the mean field approximation based on the exact result on the Toda lattice. In this model, the filling factor $\nu = 0$ (empty band), 1/2 (half filled band), and 1 (filled band) are the special points. Near $\nu = 1/2$ for example the ground state is described by the soliton lattice as described in Subsection 5.5. The electron energy spectrum consists of three disconnected bands. Furthermore, there is nothing special for $\nu = 1/3$ for example. When ν is close to 0 or 1, the order parameter becomes complex and the sliding mode exists in the ideal chain. It is not well understood why the prediction of the BDK model is so different for $\nu = 1/3$ from that of the SS model.

More recent variants of the SSH model are proposed by Rice and Mele [197, 198] for diatomic polymers and by Rice *et al.* [199] for polyyne chains. The former model allows solitons with irrational charge, whereas the latter model is the SSH model but with electrons of 4 flavors. However we shall not go into application of these models.

The existence of degenerate ground states is of central importance for the physics of trans-polyacetylene. However, some of the ideas developed for polyacetylene are applicable to the polymer chains which do not have degenerate ground states but one of the low energy states is slightly lifted from the other. In these systems solitons are necessarily confined but polarons and bipolarons (i.e. polarons with charge $\pm 2e$ or bound state of pair of charged solitons) are considered as elementary excitations [135, 136, 141]. Recently optical absorption data of polypyrrole [200] and polythiophene [201] one analysed convincingly in terms of polarons and bipolarons in these systems. We present the chemical structure of polypyrrole (i.e. pyrrole polymer) and polythiophene in Figures 12 and 13 together with their polaron configurations.

Therefore, in spite of the unique position trans-polyacetylene occupies in the conducting polymers the concepts developed for the physics of polyacetylene will find wider and wider application in physics of the conducting polymers.

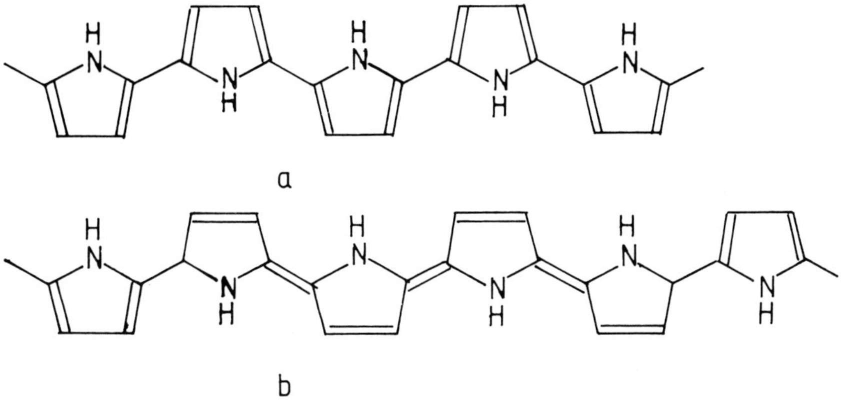

Fig. 12. The bond configurations of polypyrrole are shown. (a) the ground state, (b) the polaron (or bipolaron) state.

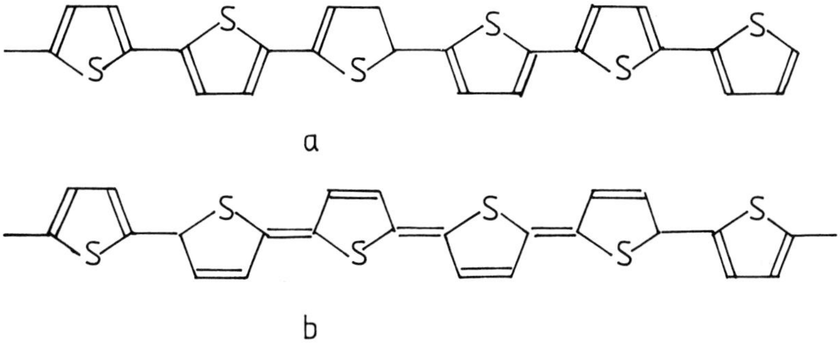

Fig. 13. The bond configurations of polythiophene are shown. (a) ground state, (b) the polaron (or bipolaron) state.

6. Sliding CDW Transport

6.1. Transport Properties

The discovery of non-Ohmic conductivity in $NbSe_3$ by Monceau *et al.* [202] opens up a new field. In the weak field limit the temperature dependence of the electric conductivity has two dips characteristic of the transition to the charge density wave (CDW) state [102]; the conductivity decreases as the quasi-particle number decreases due to the opening of the energy gap. However, Monceau *et al.* discovered that a moderate electric field $E \simeq 5$ mV/cm could eliminate these distinctive features. Now extensive studies establish that this non-Ohmic conductivity is due to the sliding of the CDW condensate. This possibility has been envisioned by Fröhlich [93] almost 30 years ago. Indeed the two features in the conductivity are associated with the two CDW transitions at $T = 144$ K and 59 K. The two CDW condensates are incommensurate but very close to the $M = 4$ commensurability [203]. Later Fleming and Grimes [204] discovered that the nonlinearity in the conductivity requires a small threshold field E_T (~ a few mV/cm). Furthermore in the non-Ohmic regime a quasi-periodic current (narrow band noise) is generated [204]. The frequency dependent conductivity is very sensitive to the frequency. For an ac field of 10 GHz (–0.5 K) the two CDW features seen in the dc conductivity almost disappear [205]. The smallness of the threshold field and the characteristic frequency rule out single particle interpretations (Zener tunneling, field ionization) for the anomaly. Furthermore independence of the CDW amplitude on the electric current as seen by X-ray even in the nonlinear regime [206] strengthens the idea that motion of the CDW condensate plays a crucial role in the nonlinear transport. More recently new systems like TaS_3 (orthorhombic and monoclinic) [207, 208], $(TaSe_4)_2I$, $(NbSe_4)_{10/3}I$ [209, 210] and the blue bronze $K_{0.3}MoO_3$ and $Rb_{0.3}MoO_3$ [211] are found to exhibit similar anomalous transport. Therefore the anomalous transport observed in the CDW states of $NbSe_3$ is generic in a class of quasi-one dimensional compounds. Since a review on this subject will be given by Monceau in Part II and since a few excellent reviews [212–214] are already available, we shall summarize the principle experimental results in this subsection. Then we shall go on to describe the theoretical models to describe the anomalous transport from a solitonic point of view.

A. *Field and Frequency Dependence of Conductivity*

The dc conductivity in $NbSe_3$ in the CDW state is found to be strongly nonlinear [202]. While later the existence of a small linear regime is established [204]. Indeed Fleming and Grimes [204] suggested the following empirical expression

$$\sigma(E) = \sigma_a + \sigma_b(E) \tag{6.1}$$

$$\sigma_b(E) = \sigma_b \theta(E - E_T)\left(1 - \frac{E}{E_T}\right) \exp(-E_0/E - E_T) \tag{6.2}$$

where σ_a is the Ohmic component and E_T is the threshold field. E_0 is another field, which is 2–4E_T for very clean $NbSe_3$ [213]. Although Equation (6.2) suggests the Zener tunneling (if we put $E_T = 0$) in a semiconductor with the energy gap 2Δ, the smallness of E_0 excludes any process associated with the quasi-particles. For example if Δ for the CDW energy gap were substituted, E_0 would be of the order of 10^4 V/cm, whereas in the cleanest $NbSe_3$ samples, the observed $E_0 \simeq 5 \times 10^{-3}$ V/cm. Furthermore, the existence of the threshold field E_T is rather difficult to understand within the tunneling model. We note also that alternative forms for $\sigma_a(E)$,

$$\sigma_b(E) = \sigma_b \theta(E - E_T) \exp(-E_0/E - E_T)$$

and

$$= \sigma_b \theta(E - E_T)(1 - E/E_T) \exp(-E_0/E)$$

produce equally good fits to the observed nonlinear conductivity as long as $E \lessapprox 3E_0$ and therefore it is not useful to argue the precise form of the nonlinear conductivity at the present time.

As already mentioned Ong *et al.* [205, 216] showed the CDW features are removed from the conductivity for $\omega = 10$ GHz. Grüner *et al.* [217] have determined the conductivity over an extended frequency range. The results may be summarized approximately as;

$$\sigma_1(\omega) - i\sigma_2(\omega) = \sigma_a + \sigma_b \frac{(\omega\eta)^2 - i(\omega\eta)}{1 + (\omega\eta)^2} \tag{6.3}$$

where $\eta = \tau\omega_0^2$.

Here Equation (6.3) is the one corresponding to an overdamped oscillator where τ is the relaxation time and ω_0 is the oscillator frequency. The experimental data are fitted with $\eta^{-1} = 3.6 \times 10^8$ Hz, although the observed frequency dependence is much broader than Equation (6.3) suggests. This result is interpreted in terms of the pinned overdamped phase modes suggested by Lee, Rice and Anderson [98]. Broadness of the frequency dependence may suggest a collection of the phase modes with different pinning frequencies and damping constants. We stress again that the smallness of both E_T and η^{-1} indicate strongly the collective origin of the E and ω dependence of the conductivity.

B. *Quasi Periodic Oscillation in Transport Current*

Perhaps the most spectacular transport phenomenon is the appearence of a quasi periodic oscillation in the conduction current in the nonlinear regime as discovered by Fleming and Grimes [204].

This oscillation is decomposed into narrow band noise and broad band noise when the voltage spectrum is Fourier analysed. Furthermore the narrow band noise consists of one fundamental and its harmonics, although sometimes a few fundamentals are observed. Subsequently Monceau *et al* [220] found that the frequency of the narrow band noise ν increases linearly with the CDW current J_{CDW}, where J_{CDW} is defined as

$$J_{\text{CDW}} = \sigma_b(E)E = J\left(1 - \frac{\sigma(O)}{\sigma(E)}\right) \tag{6.4}$$

where $\sigma(E)$ is defined in Equation (6.1). This relation suggests strongly that both the narrow band noise and the broad band noise are associated with uniform motion of the CDW condensate. The uniformly sliding CDW is described in terms of the time dependent phase ϕ;

$$\phi(x, t) = -Qvt \tag{6.5}$$

with $Q = 2k_F$. For example Equation (6.5) gives rise to the uniformly moving charge density

$$\rho_{\text{CDW}}(t) = N_0\lambda^{-1}\Delta\cos[Q(x - vt)] \tag{6.6}$$

where use is made of Equation (4.14). Equation (6.6) implies that the local charge oscillates with frequency

$$\nu = Qv \tag{6.6}$$

On the other hand substituting Equation (6.5) into Equation (4.26), we shall have the CDW current

$$J_{\text{CDW}} = -en_c(T)v \tag{6.7}$$

Eliminating v from Equations (6.6) and (6.7), we obtain

$$|J_{\text{CDW}}| = \frac{e}{Q} n_c(T)\nu \tag{6.8}$$

which is exactly what found by Monceau *et al.* [218].

More recently the temperature dependence of the linear coefficient as well as N dependence of Equation (6.8) have been studied [219, 220]. In the vicinity of $T = T_c$, it is found $n_c(T) \propto \Delta(T)$ [219] consistent with Equation (4.28). Although Monceau *et al.* [220] indicate that the experimental results for $NbSe_3$, TaS_3 (both orthorhombic and monoclinic) and $(TaSe_4)_2I$ are more consistent with Equation (6.8) but with e replaced by $2e$, we don't believe that there is any justification for such a relation. We think rather that the above analysis of Monceau *et al.* indicates the difficulty in estimating N, the density of conduction electrons, in these compounds.

C. *Impurity Effects*

As discussed by Lee, Rice, Anderson [98] the CDW is pinned either by impurities or by incommensurability potential. Therefore it is of great interest to study the origin of the threshold field E_T. In a series of experiments on $NbSe_3$ doped with Ta and Ti, Ong *et al.* [215, 216] established that both E_T and E_0 obey a y^2 relation where y is the Ta concentration. On the other hand for the Ti doped samples and the samples irradiated with high energy protons [221], E_T increases linearly with y the impurity concentration or the dosage of protons.

The observed concentration dependence of the threshold field not only indicates that the impurity pinning plays the central role in these compounds but also that the different y dependence of E_T for the Ta and Ti doping is interpreted with the concepts of the weak and the strong pinning developed by Fukuyama, Lee and Rice [40, 222]. We shall describe the theoretical concepts in the following subsection.

D. *Hysteresis, Memory Effects*

Another remarkable feature of the nonlinear conduction is the appearance of a long relaxation time of the order of millisecond. When a dc pulse is applied the onset of non-Ohmicity requires above a millisecond as first observed by Gill [223]. This relaxation time increases exponentially with energy gap 2Δ, as the temperature decreases, suggesting that the quasi-particle pair creation plays an important role in the relaxation of the CDW. When a pulse field of alternative sign is applied the conduction current exhibits the memory effect with the same relaxation time. More recently Zettl and Grüner [224] reported a 'switching' phenomenon; the jump between the Ohmic and non-Ohmic regime depends on the hysteresis in TaS_3. This implies that the underlying CDW state is not uniquely determined by the external electric field but passes through a multitude of quasi-equilibrium states before the final equilibrium configuration is reached.

6.2. Impurity Pinning and the Fukuyama Lee Rice Domain

From the observed impurity concentration dependence of the threshold field E_T [215], the impurities are clearly the dominant pinning centers. The interaction energy between electrons and impurities is given by

$$H_{\text{imp}} = \sum_i \int d^3x \, \rho(\mathbf{x}) V(\mathbf{x} - \mathbf{R}_i) \tag{6.9}$$

where $\rho(\mathbf{x}) = \Sigma_i \psi_s^+(\mathbf{x})\psi_s(\mathbf{x})$ is the local electron density and the summation is over the impurity sites $\mathbf{R}_i$. In the following we approximate the impurity potential $V(\mathbf{x}) = V_0\delta(\mathbf{x})$. Substituting for $\rho(x)$ by $\langle\rho(x)\rangle = \rho_{\text{CDW}}(x)$ given in Equation (4.14), we find

$$H_{\text{imp}} = v_{\text{imp}} \sum_i \cos(Qx_i + \phi(\mathbf{x}_i)) \tag{6.10}$$

with

$$v_{\text{imp}} = N_0 \Delta V_0 \lambda^{-1} \tag{6.11}$$

Therefore in the presence of random impurities the equilibrium configuration for $\phi(\mathbf{x})$ is determined by the phase Hamiltonian

$$\mathcal{H}(\phi) = \tfrac{1}{2} A \int d^3x \, \partial\phi c^2 \, \partial\phi + v_{\text{imp}} \sum_i \cos(Qx_i + \phi(\mathbf{x}_i)) \tag{6.12}$$

where the first term is the phase elastic energy taken from Equation (4.22) with proper generalization to the three dimensional system. Here $\mathbf{c}^2$ is the square of the tensor of the phason velocity. Since the anisotropy in the phason velocity is easily eliminated from Equation (6.12) by appropriate length scale change [40], we shall consider in the following a slightly simpler version of Equation (6.12)

$$\mathcal{H}(\phi) = \tfrac{1}{4} N_0 v_F^2 \int d^3x \, |\nabla\phi|^2 + v_{\text{imp}} \sum_i \cos(Qx_i + \phi(\mathbf{x}_i)) \tag{6.13}$$

where v_F is the Fermi velocity and N_0 is the electron density of states per spin at the Fermi surface. The Hamiltonian (6.13) has been extensively studied by Fukuyama, Lee and Rice [222, 40]. We shall summarize here the principal results. From the outset it is important to distinguish two cases: the strong pinning case and the weak pinning case. In the strong pinning case the second term dominates Equation (6.13). Then the equilibrium configurations are determined by minimizing $\phi(\mathbf{x})$ at the impurity sites;

$$\begin{aligned} \phi(\mathbf{x}_i) &= \pi - Qx_i \quad \text{modulo } (2\pi) \quad \text{for} \quad V_0 > 0 \\ &= -Qx_i \quad \text{modulo } (2\pi) \quad \text{for} \quad V_0 < 0 \end{aligned}$$

Then the rest of $\phi(\mathbf{x}_i)$ are determined to minimize the elastic energy consistent with the boundary condition Equation (6.13). When the impurity concentration is y, the pinning energy per unit volume is estimated as

$$E_{\text{pin}} \simeq \frac{\pi^2}{4} N_0 E_F^2 y^{2/3} - y|v_{\text{imp}}| \tag{6.14}$$

Here we estimate the elastic energy from

$$\frac{\partial \phi}{\partial x} \simeq \frac{\pi}{a} y^{1/3} \tag{6.15}$$

where $ay^{-1/3}$ is the average distance between two impurities and a is the atomic distance. From Equation (6.14) we can deduce that the strong pinning limit applies when

$$V_0 \geqslant \lambda E_F^2/\Delta \sim 10 E_F \tag{6.16}$$

in the case of $NbSe_3$ for example.

In the weak pinning limit on the other hand a single impurity is not strong enough to pin the local phase. In this circumstance the CDW is split into domains with linear dimension L_0. Within each domain ϕ is fairly coherent; we can define an averaged $\bar{\phi} = \langle \phi \rangle_d$. The linear dimension of the domain L_0 as well as the pinning energy per unit volume in this case is estimated as follows;

$$E_{\text{pin}} = \tfrac{1}{4} N_0 v_F^2 L_0^{-2} - |v_{\text{imp}}| y^{1/2} (a/L_0)^{3/2} \tag{6.17}$$

where the first term is estimated from

$$\frac{\partial \phi}{\partial x} \sim L_0^{-1}$$

while the second term is obtained from by averaging $\cos(Qx_i + \phi(x_i))$ within the domain

$$\langle \cos(Qx_i + \phi(x_i)) \rangle_d = \text{Re}\,\langle \exp(i(Qx_i + \bar{\phi})) \rangle_d = (y(L_0/a)^3)^{1/2} (L_0/a)^{-3} \cos \bar{\phi} \tag{6.18}$$

According to Fukuyama and Lee [222], the average can be considered as the random walk of a unit vector in the two-dimensional space with number of steps $y(L_0/a)^3$, the impurity number in the volume L_0^3. Here it is assumed that the variation in Qx_i dominates that of $\phi(x_i)$. Then minimizing Equation (6.17) for L_0, we find

$$L_0 = (N_0 v_F^2 / 3|v_{\text{imp}}|)^2 a^{-3} y^{-1} \tag{6.19}$$

and

$$E_{\text{pin}} = -\tfrac{27}{4} y^2 a^6 |v_{\text{imp}}|^4 (N_0 v_F^2)^{-3} \tag{6.20}$$

In the weak pinning limit, the pinning energy is proportional to y^2. Furthermore, Equation (6.20) gives the correct order of magnitude of the pinning voltage. Note that Equations (6.19) and (6.20) yields:

$$L_0 \simeq (\lambda\xi/3)^2 a^{-1} y^{-1} \tag{6.21}$$

$$E_{\text{pin}} \simeq -(27/4)\lambda^{-4} y^2 (a/\xi)^2 N_0 \Delta^2 \tag{6.22}$$

for $V_0 \cong E_F$ and $\xi = v_F/\Delta$ ($\sim$ 100 Å) is the BCS coherence distance for the CDW state. For $y = 0.1\%$ for example, we still have $L_0 \simeq 10\ \mu$m and $|E_{\text{pin}}| \simeq 10^{-13} E_{\text{cond}}$. The latter implies the pinning frequency $\omega_0 = 10^{-6}\Delta$ consistent with the experiments described in the preceding subsection. Indeed the isoelectronic impurities like Ta in $NbSe_3$ give rise to the weak pinning, while the non-isoelectric impurities like Ti result in the strong pinning.

Although within the present approach it is difficult to calculate the effective phase potential $V(\bar{\phi})$ or to predict the E and ω dependence of the conductivity, the impurity pinning can interpret both the y dependence and the order of magnitude of E_T correctly. Furthermore within the mean field theory Equation (6.20) predicts that the pinning frequency ω_0 diverges like $(1 - T/T_c)^{-1/2}$ near the transition temperature which is also consistent with the observed temperature dependence of E_T [225]. On the other hand at low temperatures Equation (6.20) predicts a temperature independent pinning frequency whereas the observed E_T for $NbSe_3$ increases like $\exp(-aT)$ as the temperature decreases [225].

6.3. Tunneling Model

We shall now describe the theoretical models which attempt to understand the E and ω dependence of the conductivity and the origin of quasi-periodic current in the non-Ohmic regime.

The first of them is the tunneling model, which hypothesizes that the E dependence of the conductivity is due to the quantum mechanical tunneling of some sort of the CDW mode, which is suggested by similarity of Equation (6.2) with the Zener tunneling expression. The first of such a model is the soliton-pair creation proposed by Maki [226]. In spite of several difficulties this model encounters in applying to the CDW transport in $NbSe_3$, this model is still the simplest conceptually. We believe also that this model is extremely helpful to understand Bardeen's model which is free of some of the above difficulties [227].

In order for the soliton tunneling to apply the pinning potential has to be periodic in $\phi(x)$. The commensurability potential provides such a potential, although this may not be essential.

In the presence of a periodic potential the phase dynamics is controlled by Equation (4.33). In particular the system contains the ϕ-solitons [33] with charge $Q = \pm 2\,e\,M^{-1}$, where $M = 4$ in the case of the commensurability potential.

In the presence of an electric field E along the chain direction a soliton–antisoliton pair gains a potential energy of $2eM^{-1}Ed$ where d is the distance between the pair. Making use of the instanton technique, the probability of the soliton pair production is readily obtained [226], which gives rise to a nonlinear component of the conductivity

$$\sigma_b(E) = 2(2e/M)^2 \frac{C_0}{\pi E_s} \frac{E_0}{E} \exp(-E_0/E) \tag{6.23}$$

where

$$E_0 = \pi M E_s^2/(2eC_0) \tag{6.24}$$

and $E_s = 8AC_0\omega_0 M^{-2}$ the ϕ-soliton energy.

The principal difficulty of the model is that the soliton energies needed to account for the nonlinear conductivity in $NbSe_3$ are extremely small. For example in order to account for the nonlinear conductivity below T = 144 and 59 K in $NbSe_3$, E_s has to be 0.63 and 0.15 K respectively. If this were really the case, we would not expect any anomaly in the conductivity, since the thermally excited solitons will bring forth the normal conductivity except at extremely low temperatures ($T \leqslant E_s$).

A possible way out of this difficulty is to assume that these solitons are confined due to the interchain coupling. Indeed $NbSe_3$ appears to be three dimensional rather than one dimensional. The confinement potential not only limits the density of thermally excited solitons but also gives rise to a threshold field to the tunneling process. If the confinement potential for soliton and antisoliton pair is given by

$$E_c(x) = W|x|/a \tag{6.25}$$

as in the case of polyacetylene (Equation (5.49)) where x is the distance between the pair, the probability of the pair production vanishes for $E \leqslant E_T = MW/(2ea)$, while for $E > E_T$ we shall have

$$\sigma_b(E) = 2(2e/m)^2 \frac{C_0}{\pi E_s} \frac{E_0}{E - E_T} \exp(-E_0/(E - E_T)) \tag{6.26}$$

Therefore in the presence of soliton confinement, the only strong objection to the ϕ-soliton model is how a periodic potential for ϕ can be obtained from the impurity pinning, since experimentally the soliton energy is determined by the impurity potential rather than the commensurability potential.

The Bardeen model [227, 228] does not depend on detail of $V(\phi)$, the potential for ϕ. This is both an advantage and a difficulty in his theory. Since Bardeen describes his tunneling model in this volume, we shall limit ourselves to a few remarks on his model. Indeed, if the following two conditions are satisfied, the Bardeen model can be reinterpreted in terms of the ϕ-soliton pair production: (1) The impurity potential produces a quasi-periodic potential in ϕ. (2) The commensurability index M in this potential is extremely large: $M \sim 10^3$. By choosing an enormously large M the soliton charge Q becomes extremely small (note $Q = 2e/M$), therefore in this circumstance the soliton energies required to account for the nonlinear conductivity are comfortably large (i.e. comparable to the quasi-particle energy). Although Bardeen interpreted the existence of a threshold field in terms of screening, this can be as well due to the confinement potential. Indeed from the topological point of view the soliton pair creation is rather generic as shown in Figure 14. The depinning starts as local phase around a point x changes from $\phi(x)$ to $\phi(x) \pm 2\pi/M$, as seen from some numerical analysis of the one dimensional model [229].

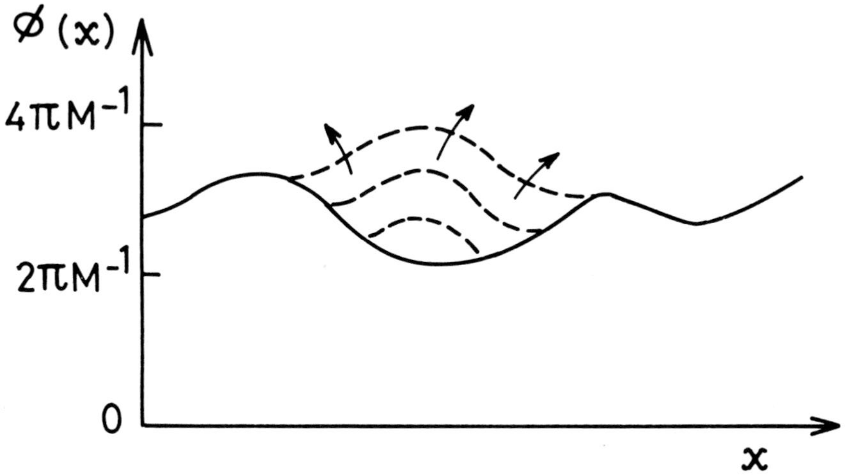

Fig. 14. The initial depinning of the phase $\phi(x)$ in the CDW is schematically shown.

Therefore the problem boils down on how to find a quasi-periodic potential for ϕ starting from the impurity potential. Perhaps a manybody approach to the Lee–Rice pinning by Klemm and Schrieffer [230] may throw light on this question. If the impurity distribution is not random but has a short range correlation, the impurity scattering might result in a quasi-periodic potential for ϕ with large M.

We shall conclude this subsection with two fundamental objections to all the tunneling models in the case of $NbSe_3$. First, to the extent that the tunneling is a purely quantum mechanical process, we expect that E_T becomes constant and independent of temperature at lower temperatures, where the pinning energy becomes temperature independent. However in the case of $NbSe_3$ and TaS_3 (orthorhombic) the observed threshold field E_T increases as the temperature is lowered [225, 206]. This suggests rather strongly that the depinning is a thermal process rather than a quantum process. Indeed in the two dimensional system, the creation of a phase vortex ring results in a similar expression as Equation (6.23);

$$\sigma_{th}(E) \propto e^{-E_0/E} \tag{6.26}$$

where

$$E_0 = \pi M E_v^2 / 2ek_B T \tag{6.27}$$

and E_v is the vortex energy per unit length [231].

Again the three dimensional coupling introduces a threshold field in Equation (6.26). This model predicts that E_0 diverges as T^{-1} at low temperatures, which is not inconsistent with observed E_0. Indeed such a T^{-1} dependence of the threshold field has been observed recently for $(NbSe_4)_{10/3}I$ by Z. Z. Wang *et al.* [232]. However, how to relate E_T to the pinning potential has yet to be worked out. Second, the number of charged pairs tunneled increases gradually with E, whereas the observed $\sigma(E)$ near $E = E_T$ indicates that the charged carrier number is independent of E near $E = E_T$ [213].

Summing up, the thermal activation process may be dominant near the threshold. However, when the CDW condensate is once set in motion it appears that we need more global description of the CDW motion, whereas quantum or thermal process are only local processes. Also the temperature dependence of E_T in blue bronzes $K_{0.30}MoO_3$ appears to be quite different from those of $NbSe_3$ and TaS_3 described above [211]. Therefore even the threshold behavior of the nonlinear conductivity may be controlled by different mechanisms depending on the quasi-one dimensional systems.

6.4. Classical Particle Model

The second model to be described is the classical particle model or washboard model. We have seen already that in the weak pinning limit the CDW condensate is split into the Fukuyama, Lee, Rice (FLR) domains. Within a FLR domain an average phase $\bar{\phi}$ is assigned. We shall concentrate on the dynamics of $\bar{\phi}$, which we shall write ϕ hereafter for simplicity. The simplest equation for ϕ may be

$$\phi_{tt} + \tau^{-1}\phi_t + \frac{\mathrm{d}V(\phi)}{\mathrm{d}\phi} = -F \tag{6.28}$$

first proposed by Grüner *et al.* [233], where

$$F = ev_F\lambda(\omega Q/2\Delta)^2\, n_s(T)E \tag{6.29}$$

and E is the electric field.

The effective phase potential $V(\phi)$ is assumed to be periodic in ϕ. The simplest choice is

$$V(\phi) = \omega_0^2 M^{-2}(1 - \cos(M\phi)) \tag{6.30}$$

where ω_0 is the pinning frequency.

In the presence of a small ac electric field with frequency ω, the induced CDW current is calculated as [233]

$$\begin{aligned} J_{\mathrm{CDW}} &= eQ^{-1}n_c(T)\frac{\partial\phi}{\partial t} \\ &= \sigma_{\mathrm{CDW}}(\omega)E_\omega \end{aligned} \tag{6.31}$$

where

$$\sigma_{\mathrm{CDW}}(\omega) = \sigma_b \frac{-i\omega\tau^{-1}}{\omega_0^2 - \omega^2 - i\omega\tau^{-1}} \tag{6.32}$$

and

$$\sigma_b = \tau e^2 v_F Q^{-1} n_s(T) \tag{6.33}$$

Comparing Equation (6.32) with the observed ac conductivity of $NbSe_3$ [217], it is concluded that the inertia term is of little importance; the first term in the left-hand side of Equation (6.28) can be neglected. Furthermore we can identify $\tau\omega_0^2 = \eta$ as defined, following Equation (6.3). Furthermore, since M in the potential (6.30) can be eliminated by rescaling ϕ completely, at least in the classical theory, we shall concentrate on

$$\tau^{-1}\phi_t + \omega_0^2 \sin\phi = -F \tag{6.34}$$

in the following.

Since Equation (6.34) is analysed in greater detail by Monceau *et al.* [218, 234], we shall summarize their results here. Equation (6.34) describes a motion of a viscous ball on a tilted washboard where the tilt angle is controlled by F. For $F < F_T = \omega_0^2$ the ball does not fall, implying $J_{\text{CDW}} = 0$. When $F > F_T$, J_{CDW} is calculated as

$$J_{\text{CDW}}(t) = eQ^{-1} n_c(T)\tau(F^2 - F_T^2)\,(F - F_T\cos(ut))^{-1} \tag{6.35}$$

with

$$u = \tfrac{1}{2}\tau\,(F^2 - F_T^2)^{1/2} \tag{6.36}$$

First of all, Equation (6.35) gives the dc current

$$J_{\text{CDW}}^{\text{dc}}(t) = eQ^{-1} n_c(T)\tau(F^2 - F_T^2)^{1/2} \tag{6.37}$$

which increases as the square root of $F - F_T$. Furthermore, Equation (6.35) has a periodic component with frequency nu. The intensities of the nth harmonics decreases as K^n with $K = F/F_T - [(F/F_T)^2 - 1]^{1/2}$.

This model predicts the well defined threshold field F_T and the non-Ohmic behavior above the threshold. Furthermore, the appearance of an oscillatory current above the threshold with fundamental frequency u is predicted. However if you look into more detail at the prediction, it is clear that even the qualitative features of the nonlinear transport are missing from the model. In particular: (1) The dc component increases as $(F - F_T)^{1/2}$ near the threshold, whereas in most cases the experiment indicates that J_{CDW} increases linearly with $F - F_T$; (2) in the limit $F \gg F_T$, σ_{CDW} given by Equation (6.37) approaches a constant value as F^{-2}, while experimentally the approach to the constant value is slower, as F^{-1}; (3) the observed higher harmonic intensities decrease much slowly with n than the prediction [234]; (4) the predicted ac components are infinitely sharp. Furthermore, there is no room for the broad band noise. The origin of these difficulties most likely lies in the drastic simplification of the model. The actual CDW condensate contains large internal degrees of freedom associated with $\phi(\mathbf{x})$. Therefore, description of the collection of $\phi(\mathbf{x})$ by a single ϕ is obviously not adequate. Furthermore, a simple periodic potential $V(\phi)$ assumed here is never derived. In spite of these shortcomings the present model established convincingly the existence of the threshold field and the sliding of the CDW for $F > F_T$, the associated non-Ohmicity of conductivity and the appearence of the periodic oscillatory current. One way to improve the classical particle model is to incorporate large degrees of freedom neglected in the model. There are several works in this direction. For example Sneddon *et al.* [235] considered the fluctuation of ϕ around $\bar{\phi}$ within a continuum model and found that the fluctuation gives rise to the nonlinear conductivity which decays like $F^{-1/2}$ in the limit $F \gg F_T$. Unfortunately, however there appears to be no experimental evidence for this slow decay term. Fisher [236] considered the behavior of the nonlinear conductivity near the threshold in analogy to a phase transition. He concluded that the characteristic length characterizing the domain size is no longer the FLR length L_0 but the coherence length, which diverges as $(F - F_T)^{-1/2}$ within the mean field approximation, which leads to $J_{\text{CDW}} \propto (F - F_T)^{3/2}$ near the threshold. Again this result appears to contradict most of the experimental results. The second approach is to consider the one dimensional version of Equation (6.13). The corresponding equation of motion is;

$$\tau^{-1}\phi_t(x) - c_0^2\phi_{xx}(x) + \omega_0^2 \sum_i \delta(x - x_i)\sin(Qx_i + \phi(x_i)) = -F \qquad (6.38)$$

where x_i's are the impurity sites. The above equation is considered as a generalization of Equation (6.34), where the position of impurities are explicitly included. This type of equation has been studied numerically by Sokoloff [237] and Pietronero and Strässler [238]. Precisely speaking the latter authors studied a modified model where the first term in the left hand side of Equation (6.38) is replaced by local ones; $\tau^{-1}\phi_t(x) \to \Sigma_i\delta(x - x_i)\tau^{-1}\phi_t(x_i)$. Both of them found that in the limit of a large number of impurities the threshold behavior of J_{CDW} is broadened in the one dimensional model. On the other hand the calculated widths of the fundamental and the higher harmonics are much broader than those observed in the nonlinear regime of $NbSe_3$. However, these are certainly improvements over the classical model. The third approach will be to incorporate the remaining degrees of freedom in terms of phase vortices or dislocations in the CDW [41]. This model will be described in the following subsection.

The most crucial difficulty of the classical particle model is the origin of a periodic potential in ϕ. In the thermodynamic limit it is known that the impurity potential averages out completely. A more elaborate formulation of the FLR theory may produce such a potential [230]. As noted by Barnes and Zawadowski [239] the higher order terms of Equation (6.9) may produce also such a potential. A more careful exploration in this direction is also desirable.

6.5. Vortex Array Model

We shall describe here a recent model put forwards by Ong *et al.* [41] to account for the narrow band noise in the nonlinear regime of the CDW transport. The model is based on the experimental fact that sources of the narrow band noise are localized and probably near the contact [240]. Furthermore in most of experiments done on $NbSe_3$ contacts formed by silver paint are the major perturbation to a thin crystal of $NbSe_3$ as shown

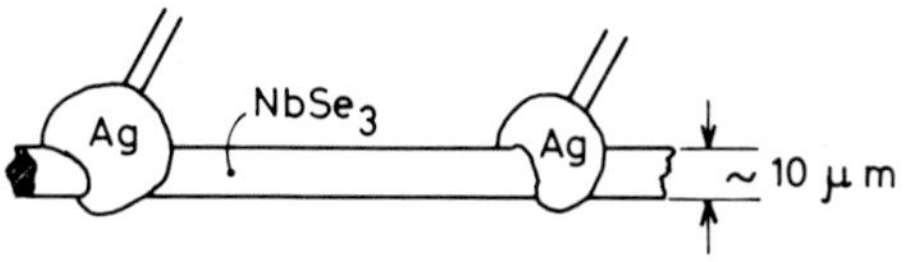

Fig. 15. A $NbSe_3$ crystal with two silver paint contacts is schematically shown.

in Figure 15. Under this circumstance the electric field under the contacts is much smaller than that in the bulk of the sample as the contacts provide bypath for the electric current. When the electric field E in the bulk is larger than E_T, the CDW between the contacts slides with velocity v; the phase of the CDW is increasing linearly with time t as $\phi = Qvt$. On the other hand since the electric field E' under a contact is much smaller the phase of the CDW under the contact is stagnant. This implies a clash between a sliding phase and a stagnant phase close under the contact. The clash is avoided, however, by introducing an array of phase vortices (or dislocations) which moves along the phase boundary perpendicular to v with velocity v_s;

$$v_s/l_v = v/\lambda_{\mathrm{CDW}} = \nu\ . \tag{6.39}$$

where l_v is the distance between two vortices in the array, $\lambda_{\mathrm{CDW}} = 2\pi Q^{-1}$ and ν is the frequency of the fundamental of the narrow band noise. To the extent that the array of vortices forms a regular lattice, the frequency of the noise agrees with the classical particle model. Making use of the Hamiltonian (6.13), the vortex energy per unit length is estimated as

$$E_v = \tfrac{1}{4} N_0 v_F^2 \int \mathrm{d}^2x\,(1/r^2) = \tfrac{1}{2}\pi N_0 v_F^2 \ln(L_0/\xi) \tag{6.40}$$

where

$$\phi = \tan^{-1}(y/x) \tag{6.41}$$

indicating a vortex at $(x, y) = (0, 0)$ is substituted, and the integral is cut off at L_0 the FLR length and ξ the coherence length. The free energy of an array of vortices is much less than the complete disruption of the CDW condensate at the phase boundary, since the ratio of two free energies $\Delta F_v/\Delta F_s = (\pi\xi/l_v)\ln(L_0/\xi)$ is always much less than unity. In the presence of an array of vortices at $x = 0$ and $y = y_i$, the extra CDW current associated with the array is calculated as follows.

Integrating Equation (4.26) around a loop enclosing the vortex array we obtain that the current difference between two sides of the array is

$$\begin{aligned}\Delta J &= J_{\mathrm{CDW}}(\mathrm{left}) - J_{\mathrm{CDW}}(\mathrm{right})\\ &= -2\pi n_c(T)eQ^{-1}\sum_i \partial y_i/\partial t\end{aligned} \tag{6.42}$$

where we substituted

$$\phi = \sum_i \tan^{-1}(y - y_i/x) \tag{6.43}$$

in Equation (4.26). Taking the time average of Equation (6.42), the dc component is nothing but the difference of the local J_{CDW} between the two sides of the vortex array. The vortex array compensates the difference of the local CDW current and achieves the conservation of the CDW current throughout the sample whenever there are a few domains with different sliding velocities. On the other hand since the motion of vortices cannot be uniform, Equation (6.42) generates an ac component as well. The present model accounts easily for the experimental fact that three fundamentals instead of one fundamental are usually observed in the four probe experiment [234]. The major advantage of the present model over the classical particle model is that we do not need to have a periodic potential for ϕ generated from the impurity potential. In the vortex model the periodic potential for an array of vortices is provided by the walls where vortices enter and exit. The narrow band noise has always width due to fluctuation around the uniform motion of vortices. It is naturally expected the width of the narrow band noise increases with the transverse cross-section of the samples. On the other hand we expect that the noise amplitude is independent of the distance between two contacts. We will show in Figure 16 the frequency spectrum of the observed current in $NbSe_3$ samples with different contact distance [241]. It is quite clear from Figure 16 that there is no appreciable change in the noise intensity for samples with contact distance between 0.07 mm and 1.6 mm, although there is a recent report contrary to the above result [242]. Indeed the predicted decrease of the noise intensity with the increasing transverse cross-section of $(TaSe_4)_2I$ has recently been reported by Mozurkewich *et al.* [243]. In any case the size and the geometry dependence of the narrow band noise is of particular importance to understand the origin of the narrow band noise.

The vortex array model also resolves a standing puzzle why the cleaner sample has stronger and sharper noise spectrum, as the model does not require a periodic potential due to the impurity potential.

From the vortex array picture the CDW condensate in the sliding regime splits into three domains, two domains at two contacts and one sliding domain in between. It is very remarkable that the CDW in the region between two contacts is monodomain, which can be seen from the absence of the second fundamental. Within the present model each domain boundary generates noise with fundamental proportional to the difference of the sliding velocities of two domains involved. This implies that the phase coherence of the CDW extends much longer than the FLR length L_0 once the CDW is set into motion.

7. Summary

We have reviewed solitons in quasi-one-dimensional systems. In the magnetic systems there is clear indication that the low temperature properties of some of these systems are dominated by solitons. In the conducting systems the notion of solitons has played a central role in understanding the electronic properties of polyacetylene and other related compounds like polythiophene and pyrrole polymer. However a further study is clearly desirable to understand the soliton dynamics. Finally in the quasi-one dimensional CDW system the phase vortices (2 dimensional solitons) will play an important role in the CDW dynamics. On the other hand, direct evidence for the ϕ solitons in any of quasi-one dimensional conductors appear to be still lacking although discommensurations are

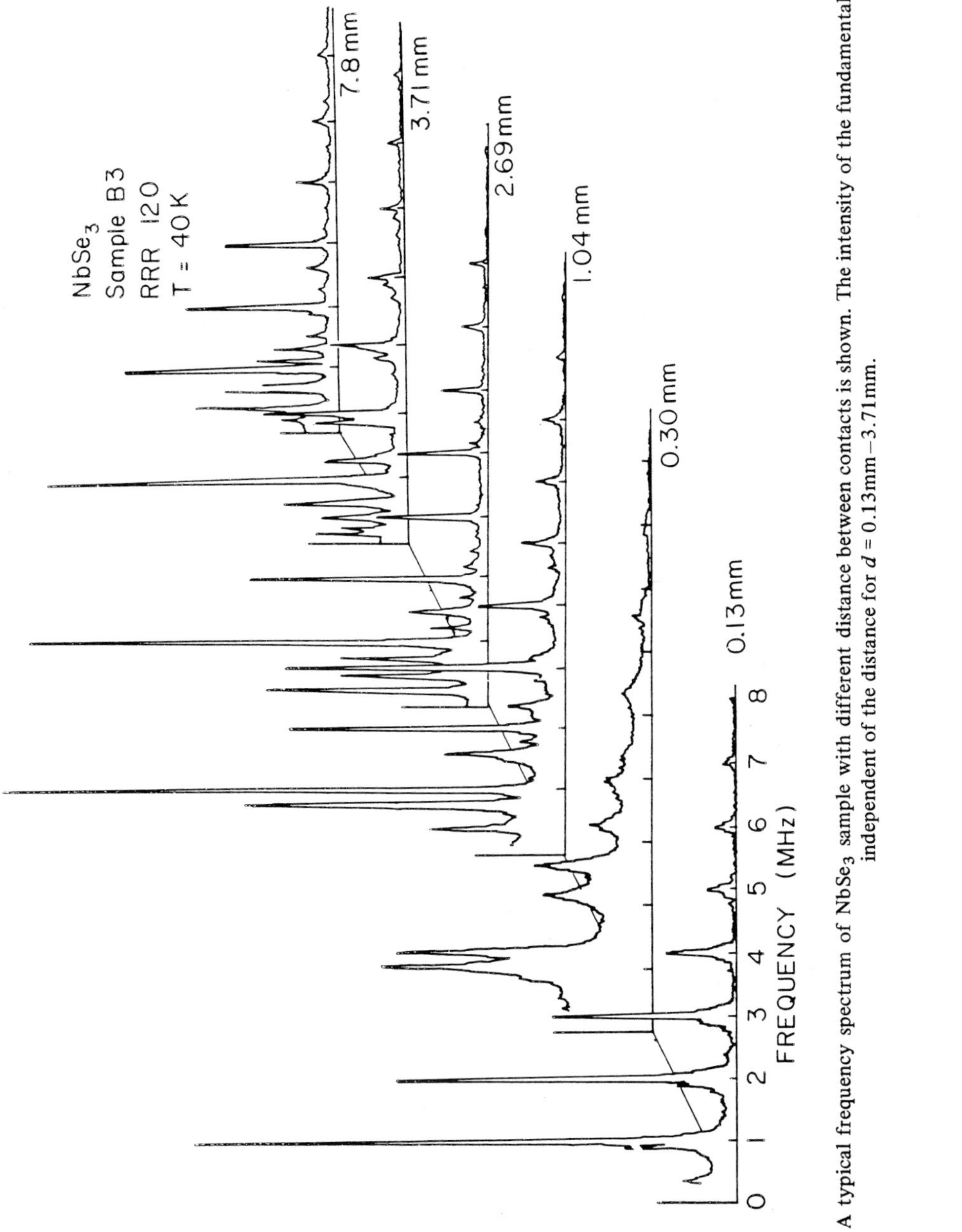

Fig. 16. A typical frequency spectrum of $NbSe_3$ sample with different distance between contacts is shown. The intensity of the fundamental is almost independent of the distance for $d = 0.13\,\text{mm} - 3.71\,\text{mm}$.

observed, which are macroscopic objects. It is hoped that this review will provide a starting point for future study of the soliton dynamics in quasi-one dimensional conductors.

Acknowledgements

I am benefited greatly from colleagues working on related subjects. Particular thanks go to Dionys Baeriswyl and Phaun Ong for numerous discussions we had on solitons in polyacetylene and on the CDW transport of $NbSe_3$. A part of this manuscript was written while I was staying at the Max Planck Institute at Stuttgart and at Aspen Center for Physics. I would like to thank both institutions for their hospitality. The present work was supported by the National Science Foundation under grant number DMR 79–16703 and DMR 82–14525.

References

1. J. Scott-Russel, *Proc. Roy. Soc. Edinburgh* 319 (1844).
2. A. Barone, F. Esposito, C. F. Maggee and A. C. Scott, *Riv. Nuovo Cimento* **1**, 227 (1971).
3. G. B. Whitham, *Linear and Nonlinear waves* (1974) John Wiley and Sons, New York.
4. A. R. Bishop and T. Schneider (eds.) *Solitons in Condensed Matter Physics* (1978) Springer Verlag, Berlin, Heidelberg, New York.
5. J. Bernasconi and T. Schneider (eds.) *Physics in One Dimensional* (1981) Springer Verlag, Berlin, Heidelberg, New York.
6. A. R. Bishop, J. A. Krumhansl and S. E. Trullinger, *Physica* **1D**, 1 (1980).
7. *J. de Physique* (Paris) **C3** (1983).
8. D. J. Kortweg and G. de Vries, *Phil. Mag.* **39**, 422 (1895).
9. A. Kochendörfer and A. Seeger, *Z. Phys.* **127**, 533 (1950).
10. J. Frenkel and T. Kontorova, *J. Phys. USSR* **1**, 139 (1939).
11. E. Fermi, J. Pasta and S. Ulam, *Los Alamos Report* LA–1940 (1955); *Collected papers of Enrico Fermi* Vol. II, p. 978 (1965) University of Chicago Press, Chicago.
12. M. Toda, *J. Phys. Soc. Japan Suppl.* **26**, 235 (1969).
13. N. J. Zabusky and M. D. Kruskal, *Phys. Rev. Lett.* **15**, 240 (1965).
14. C. S. Gardner, J. M. Greene, M. D. Kruskal and R. M. Miura, *Phys. Rev. Lett.* **19**, 1095 (1967).
15. P. D. Lax, *Comm. Pure and Appl. Math.* **21**, 1095 (1968).
16. V. E. Zakharov and A. B. Shabat, *Sov. Phys. JETP* **34**, 62 (1972).
17. M. J. Ablowitz, D. J. Kaup, A. C. Newell and H. Segur, *Studies in Appl. Math* **53**, 249 (1974).
18. A. C. Scott, F. Y. F. Chu and D. W. McLaughlin, *Proc. IEEE* **61**, 1443 (1973).
19. D. Finkelstein, *J. Math. Phys.* **7**, 1218 (1966).
20. S. Coleman, in *New Phenomena in Subnuclear Physics* ed. A. Z. Zichichi (1977) Plenum, New York.
21. G. Toulouse and M. Kleman, *J. Phys.* (Paris) **37**, L149 (1976).
22. V. P. Mineyev and G. E. Volovik, *Phys. Rev.* **B18**, 3197 (1978).
23. L. D. Landau and E. M. Lifshitz, *Statistical Physics* p. 430 (1959) Pergamon Press, London.
24. A. A. Abrikosov, *Soviet Phys. JETP* **5**, 1174 (1957).
25. J. M. Kosterlitz and D. J. Thouless, *J. Phys.* **C6**, 1181 (1973).
26. J. M. Kosterlitz and D. J. Thouless, 'Two Dimensional Physics' in *Prog. Low Temp. Phys.* vol. 7B ed. D. F. Brewer (1978) North Holland, Amsterdam.
27. W. Döring, *Z. Natur.* **31**, 373 (1948).
28. K. Maki and P. Kumar, *Phys. Rev.* **B16**, 182, 4805 (1977).
29. K. Maki and P. Kumar, *Phys. Rev.* **B17**, 1088 (1978).
30. K. Maki and Y. R. Lin-Liu, *Phys. Rev.* **B17**, 3558 (1978).
31. W. L. McMillan, *Phys. Rev.* **B14**, 1496 (1976).

32. P. Bak and V. J. Emery, *Phys. Rev. Lett.* **36**, 978 (1976).
33. M. J. Rice, A. R. Bishop, J. A. Krumhansl and S. E. Trullinger, *Phys. Rev. Lett.* **36**, 432 (1976).
34. W. P. Su, J. R. Schrieffer and A. J. Heeger, *Phys. Rev. Lett.* **42**, 1698 (1979); *Phys. Rev.* B**22**, 1101 (1980).
35. M. J. Rice, *Phys. Lett.* **71**A, 152 (1979).
36. J. A. Krumhansl and J. R. Schrieffer, *Phys. Rev.* B **11**, 3535 (1975).
37. J. K. Kjems and M. Steiner, *Phys. Rev. Lett.* **41**, 1137 (1978).
38. J. P. Boucher, L. P. Regnault, J. Rossat-Mignod, J. P. Renard, J. Bouillot and W. G. Stirling, *Sol. Stat. Comm.* **33**, 171 (1980).
39. J. P. Boucher and J. P. Renard, *Phys. Rev. Lett.* **52**, 663 (1984).
40. P. A. Lee and T. M. Rice, *Phys. Rev.* B**19**, 3970 (1979).
41. N. P. Ong, G. Verma and K. Maki, *Phys. Rev. Lett.* (in press).
42. R. F. Dashen, B. Hasslacher and A. Neveu, *Phys. Rev.* D**10**, 4114, 4130 (1974); D**11**, 3424 (1975).
43. S. Coleman, *Phys. Rev.* D**11**, 2088 (1975).
44. R. Jackiw, *Rev. Mod. Phys.* **49**, 681 (1977).
45. H. Bergknoff and H. B. Thacker, *Phys. Rev. Lett.* **42**, 135 (1979); *Phys. Rev.* D**19**, 3666 (1979).
46. H. J. Mikeska, *J. Phys.* C**11**, L29 (1978).
47. H. J. Mikeska, *J. Phys.* C**13**, 2913 (1980).
48. K. Maki, *J. Low Temp. Phys.* **41**, 327 (1980).
49. K. M. Leung, D. Hone, D. L. Mills, P. S. Riseborough and S. E. Trullinger, *Phys. Rev.* B**21**, 4017 (1980).
50. A. Seeger, H. Donth and A. Kochendörfer, *Z. Phys.* **134**, 173 (1953).
51. B. D. Josephson, *Phys. Lett.* **1**, 251 (1962); *Adv. Phys.* **14**, 419 (1965).
52. A. C. Scott, *Nuovo Cimento* **69**B, 241 (1970).
53. K. Nakajima, Y. Onodera, T. Nakamura and R. Sato, *J. Appl. Phys.* **45**, 4095 (1974).
54. J. K. Perring and T. H. R. Skyrme, *Nucl. Phys.* **31**, 550 (1962).
55. K. Maki and H. Takayama, *Phys. Rev.* B**20**, 3223 (1979).
56. J. F. Currie, J. A. Krumhansl, A. R. Bishop and S. E. Trullinger, *Phys. Rev.* B**22**, 477 (1980).
57. D. J. Scalapino, M. Sears and R. A. Ferrell, *Phys. Rev.* B**6**, 3409 (1972).
58. N. Gupta and B. Sutherland, *Phys. Rev.* A**14**, 1790 (1976).
59. H. Takayama and K. Maki, *Phys. Rev.* B**21**, 4558 (1980).
60. T. Schneider and E. Stoll, *Phys. Rev.* B**22**, 5317 (1980).
61. K. Sasaki, *Prog. Theor. Phys.* (Kyoto) **68**, 411 (1982).
62. H. Takayama and G. Sato, *J. Phys. Soc. Japan* **51**, 3120 (1982).
63. K. Sasaki and T. Tsuzuki, *J. Mag. Mag. Mat.* **31–34**, 1283 (1983).
64. T. Miyashita and K. Maki, *Phys. Rev.* B**28**, 6733 (1983).
65. J. L. Gervais and B. Sakita, *Phys. Rev.* D**11**, 2943 (1975).
66. K. Maki, *Phys. Rev.* B**24**, 3991 (1981).
67. A. Luther, *Phys. Rev.* B**14**, 2153 (1976).
68. J. D. Johnson, S. Krinsky and B. M. McCoy, *Phys. Rev.* A**8**, 2526 (1973).
69. M. Fowler and X. Zotos, *Phys. Rev.* B**24**, 2634 (1981); B**25**, 5806 (1982).
70. X. Zotos and M. Fowler, *Phys. Rev.* B**26**, 1430 (1982).
71. M. Imada, K. Hida and M. Ishikawa, *Phys. Lett.* **90**A, 79 (1982).
72. M. Fowler, *Phys. Rev.* B**26**, 2514 (1982).
73. X. Zotos, *Phys. Rev.* B**26**, 2519 (1982).
74. M. Steiner, J. Villain and C. G. Windsor, *Adv. in Phys.* **25**, 87 (1976).
75. J. Villain, *Physica* **79B**, 1 (1975).
76. N. Ishimura and H. Shiba, *Prog. Theor. Phys.* (Kyoto) **63**, 743 (1980).
77. H. Yoshizawa, K. Hirakawa, S. K. Satija and G. Shirane, *Phys. Rev.* B**23**, 2298 (1981).
78. S. E. Nagler, W. J. L. Buyers, R. L. Armstrong and B. Briat, *Phys. Rev. Lett.* **49**, 590 (1982); *Phys. Rev.* B**28**, 3873 (1983).
79. K. Maki, *Phys. Rev.* B**24**, 335 (1981).
80. L. R. Walker, R. E. Dietz, K. Andres and S. Derek, *Sol. St. Comm.* **11** 593 (1972).

81. J. P. Boucher, L. P. Regnault, J. Rossat-Mignod and J. Villain, *Sol. St. Comm.* **31**, 311 (1979).
82. L. P. Regnault, J. P. Boucher, J. Rossat-Mignod, J. R. Renard, J. Bouillot and W. G. Stirling, *J. Phys.* C**15**, 1261 (1982).
83. I. Harada, K. Sasaki and H. Shiba, *Sol. St. Comm.* **40**, 29 (1981).
84. L. Gunther and Y. Imry, *Phys. Rev. Lett.* **44**, 1225 (1980).
85. M. Büttiker and R. Landauer, *J. Phys. C.* **13**, L325 (1980).
86. M. Imada, *J. Phys. Soc. Japan* **47**, 699 (1979); **49**, 1247 (1980).
87. J. P. Boucher, H. Benner, F. Devreux, L. P. Regnault, J. Rossat-Mignod, C. Dupas, J. P. Renard, J. Bouillot and W. G. Stirling, *Phys. Rev. Lett.* **48**, 431 (1982).
88. F. Borsa, *Phys. Rev.* B**25**, 3430 (1982).
89. G. Reiter, *Phys. Rev. Lett.* **46**, 202 (1981) (*Errata* **46**, 518).
90. A. P. Ramirez and W. P. Wolf, *Phys. Rev. Lett.* **49**, 227 (1982).
91. X. Zotos, *J. de Phys.* (Paris) C3–477 (1983).
92. R. E. Peierls, *Quantum Theory of Solids* (1955), Oxford University Press, Oxford.
93. H. Fröhlich, *Proc. Roy. Soc.* A**223**, 296 (1954).
94. A. B. Migdal, *Sov. Phys. JETP* **7**, 996 (1958).
95. W. Kohn, *Phys. Rev. Lett.* **2**, 393 (1959).
96. J. T. Devreese, R. P. Evard and V. E. van Doren (eds.) *Highly Conducting One-Dimensional Solids* (1979) Plenum Press, New York.
97. P. A. Lee, T. M. Rice, and P. W. Anderson, *Phys. Rev. Lett.* **31**, 462 (1973).
98. P. A. Lee, T. M. Rice, P. W. Anderson, *Sol. Stat. Commun.* **14**, 703 (1974).
99. S. A. Brazovskii and I. E. Dzyaloshinskii, *Sov. Phys. JETP* **44**, 1233 (1976).
100. J. Bardeen, L. N. Cooper, and J. R. Schrieffer, *Phys. Rev.* **108**, 1175 (1957).
101. K. Maki and M. Sakurai, *Prog. Theor. Phys.* **47**, 1110 (1972).
102. J. Zittartz, *Phys. Rev.* **165**, 605 (1568).
103. B. Mühlschlegel, *Z. für Phys.* **155**, 313 (1959).
104. W. P. Su and J. R. Schrieffer, *Phys. Rev. Lett.* **46**, 738 (1981).
105. T. Takoshima, M. Ido, K. Tsutsumi, T. Sambongi, S. Homma, K. Yamaya and Y. Abe, *Sol. St. Comm.* **35**, 911 (1980).
106. A. L. Fetter and M. J. Stephen, *Phys. Rev.* **168**, 475 (1968).
107. B. Sutherland, *Phys. Rev.* A**8**, 2514 (1973).
108. P. Bak, *Phys. Rev. Lett.* **48**, 692 (1982).
109. M. Weger and B. Horovitz, *Sol. St. Comm.* **43**, 583 (1982).
110. A. J. Heeger and A. G. MacDiarmid in [5].
111. W. P. Su, S. Kivelson and J. R. Schrieffer in [5].
112. D. Baeriswyl, G. Harbeke, H. Kiess, and W. Mayer in *Electronic Properties of Polymers* ed. J. Mort and G. Pfister (1982) John Wiley & Sons, New York.
113. W. P. Su in *Handbook on Conducting Polymers* ed. Terje Skotheim, Marcel Dekker, New York, to be published.
114. A. J. Heeger and J. R. Schrieffer, *Rev. Mod. Phys.*, to be published.
115. I. B. Goldberg, H. R. Crowe, P. R. Newman, A. J. Heeger, and A. G. MacDiarmid, *J. Chem. Phys.* **70**, 1132 (1979).
116. Y. Tomkiewicz, T. D. Schultz, H. B. Brom, T. C. Clarke, and G. Street, *Phys. Rev. Lett.* **43**, 1532 (1979).
117. B. Francois, M. Bernard, and J. J. André, *J. Chem. Phys.* **75**, 4142 (1981).
118. B. R. Weinberger, E. Ehrenfreund, A. Pron, A. J. Heeger, and A. G. MacDiarmid, *J. Chem. Phys.* **72**, 4749 (1980).
119. R. Jackiw and C. Rebbi, *Phys. Rev.* D**13**, 3398 (1976).
120. J. A. Pople and S. H. Walmsley, *Mol. Phys.* **5**, 15 (1962).
121. W. P. Su, *Sol. St. Comm.* **35**, 899 (1980).
122. S. A. Brazovskii, *Sov. Phys. JETP Lett.* **28**, 606 (1979); *Sov. Phys. JETP* **51**, 342 (1980).
123. H. Takayama, Y. R. Lin-Liu, and K. Maki, *Phys. Rev.* B**21**, 2388 (1980).
124. B. Horovitz, *Phys. Rev.* B**22**, 1101 (1980).
125. P. G. de Gennes, *Superconductivity of Metals and Alloys* (1966), Benjamin Inc., New York.

126. K. Maki and M. Nakahara, *Phys. Rev.* B**23**, 5005 (1981).
127. J. T. Gammel and J. A. Krumhansl, *Phys. Rev.* B**24**, 1035 (1981).
128. N. Suzuki, M. Ozaki, S. Etemad, A. J. Heeger and A. G. MacDiarmid, *Phys. Rev. Lett.* **45**, 1483 (1980).
129. A. Feldblum, J. H. Kaufman, S. Etemad, A. J. Heeger, T.-C. Chung and A. G. MacDiarmid, *Phys. Rev.* B**26**, 815 (1982).
130. S. Kivelson, T.-K. Lee, Y. R. Lin-Liu, I. Peschel, and Yu Lu, *Phys. Rev.* **B25**, 4173 (1982).
131. S. A. Brazovskii, S. Gordyunin, and N. N. Kirova, *Soviet Phys. JETP Lett.* **31**, 456 (1980).
132. B. Horovitz, *Phys. Rev. Lett.* **46**, 742 (1981).
133. M. Nakahara and K. Maki, *Phys. Rev.* B**24**, 1045 (1981).
134. W. P. Su and J. R. Schrieffer, *Proc. Natl. Acad. Sci USA* 77, 5626 (1980).
135. S. A. Brazovskii and N. Kirova, *Sov. Phys. JETP Lett.* **33**, 6 (1981).
136. D. K. Campbell and A. R. Bishop, *Phys. Rev.* B**24**, 4859 (1981) and *Nucl. Phys.* B**200**, 297 (1982).
137. R. Dashen, B. Hasslacher and A. Neveau, *Phys. Rev.* D**12**, 2443 (1975).
138. Y. Onodera and S. Okuno, *J. Phys. Soc. Japan* **52**, 2478 (1983).
139. S. Etemad, A. Feldblum, A. G. MacDiarmid, T. C. Chung and A. J. Heeger, *J. Phys.* (Paris) **C–3**, 413 (1983).
140. K. Fesser, A. R. Bishop and D. K. Campbell, *Phys. Rev.* B**27**, 4804 (1983).
141. J. L. Brédas, R. R. Chance and R. Silbey, *Phys. Rev.* B**26**, 5843 (1982).
142. M. Nechtschein, F. Devreux, R. L. Greene, T. Clarke, and G. B. Street, *Phys. Rev. Lett.* **44**, 356 (1980).
143. W. G. Clark, K. Glover, G. Mozurkewich, C. T. Murayama, J. Sanny, S. Etemad and M. Maxfield. *J. de Phys.* (Paris) **C–3**, 293 (1983).
144. K. Maki, *Phys. Rev.* B**26**, 2181; 2187 and 4539 (1983).
145. N. S. Shiren, Y. Tomkiewicz, T. G. Kazyaka, A. R. Taranko, H. Thomann, L. Dalton, and T. C. Clarke, *Sol. St. Comm.* **44**, 1157 (1982).
146. H. Shirakawa, T. Ito and S. Ikeda, *Makromol. Chem.* **179**, 1565 (1978).
147. M. Schwoerer, U. Lauterbach, W. Müller, and G. Wegner, *Chem. Phys. Lett.* **69**, 359 (1980).
148. S. Ikehata, J. Kaufer, T. Woerner, A. Pron, M. A. Druy, A. Sivak, A. J. Heeger, and A. G. MacDiarmid, *Phys. Rev. Lett.* **45**, 1123 (1980).
149. R. Fincher Jr., M. Ozaki, M. Tanaka, D. Peebles, L. Lauchlan, A. J. Heeger, and A. G. MacDiarmid, *Phys. Rev.* B**20**, 1589 (1979).
150. J. F. Rabolt, T. C. Clarke, and G. B. Street, *J. Chem. Phys.* **71**, 4614 (1979).
151. C. R. Fincher, M. M. Ozaki, A. J. Heeger and A. G. MacDiarmid, *Phys. Rev.* B**19**, 4140 (1979).
152. S. Etemad, A. Pron, A. J. Heeger, and A. G. MacDiarmid, E. J. Mele and M. J. Rice, *Phys. Rev.* B**23** 5137 (1981).
153. E. J. Mele and M. J. Rice, *Phys. Rev. Lett.* **45**, 926 (1980).
154. B. Horovitz, *Sol. Stat. Commun.* **41**, 729 (1982).
155. C. K. Chiang, C. R. Fincher Y. W. Park, A. J. Heeger, H. Shirakawa, E. J. Louis, S. C. Gau, and A. G. MacDiarmid, *Phys. Rev. Lett.* **39**, 1098 (1977).
156. Y. W. Park, A. J. Heeger, M. A. Druy, and A. G. MacDiarmid, *J. Chem. Phys.* **73**, 946 (1980).
157. S. Kivelson, *Phys. Rev. Lett.* **46**, 1344 (1981); *Phys. Rev.* B**25**, 3798 (1982).
158. A. J. Epstein, H. Rommelman, M. Abkowitz and H. W. Gibson, *Phys. Rev. Lett.* **47**, 1549 (1981).
159. L. S. Lauchlan, S. Etemad, T.-C. Chung, A. J. Heeger, and A. G. MacDiarmid, *Phys. Rev.* B**24**, 3701 (1981).
160. J. P. Sethna and S. Kivelson, *Phys. Rev.* B**26**, 3513 (1982).
161. S. Zhao-bin and Yu Lu, *Phys. Rev.* B**27**, 5199 (1983).
162. J. Orenstein and G. Baker, *Phys. Rev. Lett.* **49**, 1043 (1982).
163. Z. Vardeny, J. Strait, D. Moses, T.-C. Chung, and A. J. Heeger, *Phys. Rev. Lett.* **49**, 1657 (1982).
164. C. V. Shank, R. Yen, R. L. Fork, J. Orenstein, and G. L. Baker, *Phys. Rev. Lett.* **49**, 1660 (1982).

165. Z. Vardeny, J. Orenstein, and G. L. Baker, *Phys. Rev. Lett.* **50**, 2032 (1983).
166. G. B. Blanchet, C. R. Fincher, T.-C. Chung, and A. J. Heeger, *Phys. Rev. Lett.* **50**, 1938 (1983).
167. G. B. Blanchet, C. R. Fincher and A. J. Heeger, *Phys. Rev. Lett.* **51**, 2132 (1983).
168. J. D. Flood and A. J. Heeger, *J. Phys.* (Paris) C3–397 (1983).
169. R. Ball, W. P. Su, and J. R. Schrieffer, *J. Phys.* (Paris) C3–429 (1983).
170. D. Baeriswyl, *J. de Phys.* C3–381 (1983).
171. M. Nakahara and K. Maki, *Phys. Rev.* B**25**, 7789 (1982).
172. W. P. Su, *Sol. St. Comm.* **42**, 297 (1982).
173. J. E. Hirsch and E. Fradkin, *Phys. Rev. Lett.* **50**, 420 (1982).
174. E. Fradkin and J. E. Hirsch, *Phys. Rev.* B**27**, 1680 (1983).
175. M. Karowski and H. J. Thun, *Nucl. Phys.* B**190**, 61 (1981).
176. H. Fukutome and M. Sasai, *Prog. Theor. Phys.* **67**, 41 (1982); *ibid.* **69**, 1, 373 (1983).
177. K. Takano, T. Nakano and H. Fukuyama, *J. Phys. Soc. Japan* **51**, 336 (1982).
178. C. Aslangul and D. Saint-James, *J. Phys.* (Paris) **44**, 953 (1983).
179. S. Mazumdar and S. N. Dixit, *Phys. Rev. Lett.* **51**, 292 (1983).
180. J. E. Hirsch, *Phys. Rev. Lett.* **51**, 296 (1983).
181. H. Thomann, L. R. Dalton, Y. Tomkiewicz, N. S. Shiren and T. C. Clarke, *Phys. Rev. Lett.* **50**, 553 (1983).
182. K. R. Subbaswamy and M. Grabowski, *Phys. Rev.* B**24**, 2168 (1981).
183. A. Heeger and J. R. Schrieffer, preprint.
184. C. R. Fincher, C. E. Chen, A. J. Heeger, A. G. MacDiarmid, and J. B. Hastings, *Phys. Rev. Lett.* **48**, 100 (1982).
185. D. Baeriswyl and K. Maki, *Phys. Rev.* B**28**, 2068 (1983).
186. P. M. Grant and I. P. Batra, *Sol. St Comm.* **29**, 225 (1979); *J. Phys.* (Paris) C3–437 (1983).
187. D. Moses, A. Feldblum, E. Ehrenfreund, A. J. Heeger, T.-C. Chung, and A. G. MacDiarmid, *Phys. Rev.* B**26**, 3361 (1982).
188. R. H. Baughman and G. Moss, *J. Chem. Phys.* **77**, 6321 (1982).
189. C. S. Yannoni and T. C. Clarke, *Phys. Rev. Lett.* **51**, 1191 (1983).
190. K. Maki, *Synthetic Metals* **9**, 185 (1984).
191. Y. R. Lin-Liu and K. Maki, *Phys. Rev.* B**22**, 5754 (1980).
192. M. J. Rice, *Phys. Rev. Lett.* **51**, 142 (1983).
193. J. Hara, T. Nakano, and H. Fukuyama, *J. Phys. Soc. Japan* **51**, 34 (1982).
194. J. T. Gammel and J. A. Krumhansl, *Phys. Rev.* B**27**, 7659 (1983).
195. B. Sutherland, *Phys. Rev.* B**27**, 7209 (1983).
196. S. A. Brazovskii, N. E. Dzyaloshinskii and I. M. Krichever, *Sov. Phys. JETP* **56**, 212 (1982).
197. M. J. Rice and E. J. Mele, *Phys. Rev. Lett.* **49**, 1455 (1982).
198. B. Horovitz and B. Schaub, *Phys. Rev. Lett.* **50**, 1942 (1983).
199. M. J. Rice, A. R. Bishop and D. K. Campbell, *Phys. Rev. Lett.* **51**, 2136 (1963).
200. J. C. Scott, J. L. Brédas, K. Yakushi, P. Pfluger and G. B. Street, *Synthetic Metals* (in press); J. L. Brédas, J. C. Scott, K. Yakushi and G. B. Street, preprint.
201. M. Kobayashi, J. Chen, T.-C. Chung, F. Moraes, A. J. Heeger and F. Wudl, preprint. T.-C. Chung, H. H. Kaufman, A. J. Heeger, and F. Wudle, preprint.
202. P. Monceau, N. P. Ong, A. M. Portis, A. Meerschaut, and J. Rouxel, *Phys. Rev. Lett.* **37**, 602 (1977).
203. K. Tsutsumi, T. Takagaki, M. Yamamoto, Y. Shiozaki, M. Ido, T. Sambongi, K. Yamaya, and Y. Abe, *Phys. Rev. Lett.* **39**, 1675 (1977).
204. R. M. Fleming and C. C. Grimes, *Phys. Rev. Lett.* **42**, 1423 (1979).
205. N. P. Ong and P. Monceau, *Phys. Rev.* B**16**, 3443 (1977).
206. R. M. Fleming, D. C. Moncton and D. B. McWhan, *Phys. Rev.* B**18**, 5560 (1978).
207. A. H. Thompson, A. Zettl and G. Grüner, *Phys. Rev. Lett.* **47**, 64 (1981); G. Grüner, A. Zettl, W. G. Clark, and A. H. Thompson, *Phys. Rev.* B**23**, 6813 (1981).
208. A. Meerschaut, J. Rouxel, P. Haen, P. Monceau and M. Nunez-Regueiro, *J. Phys. Lett.* **40**, L157 (1979).

209. Z. Z. Wang, M. C. Saint-Lager, P. Monceau, M. Renard, P. Gressier, A. Meerschaut, L. Guemas, and J. Rouxel, *Sol. St. Comm.* **46**, 325 (1983).
210. M. Maki, M. Kaizer, A. Zettl, and G. Grüner, *Sol. St. Comm.* **46**, 497 (1983).
211. J. Dumas, C. Schlenker, J. Marcus and R. Buder, *Phys. Rev. Lett.* **50**, 757 (1983); J. Dumas and C. Schlenker, *Sol. St. Comm.* **45**, 885 (1983).
212. R. M. Fleming, in [5].
213. N. P. Ong, *Can J. Phys.* **60**, 757 (1982).
214. G. Grüner, *Physica* **8D**, 1 (1983).
215. J. W. Brill, N. P. Ong, J. C. Eckert, J. W. Savage, S. K. Khanna and R. B. Somoano, *Phys. Rev.* **B23**, 1517 (1981).
216. N. P. Ong, J. W. Brill, J. C. Eckert, J. W. Savage, S. K. Khanna, and R. B. Somoano, *Phys. Rev. Lett.* **42**, 811 (1979).
217. G. Grüner, L. C. Tippie, J. Sanny, and W. G. Clark, *Phys. Rev. Lett.* **45**, 935 (1980).
218. P. Monceau, J. Richard and J. Renard, *Phys. Rev. Lett.* **45**, 43 (1980); *Phys. Rev.* **B25**, 931 (1982).
219. J. Bardeen, E. BenJacob, A. Zettl, and G. Grüner, *Phys. Rev. Lett.* **49**, 493 (1982).
220. P. Monceau, M. Renard, J. Richard, M. C. Saint-Lager, H. Salva, and Z. Z. Wong, *Phys. Rev.* **B28**, 1646 (1983).
221. W. W. Fuller, P. M. Chaikin, and N. P. Ong, *Sol. St. Comm.* **39**, 547 (1981); W. W. Fuller, G. Grüner, P. M. Chaikin, and N. P. Ong, *Phys. Rev.* **B23**, 6259 (1981).
222. H. Fukuyama and P. A. Lee, *Phys. Rev.* **B17**, 535 (1978).
223. J. C. Gill, *Sol. St. Comm.* **39**, 1203 (1981).
224. A. Zettl and G. Grüner, *Phys. Rev.* **B26**, 2298 (1982).
225. N. P. Ong, *Phys. Rev.* **B17**, 3243 (1978).
226. K. Maki, *Phys. Rev. Lett.* **39**, 46 (1977); *Phys. Rev.* **B18**, 1641 (1978).
227. J. Bardeen, *Phys. Rev. Lett.* **42**, 1498 (1979); **46**, 1978 (1980).
228. J. Bardeen, Lectures for International School of Physics 'Enrico Fermi', Varenna, Italy, July 4–6 (1983).
229. P. B. Littlewood and T. M. Rice, *Phys. Rev. Lett.* **48**, 44 (1982).
230. R. A. Klemm and J. R. Schrieffer, *Phys. Rev. Lett.* **51**, 47 (1983).
231. K. Maki, *Phys. Lett.* **70A**, 449 (1979).
232. Z. Z. Wang, P. Monceau, M. Renard, P. Gressier, L. Guemas, A. Meerschaut, *Sol. St. Comm.* **47**, 439 (1983).
233. G. Grüner, A. Zawadowski and P. M. Chaikin, *Phys. Rev. Lett.* **46**, 511 (1981).
234. J. Richard, P. Monceau, and M. Renard, *Phys. Rev.* **B25**, 948 (1982).
235. L. Sneddon, M. C. Cross and D. S. Fisher, *Phys. Rev. Lett.* **49**, 292 (1982).
236. D. S. Fisher, *Phys. Rev. Lett.* **50**, 1486 (1983).
237. J. B. Sokoloff, *Phys. Rev.* **B23**, 1992 (1981).
238. L. Pietronero and S. Strässler, *Phys. Rev.* **B28**, 5863 (1983).
239. S. E. Barnes and A. Zawakowski, *Phys. Rev. Lett.* **51**, 1003 (1983).
240. N. P. Ong and G. Verma, *Phys. Rev.* **B27**, 4495 (1983).
241. N. P. Ong and G. Verma, Proceedings of the international symposium on 'Nonlinear Transport and Related Phenomena in Inorganic Quasi-One Dimensional Conductors' Sapporo, Japan, October 20–22 (1983).
242. G. Mozurkewich and G. Grüner, *Phys. Rev. Lett.* **51**, 2206 (1983).
243. G. Mozurkewich, M. Maki and G. Grüner, *Sol. St. Comm.* **48**, 453 (1983).

THEORY OF THE SUPERCONDUCTING PROPERTIES OF QUASI-ONE-DIMENSIONAL MATERIALS

RICHARD A. KLEMM

Corporate Research Science Laboratories,
Exxon Research and Engineering Company,
Route 22 East
Annandale, NJ 08801,
U.S.A.

1. Introduction

In recent years, there has been a continuing interest in the superconducting properties of materials that are nearly one-dimensional in their electronic properties [1–22]. A few materials have been found which exhibit strong anisotropy in their electron conduction, with one axis being a much better conduction direction than the others in the normal state. These materials have been of interest due to the strong anisotropies in the superconducting properties they exhibit, and the hope that materials of sufficiently strong superconducting anisotropy might exhibit novel behavior has been growing, as better crystals of highly anisotropic superconductors have been made.

1.1. General Features of a Quasi-One-Dimensional Superconductor

Let us first define what is meant by a quasi-one-dimensional superconductor. First of all, it is a material that is metallic in the normal state, and exhibits a bulk superconducting transition at the critical temperature $T_c > 0$. Below T_c, the material is in the superconducting phase, and exhibits certain properties characteristic of all bulk superconductors: a Meissner effect in which the magnetic flux is excluded for fields less than the critical field H_c (or in the case of Type II superconductors, below the lower critical field H_{c1}), a supercurrent, and an upper critical field H_{c2} at which the superconductivity is destroyed [23–24]. In real systems, it may not always be possible to observe all of these phenomena conclusively, but they must exist in principle.

Secondly, the properties of a quasi-one-dimensional superconductor are very anisotropic. In the normal state, the conductivity in one direction is much larger than in the two other crystal directions. In the superconducting state, the critical fields are anisotropic, with H_{c2} in the highest conduction direction exceeding its value in the other two directions. This anisotropy is similar to that observed in quasi two-dimensional, or layered, superconductors, in which there is a highly conducting plane and hence H_{c2} in that plane exceeds its value normal to the highly conducting plane [25–27]. In addition, the lower critical fields for highly anisotropic materials should exhibit a kink in their angular dependence, as we shall see in Section 3.3.

Thirdly, one must make the distinction between a conventional anisotropic 'bulk' superconductor and a quasi-one-dimensional superconductor. Although both materials

P. Monceau (ed.), Electronic Properties of Inorganic Quasi-One-Dimensional Materials, I, 195–241.

exhibit superconducting order throughout the crystal, an anisotropic bulk superconductor should exhibit an anisotropy of H_{c2} that is relatively temperature independent below T_c, as its temperature dependence in a given field direction is essentially that expected for an isotropic superconductor. A quasi-one-dimensional superconductor, on the other hand, should exhibit 'dimensional crossover' in the temperature dependence of the upper critical field parallel to the highly conducting direction, as in the case of certain intercalated layered compounds and Nb–Ge layered composites [26–27]. In addition, novel behavior in the temperature dependence of H parallel to the conducting axis is expected, as will be discussed later.

It should be noted that the above properties distinguish a quasi-one-dimensional superconductor from two other systems of interest: a purely one-dimensional 'superconductor', and a thin wire, which for widths less than the coherence length exhibits one-dimensional superconducting fluctuations. A purely one-dimensional conductor can, for attractive effective electron–electron interactions, exhibit a tendency for superconducting ordering at low temperature, but true long range order only sets in at $T = 0$ [28]. A thin wire, on the other hand, may be made thick enough to exhibit true long range order. However, unless it is made of highly anisotropic material, it only exhibits critical field anisotropies due to its geometry, and would not exhibit the anomalies in the temperature dependence of H_{c2} and the angular dependence of H_{c1} that are predicted to occur in a quasi-one-dimensional superconductor [29]. This case will be examined in detail in Section 4.2.

To date, it has not been clearly established that any existing materials are truly quasi-one-dimensional superconductors. There have been a number of superconducting materials of strong filamentary anisotropy that have been made and studied, but none of them have exhibited the anomalies in the temperature dependence of H_{c2} that would characterize a truly quasi-one-dimensional superconductor [1–22]. In that sense, the existing materials are perhaps better described as bulk superconductors with strong filamentary anisotropy. By analogy with the temperature dependence of H_{c2} in the layered compounds, the superconductors with filamentary anisotropy such as $(SNBr_{0.4})_x$, $TaSe_3$, and $Tl_2Mo_6Se_6$ are more like the rather anisotropic unintercalated layered materials TaS_2 and $NbSe_2$ than the intercalated ones such as $TaS_2(\text{pyridine})_{1/2}$ [5, 8, 20, 26]. On the other hand, the magnitude of the H_{c2} anisotropy in the latter two materials is similar to that observed in intercalated materials. Hence, it is possible that a more detailed study of the temperature dependence of H_{c2} in these materials might establish truly quasi-one-dimensional behavior. It is also possible in principle that intercalation compounds of these materials might be made. If so, such materials should exhibit the predicted anomalies in the temperature dependence of H_{c2}. Nevertheless, the existing materials are sufficiently anisotropic that the angular dependence of the lower critical field should exhibit a cusp, as has been seen in the unintercalated layered compound $NbSe_2$ [30].

In addition, there is a class of materials which appear to exhibit a great deal of disorder. In particular, $(SN)_x$ appears to be made of many strands or filaments [1]. Also, the samples of $NbSe_3$ which appear to exhibit superconductivity are also strand-like [3, 22]. As we shall see in Section 4, this type of disorder would tend to disguise any anomalies in the critical field behavior characteristic of a true quasi-one-dimensional superconductor.

1.2. PRELIMINARY THEORETICAL CONSIDERATIONS

Before we consider the properties of quasi-one-dimensional superconductors, we will first discuss the occurrence of superconductivity in such materials. There are two basic theoretical approaches to treating filamentary materials. The simplest approach is to assume that the material is similar to ordinary bulk materials except that the Fermi surface is anisotropic. This is often called the 'anisotropic mass' model, as the low energy single electron states are taken to have the form,

$$\epsilon(k_x, k_y, k_z) = \hbar^2 (k_x^2/m_x + k_y^2/m_y + k_z^2/m_z)/2, \tag{1.1}$$

where the three crystal axes are taken to be in the $\hat{x}$, $\hat{y}$, and $\hat{z}$ directions. The advantage of this approach is that it is possible to include any interaction that can be treated in an ordinary isotropic superconductor; among those of interest are Pauli paramagnetism, which can be responsible for the destruction of superconductivity at large magnetic fields, and impurity scattering such as spin-orbit scattering, which can tend to counteract the Pauli pair-breaking under appropriate circumstances. On the other hand, the above approach does not explicitly take the quasi-one-dimensionality into account, as the system is assumed to be inherently three-dimensional in nature [31].

The other basic approach is to treat the system as a lattice of weakly coupled one-dimensional metallic chains. In this model, the single electron states are assumed to be free particle-like along the chains, and of the tight-binding form between chains,

$$\epsilon(k_x, k_y, k_z) = \hbar^2 k_z^2/2m + J_x \cos(k_x a) + J_y \cos(k_y b), \tag{1.2}$$

where J_x and J_y are the single particle energies for tunneling from one chain to the next in the x, y directions, and a and b are the respective lattice constants. The advantage of this approach is that it explicitly treats the quasi-one-dimensionality of the system, and hence the electron–electron interactions along the chain can be treated much better than in a three-dimensional system. In addition, the long-wavelength excitations are of the anisotropic mass form, so that this approach can treat both dimensionality regimes. Unfortunately, this model has the disadvantages that it is difficult to treat the impurity scattering for the technical reason that crossed impurity-averaging diagrams cannot be neglected [32, 33].

In addition, for very small interchain tunneling energies, the model described by Equation (1.2) is complicated by the competition of four types of pairing excitations: in addition to ordinary singlet pairing of two electrons (SS), there can be triplet superconducting pairing (TS), and electron–hole pairings of either the spin-density wave (SDW) or charge-density wave (CDW) types, depending upon the signs of the effective interactions between the charges and spins of the electrons [28]. Basically, when the effective interaction between the charges of two electrons is attractive, the electrons will tend to form pairs. Whether they form singlet or triplet pairs depends upon the sign of the effective interaction between their spins. By particle–hole symmetry, a repulsive interaction between electrons corresponds to an attractive interaction between a particle and a hole. Hence, particle–hole pairings are favored for repulsive interactions between the charges of the electrons. Attractive electronic spin–spin interactions result in the favoring of SDW excitations, and repulsive electronic spin–spin interactions favor CDW excitations. This situation is pictured in Figure 1. It should be emphasized that in a

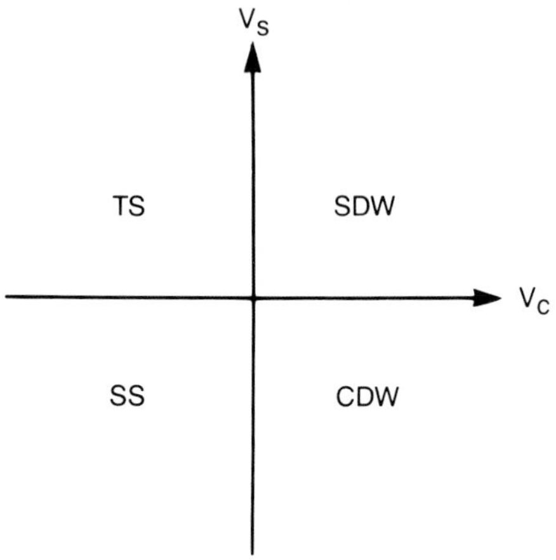

Fig. 1. The regions of different dominant two-body response functions at low temperature for a purely one-dimensional metal are shown. V_C and V_S are the effective interactions between the charges and spins of the bodies, respectively. The four types of ordering are singlet and triplet superconductivity (SS and TS), and spin- and charge-density wave (SDW and CDW) ordering. A more detailed description is given in reference [28].

one-dimensional electron gas, there is no long-range order; these regions in the phase diagram (Figure 1) merely indicate which types of two-particle excitations become important at low temperatures, as for $T > 0$ the one-dimensional system is a normal metal.

Although the phase diagram pictured in Figure 1 is that of a one-dimensional system, it also corresponds to three-dimensional metals that exhibit 'nesting' of the Fermi surface, as pictured in Figure 2. A Fermi surface exhibits 'nesting' if there is a particular wave

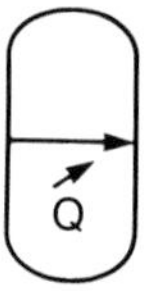

Fig. 2. An idealized Fermi surface cross-section exhibiting nesting over a finite fraction of its area is pictured.

vector **Q** that spans a finite fraction of the Fermi surface. In practice, it is not really necessary for a single wave vector to exactly span a finite fraction of the Fermi surface; a near-spanning of the Fermi surface over a few percent of the Fermi surface is usually sufficient for particle–hole excitations to be important, provided that the interactions are of the appropriate sign [34–35].

As is indicated in Figure 1, there are regions in the phase diagram in which more than one type of pairing can be important. In particular, in the lower half of the diagram, there is a region in which both SS and CDW excitations are important. Assuming that interchain interactions can drive the CDW excitations in this region to a long-range CDW phase transition, it may still be possible for a superconducting transition to occur at a lower temperature, if the CDW transition creates a gap over only part of the Fermi surface (i.e., the region of 'nesting'), leaving a finite fraction of the Fermi surface available for superconducting pairing. This phenomenom is well-known in the layered compounds $NbSe_2$ and TaS_2 [34–36], and appears to play a role in the filamentary material $NbSe_3$ as well. In $NbSe_3$, there are two incommensurate CDWs that form at high temperatures [37], and at low temperatures the system appears to exhibit evidence for a tendency towards superconductivity [3]. In addition, in the region in which TS excitations predominate, SS excitations are also important. In the absence of impurities, one would expect that interchain tunneling would prefer to drive a triplet rather than a superconducting transition. However, as triplet pairing is suppressed by normal as well as magnetic impurities, unless the crystal were made extremely pure, it is doubtful that a triplet transition would occur. On the other hand, singlet pairing is not substantially suppressed by normal impurities; hence in this region of the phase diagram, it is most likely that if a superconducting transition occurs, it is of the conventional BCS [38], or singlet, type.

Finally, we note that a lattice of coupled chains, a CDW transition can be driven by either interchain Coulomb interactions (i.e., interchain backscattering) or by interchain tunneling. Since superconducting pairing is not enhanced by interchain electron-electron interactions (unless interchain pairing is important), a superconducting transition can only be driven by interchain tunneling [32].

2. Anisotropic Mass Model

The simplest model of an anisotropic superconductor is represented by the standard Ginzburg–Landau (GL) free energy, modified to allow for anisotropy in the elastic energy of the order parameter [23, 39, 40]

$$F = \int d^3 r \left\{ \alpha |\psi|^2 + \beta |\psi|^4/2 + b^2/8\pi + \sum_{\mu} |(-i\partial_\mu + 2ea_\mu)\psi|^2/(2m_\mu) \right\}, \quad (2.1)$$

where $\mathbf{b}(r) = \nabla \times \mathbf{a}$ is the microscopic magnetic induction, $\mathbf{a}$ is the vector potential, e is the magnitude of the electronic charge, α and β are the usual Ginzburg–Landau parameters, $\psi(r)$ is the order parameter, $\partial_\mu \equiv \partial/\partial x_\mu$, x_μ is x, y, z for $\mu = 1, 2, 3$, respectively, and the integration extends over the volume V of the sample. We use the natural units $\hbar = c = k_B = 1$.

Equation (2.1) may be derived from microscopic theory, using the single particle states represented by Equation (1.1) [24]. It is valid for temperatures close to T_c ($|T - T_c|/T_c \ll 1$), and for fields small with respect to the Pauli-limiting field $H_p = \sqrt{2}T_c/\mu_B$, where μ_B is the Bohr magneton [41]. The third term in Equation (2.1) is the magnetic energy term, which is not important for calculating H_{c2}, but is important for calculating the other critical fields. In the absence of a field, the order parameter inside the sample may be taken to be the constant that minimizes the free energy, $\psi(r) = 0$ for $\alpha, \beta > 0$, and $\psi(r) = \psi_0 = (-\alpha/\beta)^{1/2}$ in equilibrium for $\alpha < 0, \beta > 0$. We always must have $\beta > 0$ for the superconducting state to be stable. This is pictured in Figure 3. Since

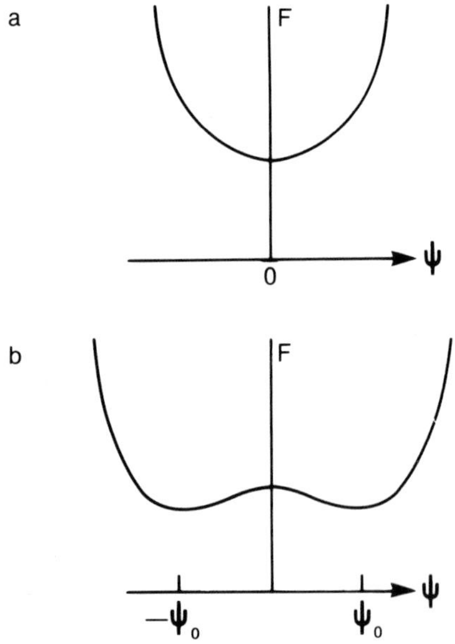

Fig. 3. Shown are schematic plots of the Ginzburg–Landau free energy for a spatially constant order parameter ψ. In (a), $T > T_c$. In (b), $T < T_c$, giving minima at $\pm\psi_0$.

$\psi(r)$ is the order parameter, we expect its mean-field value to be zero above T_c and non-zero below T_c. Assuming that the Ginzburg–Landau parameters α and β can be expanded in powers of $T - T_c$, this implies $\alpha \cong \alpha_0 (T - T_c)$ and $\beta \cong$ constant >0 near to T_c. This temperature dependence of the Ginzburg–Landau parameters is consistent with that obtained from microscopic theory near to T_c [24].

2.1. THE UPPER CRITICAL FIELD

The upper critical field is the largest field for which superconductivity can exist. In a homogeneous material, it is thus also the field at which the superconductivity is destroyed. It may be calculated by treating the magnetic field as uniform, and assuming ψ_0 to be very small. These assumptions are valid since the superconductivity is destroyed everywhere, and at the temperature for which a given field destroys the superconductivity, the order parameter vanishes. Hence, the upper critical field is a thermodynamic property of the specimen [23].

We assume the magnetic field (which at H_{c2} is equal to the magnetic induction) is constant in magnitude and directed arbitrarily with respect to the crystal axes,

$$\mathbf{b} = b(\sin\theta\cos\phi, \sin\theta\sin\phi, \cos\theta). \tag{2.2}$$

We now express all quantities in dimensionless form. First, we define a coherence length $\xi(T)$ and a penetration depth $\lambda(T)$ by [39]

$$\xi(T) = (-2m\alpha)^{-1/2} \tag{2.3}$$

and

$$\lambda(T) = (m/16\pi e^2 \psi_0^2)^{1/2}, \tag{2.4}$$

where

$$m = (m_1 m_2 m_3)^{1/3}. \tag{2.5}$$

We then use the conventional normalization [23] in which length is measured in units of λ, magnetic field in units of $\sqrt{2}H = \Phi_0/(2\pi\xi\lambda)$, vector potential in units of $\lambda\sqrt{2}H = \Phi_0/(2\pi\xi)$, and energy density in units of $H^2/(4\pi) = \Phi_0^2/(32\pi^3\xi^2\lambda^2) = \alpha^2/\beta$, where the flux quantum $\Phi_0 = \pi/e$. The reduced free energy is then given by

$$F = \int d^3r \left\{ -|f|^2 + |f|^4/2 + b^2 + \sum_\mu (m/m_\mu)|(-i\partial_\mu/\kappa + a_\mu)f|^2 \right\}, \tag{2.6}$$

where $f = \psi/\psi_0$ is the reduced order parameter and $\kappa = \lambda/\xi$ is the effective Ginzburg–Landau parameter. The cross-term in the last term in Equation (2.6) may be eliminated by making a gauge transformation,

$$f = f_0 \exp(i\gamma) \tag{2.7}$$

and choose a to eliminate the superfluid velocity,

$$\mathbf{a}_0 = \mathbf{a} + \nabla\gamma/\kappa. \tag{2.8}$$

The free energy may now be written as

$$F = \int d^3r \left\{ -f_0^2 + f_0^4/2 + \sum_\mu (m/m_\mu)[(\partial_\mu f_0)^2/\kappa^2 + a_0^2 f_0^2] + b^2 \right\}. \tag{2.9}$$

Equation (2.9) and $\mathbf{b} = \text{curl}\,(\mathbf{a}_0)$ may now be used to calculate the upper and lower critical fields. The lower critical field will be found in Section 2.2. Because of its simplicity, the upper critical field may be found in a number of ways. The procedure that follows is probably the most straightforward.

The magnetic vector potential $\mathbf{a}_0$ is taken to be of the form,

$$\mathbf{a}_0 = b\,\{z \sin\theta \sin\phi - y\cos\theta, x\cos\theta - z\sin\theta\cos\phi, y\sin\theta\cos\phi - \\ - x\sin\theta\sin\phi\}/2, \tag{2.10}$$

which clearly satisfies $\mathbf{b} = \text{curl}(\mathbf{a}_0)$. Other forms for $\mathbf{a}_0$ satisfying this requirement are of course also acceptable, but the above form is chosen because of its symmetry. The upper critical field may now be calculated by minimizing the free energy as given by Equation (2.9) with respect to variations in ψ, neglecting the resulting term of order ψ^3 which is vanishingly small,

$$\sum_{\mu} \{-\partial_\mu{}^2 f_0/\kappa^2 + a_0^2 f_0\}/m_\mu - f_0/m = 0. \tag{2.11}$$

Equation (2.11) may be transformed into a one-dimensional harmonic oscillator form by first performing an anisotropic scale transformation to equalize the coefficients of the derivative terms, and then rotating the axes so that one of them is parallel to **b**. In performing these transformation, the Maxwell relation $\mathbf{b} = \text{curl}\ \mathbf{a}_0$ must be preserved. At $b = H_{c2}$, **b** is independent of position in the specimen, and the Maxwell relation will automatically be satisfied.

We first transform the scale of the system by

$$x_\mu = x'_\mu (m/m_\mu)^{1/2}. \tag{2.12}$$

This changes Equation (2.11) to the form

$$-\nabla^2 f_0/\kappa^2 + \{(b'_y z' - b'_z y')^2/4 + (b'_z x' - b'_x z')^2/4 + \\ + (b'_x y' - b'_y x')^2/4\} f_0 - f_0 = 0, \tag{2.13}$$

where the scaled components of the magnetic induction b' are given by

$$b'_\mu = b_\mu (m_\mu/m)^{1/2}, \tag{2.14}$$

and where the components of b are given by Equation (2.2). We then rotate about the z' axis by an angle $-\phi'$,

$$\begin{aligned} x' &= x'' \cos\phi' - y'' \sin\phi' \\ y' &= y'' \cos\phi' + x'' \sin\phi' \\ z' &= z''. \end{aligned} \tag{2.15}$$

The phase ϕ' is then chosen to eliminate the term proportional to $x''y''$. One such possible choice is

$$\tan\phi' = b'_y/b'_x. \tag{2.16}$$

This choice of ϕ' also eliminates the term proportional to $y''z''$. The axes are now rotated by an angle $-\theta'$ about the y'' axis,

$$\begin{aligned} x &= x'' \cos\theta' + z'' \sin\theta' \\ z &= z'' \cos\theta' - x'' \sin\theta' \\ y &= y''. \end{aligned} \tag{2.17}$$

The phase θ' is chosen to eliminate the resulting xz term, which implies that

$$\cos\theta' = b'_z/[b\alpha(\theta, \phi)], \tag{2.18}$$

where the anisotropy factor $\alpha(\theta,\phi)$ is given by

$$\alpha(\theta,\phi) = \left[\sum_\mu b'^2_\mu\right]^{1/2}/b = \left[\sum_\mu (m_\mu/m)(\hat{e}_\mu\cdot\hat{b})^2\right]^{1/2}, \tag{2.19}$$

with the $\hat{e}_\mu\cdot\hat{b}$ being the direction cosines of b as given by Equation (2.2). This choice of the phase θ' diagonalizes the Ginzburg–Landau equation,

$$-\nabla^2 f_0/\kappa^2 + \alpha^2(\theta,\phi)(x^2+y^2)f_0/4 = f_0. \tag{2.20}$$

Clearly, this is the Schroedinger equation for an harmonic oscillator with cylindrical symmetry, and may readily be solved. The upper critical field is the largest field for which superconductivity can persist, and it is easily seen that in this case that it is obtained from the lowest eigenvalue of Equation (2.20),

$$b = \kappa/\alpha(\theta,\phi). \tag{2.21}$$

In Figure 4, we have plotted the anisotropy of H_{c2} as given by Equation (2.21) for the

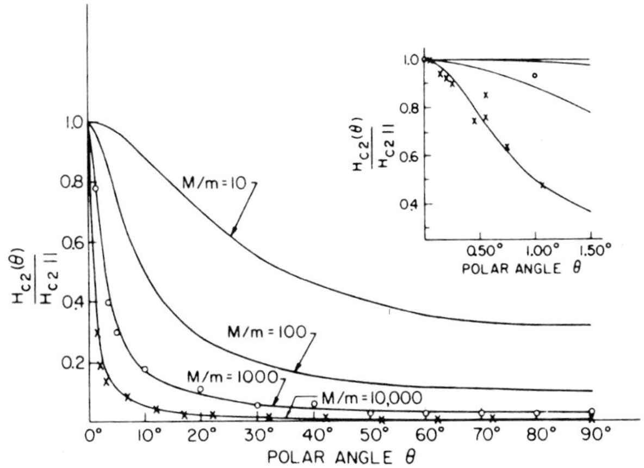

Fig. 4. Polar dependence of the upper critical field for various effective mass ratios [50]. Plotted also are the data points × for $(SN)_x$ at T = 0.255 K [2], and the data points ° for $TaSe_3$ at T = 1.73 K [5].

field perpendicular to one of the crystal axis directions, for four different effective mass anisotropies. For comparison, data on $(SN)_x$ and $TaSe_3$ have been included. In Figures 5 and 6, the angular dependence of H_{c2} as given by Equation (2.21) is compared with data on $TaSe_3$ for two different azimuths. Finally, in Figure 7, the angular dependence of H_{c2} is compared with data on $Tl_2Mo_6Se_6$. The agreement is generally quite good.

The temperature dependence of H_{c2} as given by Equation (2.21) is linear, since κ is roughly temperature independent, and the magnetic field is measured in the standard fashion, indicated previously. As the GL theory is valid for small values of the magnetic field and $(T_c - T)$, we expect deviations from linearity at low temperatures. At low

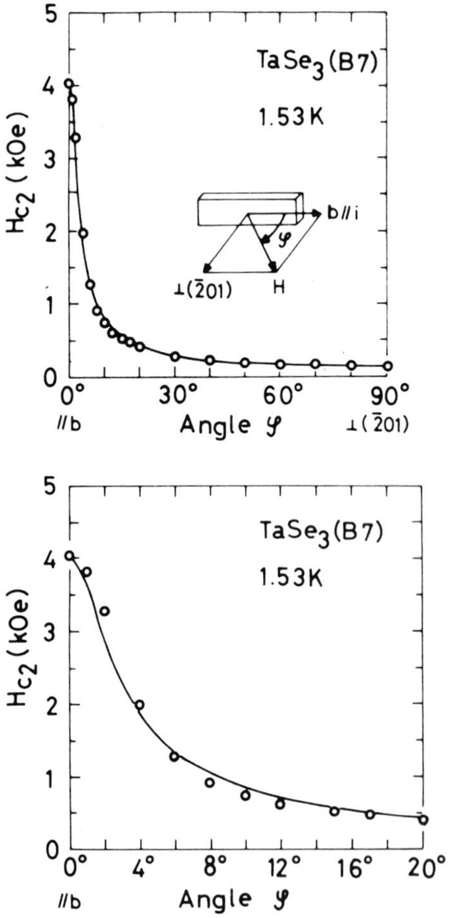

Fig. 5. Angular dependence of the upper critical field in $TaSe_3$ at 1.53 K [5]. The angle ϕ in this figure is the angle from the b axis in the plane including the b axis and the normal to the (201) plane. The solid line is the anisotropic mass formula (Equation (2.21)) with an effective mass anisotropy of 700.

temperatures, the temperature dependence of H_{c2} for an isotropic superconductor flattens out, as predicted by Abrikosov and Gor'kov [42]. In Figure 8, the temperature dependence of H_{c2} for the parallel and perpendicular directions in the doped sample $Nb_{1-x}Ta_xSe_3$ is shown. The linear temperature dependence near to T_c is as expected, and the Abrikosov–Gor'kov theory fits quantitatively. In Figure 9, H_{c2} for the direction perpendicular to the filament is plotted for $(SN)_x$ and $(SNBr_{0.4})_x$. The brominated sample has the expected temperature dependence for a bulk material, exhibiting the linear temperature dependence near to T_c, and bending over at low temperatures. The sample of $(SN)_x$ studied, however, has a very different temperature dependence. This upward curvature of H_{c2} is not well understood (it is apparently not 'dimensional cross-over', which will be discussed in Section 3), but it is similar to the upward curvature

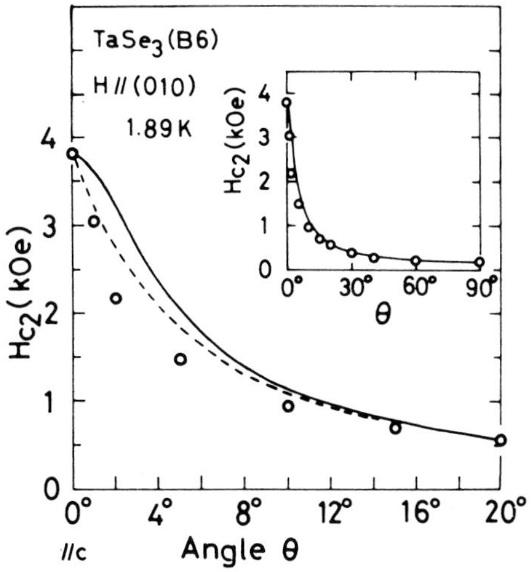

Fig. 6. Polar (θ) dependence of H_{c2} in $TaSe_3$ at 1.53 K. The angle θ is the angle between the b and c axes. The solid line is the anisotropic mass formula with a mass anisotropy of about 200. The dashed curve is a fit to the Tinkham thin film formula [57]. The data is for a different sample from that in Figure 5 [5].

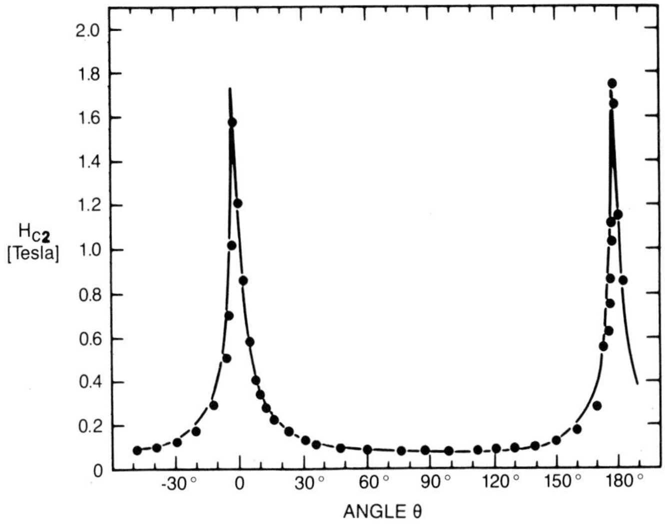

Fig. 7. Angular dependence of H_{c2} for $Tl_2Mo_6Se_6$ (after ref. [20]).

observed in the layered superconductors for the field perpendicular to the layers [26]. This anomalous temperature dependence near to T_c seems to be a general feature of many, but not all highly anisotropic materials.

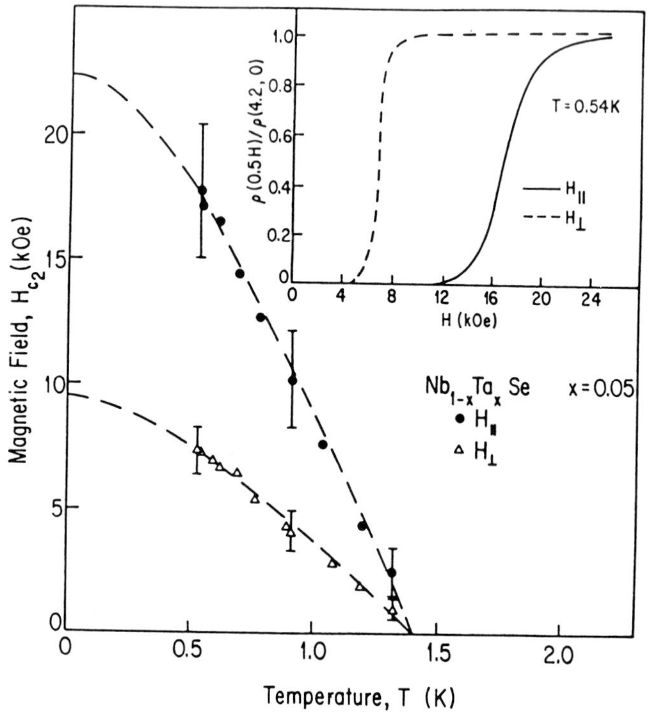

Fig. 8. Temperature dependence of H_{c2} in the doped sample $Nb_{1-x}Ta_xSe_3$ for $x = 0.05$, measured resistively [10]. The dashed curves are fits to the Abrikosov–Gorkov formula [42].

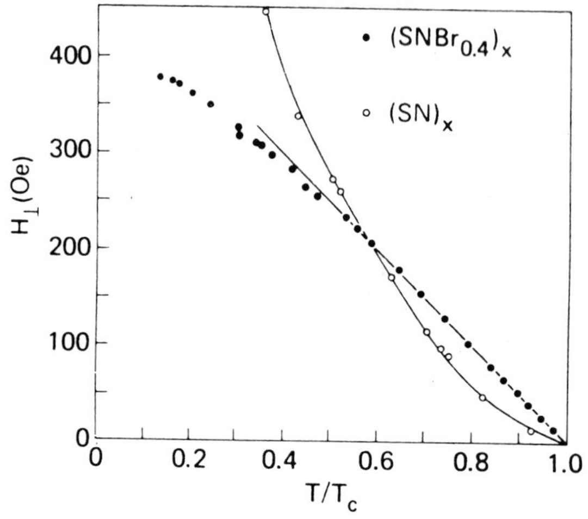

Fig. 9. Temperature dependence of the perpendicular H_{c2} in $(SN)_x$ and $(SNBr_{0.4})_x$. The data are for a single crystal before and after bromination, and the solid lines are guides to the eye [14].

2.2. THE LOWER CRITICAL FIELD

We now focus upon the lower critical field in the anisotropic mass model. Since below the upper critical field H_{c2}, the magnetic induction is not a constant, and the order parameter is not vanishingly small, one must keep the spatial dependence of the magnetic induction and the non-linear term in the free energy. Since the lower critical field is the field at which a single magnetic vortex enters the superconductor, we expect the magnetic induction to be largest in magnitude in the vortex core, and to vanish far from the vortex core. The order parameter, on the other hand, vanishes at the center of the vortex core, and reaches the constant value ψ_0 far from the vortex core. In general, the magnetic induction just outside the core will not be constant in direction. However, it is difficult to solve for the magnetic induction if both its magnitude and direction are allowed to vary in the sample. It is possible to solve for H_{c1} analytically if one assumes that only the magnitude of the magnetic induction can vary in the sample; i.e., the direction of the magnetic induction is taken to be a constant [39],

$$\mathbf{b}(r) = b(r)(\sin\theta\cos\phi, \sin\theta\sin\phi, \cos\theta). \tag{2.22}$$

We then transform the free energy in exactly the same way as the linearized Ginzburg–Landau equation was previously transformed. After performing the scale and rotation transformation [Equations (2.12), (2.15) and (2.17)], the free energy is reduced to

$$F = \int d^3r \, \{-f_0^2 + f_0^4/2 + (\nabla f_0)^2/\kappa^2 + a_0^2 f_0^2 + b^2/\alpha^2(\theta, \phi)\}. \tag{2.23}$$

As we might expect from the calculation of H_{c2}, which only depends upon the single parameter κ/α, the free energy can be further simplified by performing an isotropic scale transformation.

$$\begin{aligned} x_\mu &\to x_\mu/\alpha \\ \partial_\mu &\to \alpha\partial_\mu \\ a_{0\mu} &\to a_{0\mu} \\ b_\mu &\to \alpha b_\mu . \end{aligned} \tag{2.24}$$

This transformation satisfies Maxwell's equation, as required. The free energy becomes

$$F = \alpha^{-3} \int d^3r \, [-f_0^2 + f_0^4/2 + b^2 + a_0^2 f_0^2 + \tilde{\kappa}^{-2}(\nabla f_0)^2], \tag{2.25}$$

where

$$\tilde{\kappa} = \kappa/\alpha. \tag{2.26}$$

The free energy is now just a function of $\tilde{\kappa}$ in this approximation, and the thermodynamic properties can thus be calculated. As we found previously, the upper critical field is given by $\tilde{\kappa}$, as should be expected since at H_{c2}, the direction and magnitude of the magnetic induction are constant. The lower critical field $\mathbf{H}_{c1}$, however, is not parallel to the magnetic induction except in an isotropic system, since in thermodynamic equilibrium the magnetic field $\mathbf{H}(\mathbf{B})$ is given by [23]

$$\mathbf{H}(\mathbf{B}) = \nabla_{\mathbf{B}}\mathbf{F}(\mathbf{B})/2, \tag{2.27}$$

where H and B are the macroscopic field and induction, respectively, and $\mathbf{F}(\mathbf{B}) = F/V + 1/2$ is the Helmholtz free energy per unit volume of the superconducting relative to the Meissner state. As the Helmholtz free energy is the same as for an isotropic material with the replacement $\kappa \to \tilde{\kappa}$, the component of $\mathbf{H}_{c1}$ parallel to $\mathbf{B}$ will have the same functional dependence upon $\tilde{\kappa}$ as an isotropic material has upon κ. In the large $\tilde{\kappa}$ limit [43],

$$H_{\parallel \mathbf{B}} \cong (2\tilde{\kappa})^{-1}(\ln \tilde{\kappa} + 0.497). \tag{2.28}$$

From Equation (2.26), we obtain in the limit $B \to 0$,

$$\mathbf{H}_{c1}(\theta, \phi) = [\hat{B} + \hat{\theta}\,\partial/\partial\theta + \hat{\phi}(\sin\theta)^{-1}\,\partial/\partial\phi]\, H_{\parallel \mathbf{B}}(\theta, \phi). \tag{2.29}$$

In Figure 10, we have plotted the parallel and perpendicular components of $\mathbf{H}_{c1}$, along

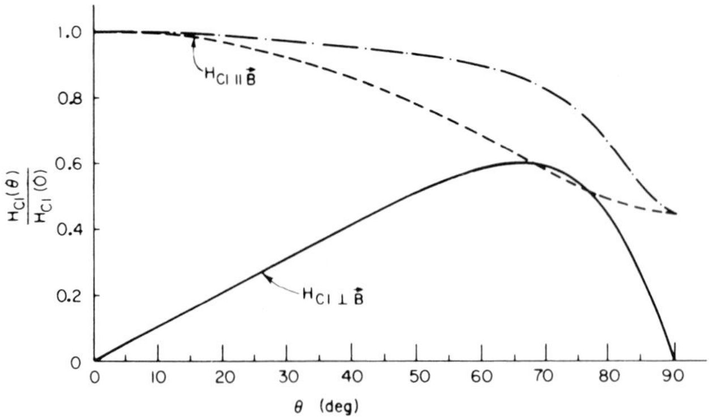

Fig. 10. Shown are the components of H_{c1} parallel and perpendicular to $\mathbf{B}$, and the magnitude of $\mathbf{H}_{c1}$, as a function of the angle θ that $\mathbf{B}$ makes with the normal to the filament ($\hat{z}$) axis, for $\kappa = 10$ and effective mass anisotropy $\epsilon = 0.1$ [39].

with its magnitude, for an anisotropic material of uniaxial symmetry with effective mass ratio $\epsilon = 0.1$, as a function of θ, the angle $\mathbf{B}$ makes with the normal to the filament ($\hat{z}$) axis. The curve for the magnitude of $\mathbf{H}_{c1}$ is redrawn in Figure 11, where we have plotted the dependence of H_{c1} on the angle θ that the field $\mathbf{H}$ makes with the normal to the filament axis. In this figure, curves for $\epsilon = 0.01$ and 0.001 are also shown. For those two curves, there is a region where the free energy is multiple-valued. The value of H_{c1} that is favored is the lowest value, and thus the angular dependence of H_{c1} will exhibit

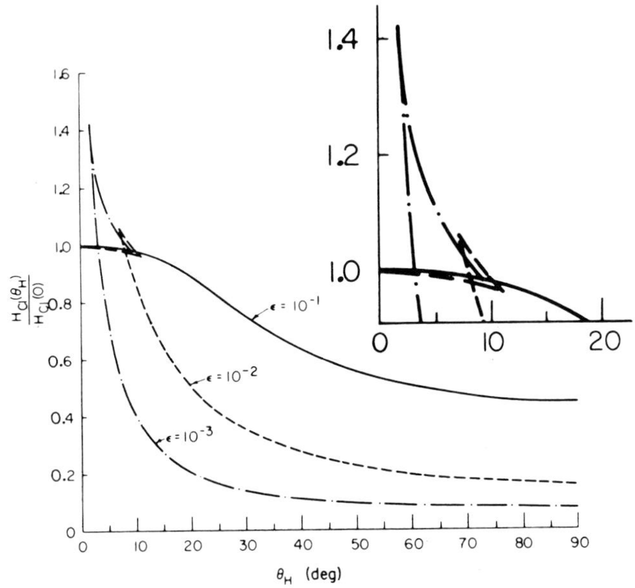

Fig. 11. The lower critical field H_{c1} for materials with filamentary symmetry with $\kappa = 10$, and effective mass ratios $\epsilon = 0.1, 0.01$, and 0.001, as a function of θ_H, the angle **H** makes with the normal to the filament ($\hat{z}$) axis. The inset is an enlargement of the upper-left-hand portion of the figure [39].

a kink for sufficient anisotropy [39, 44]. We note that the lower critical field in a real material is dependent upon the crystal shape, due to demagnetization effects [45]. Since nearly all of the quasi-one-dimensional superconductors are filamentary in shape, they can be approximated by elongated ellipsoids. In Figure 12, we have plotted the dependence of H_{c1} upon the external angle θ_e for the same uniaxial anisotropies as in Figure 11, but for a material that is much longer than it is wide, as is often the case experimentally [45]. We note that for the external field parallel to the long axis, there is no demagnetization correction. For the field perpendicular to the long axis, the demagnetization factor changes H_{c1} by a relative factor of 2.

Although there have not yet been any experiments on the angular dependence of the lower critical field in inorganic quasi-one-dimensional superconductors, there have been measurements on closely related materials. Schwenk, Andres, and Wudl [46] have recently measured the angular dependence of the lower critical field in the organic quasi-one-dimensional superconductor $(TMTSF)_2ClO_4$ in the basal plane, where demagnetization effects are not very pronounced. Their results are shown in Figure 13, and compared with the expected angular dependence calculated from the anisotropic mass theory assuming (as above) the direction of **B** to be a constant. Considering the difficulty in measuring H_{c1} precisely, the agreement appears to be reasonably good. There does not appear to be any obvious evidence of the kink in the angular dependence of H_{c1} predicted in the above discussion, but the experiments do not rule it out, either.

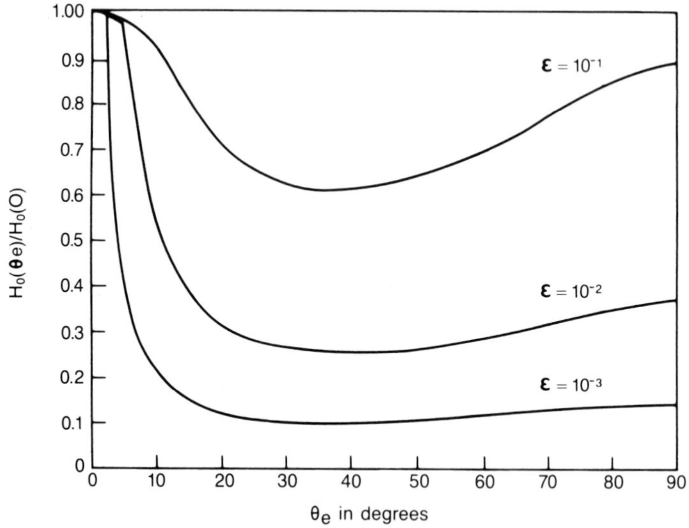

Fig. 12. The lower critical field H_{c1} for materials with filamentary symmetry with the length/width → ∞, as a function of θ_e, the angle the applied magnetic field makes with the normal to the filament ($\hat{z}$) axis, for various anisotropies.

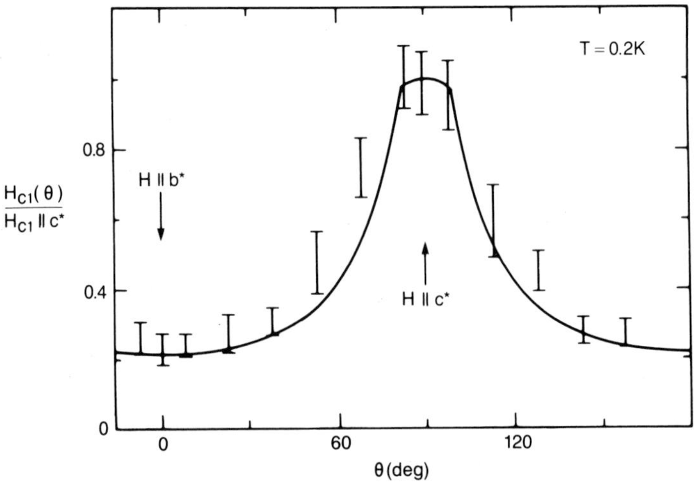

Fig. 13. The lower critical field for $(TMTSF)_2ClO_4$ in the basal plane, as a function of the angle the applied magnetic field makes with the $\hat{b}^*$ axis [46]. The solid line is a fit to the Klemm–Clem formula, given by Equation (2.28 and 2.29).

In addition, there have been measurements on the related material $NbSe_2$ by Denhoff and Gygax [30]. This material has a strong, uniaxial anisotropy, but is layered in structure. Although their measurements did not agree quantitatively with the above theory, they did observe a kink in the angular dependence of H_{c1}. They were able to quantitatively fit their data by assuming that the direction of **B** was either parallel or perpendicular to the layers, and treating the value of H_{c1} in the parallel direction as an adjustable parameter. However, as their restriction upon the direction of **B** is even more stringent than in the above discussion, it must correspond to a higher free energy. Although their experiments were very carefully performed, there appeared to be a great deal of flux pinning, which was most pronounced for an applied field in the direction of the kink [30].

In the above discussion, it was assumed that the direction of the magnetic induction **B** was a constant throughout the sample. This assumption has been questioned by Kogan [47] and by Kogan and Clem [48], who showed that the bulk free energy is minimized if the direction of **B** is not a constant. Exactly at H_{c2}, **B** is a constant in the sample, both in magnitude and in direction. Although the above discussion allows for the magnitude of **B** to vary locally in the sample (as it surely must near H_{c2}, where a single vortex has entered the sample, and the magnitude of **B** must decrease substantially as one moves away from the vortex core), it assumes that the direction of **B** is also constant in the sample. This assumption is strictly speaking only valid in the following limits [48]: (1) at H_{c2}; (2) when the vortex is parallel to a crystal axis; and (3) for $\tilde{\kappa} \gg 1$. At H_{c2}, the order parameter is a (small) constant, and the field is uniform, as we assumed above. However, Kogan and Clem showed that just below H_{c2}, there is a current parallel to the vortex axis in an anisotropic material, unless the vortex axis is parallel to a crystal axis. Unfortunately, they were not able to obtain the angular dependence of the magnetization near to H_{c2} in closed form [48].

Let us examine the conditions under which the vortex axis will be essentially parallel to a crystal axis. Assuming the direction of **B** to be a constant, as we have above, the kink in the angular dependence of H_{c1} predicted for highly anisotropic materials is due to a switching of the direction of **B** from being nearly parallel to one crystal axis to an angle nearly parallel to another crystal-axis. That is, the energy is lowered if the vortices lie parallel to the most highly conducting axis. It costs energy, however, if the vortices differ substantially from the direction of **H**. Hence, for applied fields nearly perpendicular to the most conducting axis, the vortices are forced to lie nearly parallel to **H**. However, as the direction of **H** is rotated towards the most conducting direction, the direction of **B** is always closer to the most conducting direction, and at a critical angle, jumps to an angle very close to the conducting axis direction. The position of this critical angle depends upon the anisotropy of the material. As can be seen from Figures 11 and 12, for anisotropy factor $\epsilon = 0.1$, there is no critical angle if the direction of **B** is restricted to being a constant. For very high anisotropies, however, the critical angle for **H** is very close to perpendicular to the most conducting axis. That is, **B** lies essentially parallel to the most conducting axis unless the applied field is very close to perpendicular to that direction, in which case **B** switches suddenly to the perpendicular direction.

The parameter $\tilde{\kappa}$ measures the stiffness of the vortices. For $\tilde{\kappa}$ small, it does not cost much energy to bend the vortex lines. For $\tilde{\kappa} \gg 1$, however, it costs a great deal of energy to bend the vortex lines. In the above discussion, it was found that $\tilde{\kappa}$ was a strong function of the angle of **B**. In the fits to the experiments that we mentioned, however, the minimum values of $\tilde{\kappa}$ were on the order of 10. This minimum value occurs for the vortex near to perpendicular to the most conducting axis. For angles near to the conducting axis, $\tilde{\kappa}$ is considerably larger, at least for highly anisotropic materials. Hence, for these angles, **B** is nearly parallel to the most conducting axis, and hence nearly constant in direction. For applied fields near to perpendicular to the most conducting axis, the non-uniformity of the direction of **B** is the most pronounced.

Finally, the temperature dependence of the lower critical field in the GL model is linear, as for H_{c2}. Figure 14 exhibits data for the perpendicular 'lower critical field' in

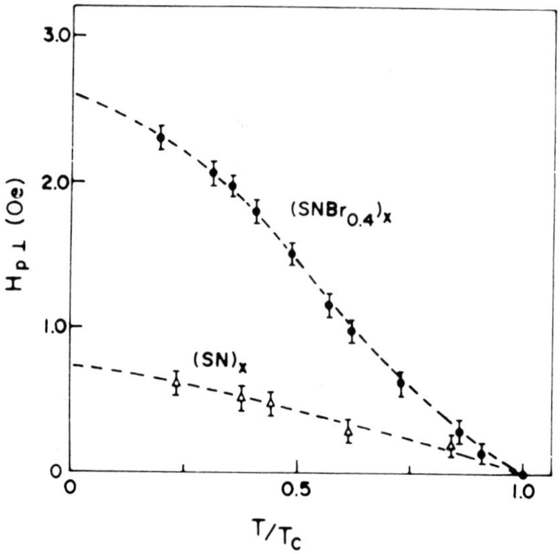

Fig. 14. Lower critical field as determined from the maxima in the magnetization curves for $(SN)_x$ and $(SNBr_{0.4})_x$ for fields in the direction perpendicular to the filaments, as a function of temperature. From [8].

$(SN)_x$ and in $(SNBr_{0.4})_x$, as determined from the maximum in the magnetication curves (actually, the appropriate definition of H_{c1} is the field at which the magnetization curve first departs from linearity, but in these materials such well-defined points were difficult to ascertain [8]. Hence, the points plotted are really upper limits upon H_{c1}).

3. Josephson-Coupled Chain Model

3.1. THE MODEL

The model for the quasi-one-dimensional conductors that explicitly treats the lower dimensionality of the system arises from a single particle energy of the form given by Equation (1.2). That is, the crystal is assumed to be made up of a rigid lattice of chains, and that the electrons propagate from chain to chain by single particle tunneling. By analogy with the layered superconductors, the superconducting properties of these systems near to T_c and for relatively small fields can be understood in terms of a Josephson-coupled model. That is, the order parameters on adjacent chains are coupled by Josephson tunneling, as in the Lawrence–Doniach theory [49] of coupling of the order parameters on adjacent layers of a layered superconductor [25, 47].

In a layered superconductor, when a magnetic field is applied perpendicular to the layers, the Cooper pairs form Landau orbits within a given layer, and hence do not feel any interlayer phase difference. Thus, the vortex currents circulate freely within the layers (Figure 15a), and the normal vortex cores penetrate all of the layers, causing the

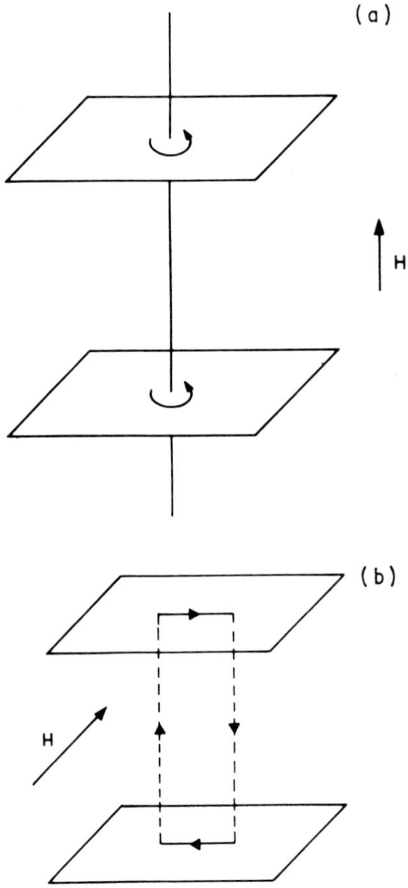

Fig. 15. (a). Field applied perpendicular to the layers in a layered superconductor. The vortex currents circulate within the layers. (b) Field parallel to the layers in a layered superconductor. The vortex currents partially circulate within the layers and partially Josephson tunnel from layer to layer. From [50].

perpendicular upper critical field H_{c2} to be essentially that of a bulk superconductor. However, when the magnetic field is applied parallel to the layers, the vortex currents complete orbits by both circulating within and tunneling between two different layers, as shown in Figure 15b. The system behaves as a series of coupled Josephson junctions.

There is a dimensional-crossover temperature T^*, at which the vortex currents propagate only between nearest-neighbor planes, the normal cores of the vortices (with diameter $\xi_\perp(T^*)/\sqrt{2}$ at the temperature T^* in the perpendicular direction) fitting between the adjacent layers, a distance s apart. Below T^*, the layers decouple, and the orbital pair-breaking is removed as a mechanism for the destruction of superconductivity in a layer. The layers then all behave as a thin film (of microscopic thickness) in a parallel field, which for a small thickness, can have a much higher H_{c2} than does a bulk superconductor; H_{c2} will be determined by other mechanisms such as Pauli paramagnetism and spin-orbit scattering.

Analogously for a quasi-one-dimensional superconductor consisting of a lattice of coupled superconducting chains, when the magnetic field is applied perpendicular to the chains, the vortex currents must tunnel at least twice from chain to chain in order to complete one orbit, as pictured in Figure 16a. We thus expect that H_{c2} for a quasi-one-

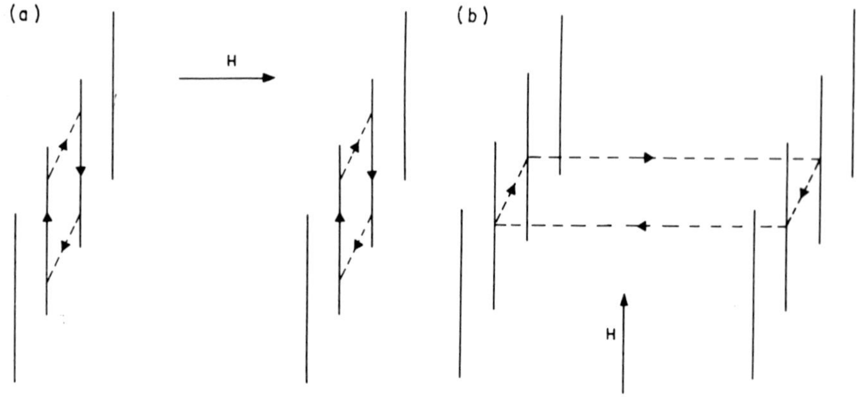

Fig. 16. (a) Field applied perpendicular to the chains in a filamentary superconductor. (b) Field applied parallel to the chains. From [50].

dimensional superconductor will behave just as H_{c2} for a layered superconductor. In particular, below a dimensional-crossover temperature T^*, the chains decouple, and H_{c2} diverges in the Ginzburg–Landau theory. It turns out, however, that this decoupling temperature T^* is a strong function of azimuthal angle ϕ. When the field is parallel to either chain lattice direction ($\phi = 0$ or $\pi/2$), only the vortex core diameter perpendicular to this direction need become smaller than the lattice dimension. When the field is not imposed along a crystal lattice direction, the vortex core crossection must fit between the chain lattice at that particular azimuthal angle, which implies that the vortex core crossection must be smaller than for a field in either direction of the crystal lattice directions; hence the decoupling temperature T^* for a general azimuthal angle is less than for the two lattice directions.

For a field parallel to the chains, the vortex currents propagate entirely by tunneling, as pictured in Figure 16b. The system behaves as a two-dimensional lattice of coupled Josephson junctions – a SQUID (superconducting quantum interference device) grid. As long as the chains are coupled, they cannot discriminate between different integral numbers of flux quanta through each SQUID, resulting formally in a periodic $H_{c2,\parallel}$ at any temperature. As for the field in the perpendicular direction, there is a dimensional-crossover temperature T^* at which the minimum vortex current path is obtained (the vortex core fitting entirely between neighboring chains). In addition, for temperatures above T^*, it turns out that it is energetically more favorable for the vortex lattice to fit commensurately with the chain lattice, so that cusps in the $H_{c2,\parallel}(T)$ curve result at temperatures corresponding to a commensurate fitting of the lattice; these cusps are most pronounced for a commensurability ratio of two small integers.

Let us now consider the Ginzburg–Landau model for a rectangular ($a \times b$) lattice of superconducting chains parallel to the z direction. We ignore the finite chain thickness, the interaction of the magnetic field with the electron spins, and the impurities. The free energy in gauge-invariant form is [53]

$$\begin{aligned} F = \int \mathrm{d}z\, ab \sum_{ij} \{ & \alpha |\psi_j^i(z)|^2 + \beta |\psi_j^i(z)|^4/2 + \\ & + |(i\partial/\partial z + 2eA_z(\mathbf{r}))\psi_j^i(z)|^2/2m + \\ & + \zeta_x |\psi_j^{i+1}(z) \exp\left(-2ie \int_{ia}^{(i+1)a} \mathrm{d}x\, A_x(\mathbf{r})\right) - \psi_j^i(z)|^2 + \\ & + \zeta_y |\psi_{j+1}^i(z) \exp\left(-2ie \int_{jb}^{(j+1)b} \mathrm{d}y\, A_y(r)\right) - \psi_j^i(z)|^2 + \\ & + |\mathbf{H}(\mathbf{r}) - \mathbf{H}_a|^2/(8\pi)\}, \end{aligned} \tag{3.1}$$

where $\alpha = -(2m\xi^2(T))^{-1} = (2m\xi^2(0))^{-1}(T - T_c)/T_c$ is the usual Ginzburg–Landau parameter, and $\zeta_x = (2M_x a^2)^{-1}$, $\zeta_y = (2M_y b^2)^{-1}$ are the interchain coupling parameters due to the Josephson tunneling of electron pairs, written in terms of effective masses such that in the limits a, $b \to 0$, F reduces to the anisotropic mass form. The magnetic field is taken to be in the direction (θ, ϕ) as given by Equation (2.2). The vector potential for this calculation is taken to be of the form [51],

$$\mathbf{A} = H_a(z \sin\theta \sin\phi,\ x\cos\theta - z\sin\theta\cos\phi,\ 0). \tag{3.2}$$

Note that the symmetry of the problem is different than in the anisotropic mass model, so it is simpler in practice to separate out the z axis from the others. Other forms for $\mathbf{A}$ are certainly possible, but this one is particularly useful for $\theta = 0$ and $\pi/2$, the two predominant angles of interest. We take the applied field $H_a = H$.

3.2. THE FIELD PERPENDICULAR TO THE CHAINS

Let us first consider the case of $H_{c2}(\theta = \pi/2)$. By varying the free energy and neglecting the term cubic in ψ (since $\psi \to 0$ as $H \to H_{c2}$), we obtain

$$\alpha\psi_l^k - (2m)^{-1}\,\mathrm{d}^2\,\psi_l^k/\mathrm{d}z^2 +$$
$$+ \zeta_x\,\{2\psi_l^k - \exp(2ihaz\sin\phi/d^2)\psi_l^{k-1} - \exp(-2ihaz\sin\phi/d^2)\psi_l^{k+1}\} +$$
$$+ \zeta_y\,\{2\psi_l^k - \exp(2ihbz\cos\phi/d^2)\psi_{l-1}^k - \exp(-2ihbz\cos\phi/d^2)\psi_{l+1}^k\} = 0, \tag{3.3}$$

where $h \equiv eHd^2$ is the reduced field, and $d \equiv (a^2\cos^2\phi + b^2\sin^2\phi)^{1/2}$ is a characteristic length. We note that d does not enter any final expressions for H_{c2}. Fourier transformation of Equation (3.3) yields

$$-(2m)^{-1}\,\mathrm{d}^2\,\psi/\mathrm{d}z^2 + 2\,\zeta_x\,\{1 - \cos(2haz\sin\phi/d^2)\}\psi +$$
$$+ 2\zeta_y\,\{1 - \cos(2hbz\cos\phi/d^2)\}\psi = -\alpha\psi = E\psi, \tag{3.4}$$

where we have set the perpendicular Fourier wavevectors Q_x and Q_y equal to 0 in order to yield the lowest eigenvalue in Equation (3.4), corresponding to the largest h and hence the largest field H_{c2} for which superconductivity can persist. Equation (3.4) is in the form of the two-term Hill equation [52] and, for general ϕ, is complicated, as the two periodic potentials are incommensurate with respect to each other. Three values of ϕ lead immediatedly to a solution, however: $\phi = 0$, $\pi/2$, and $\tan^{-1}(b/a)$, for which the equation reduces to the Mathieu equation [53],

$$-(2m)^{-1}\,\mathrm{d}^2\,\psi/\mathrm{d}z^2 + \gamma(1 - \cos qz)\psi = E\psi = -\alpha\psi, \tag{3.5}$$

where for $\phi = 0$, $\gamma = 2\zeta_y$, $q = 2hb/d^2$; for $\phi = \pi/2$, $\gamma = 2\zeta_x$, $q = 2ha/d^2$; and for $\phi = \tan^{-1}(b/a)$, $\gamma = 2(\zeta_x + \zeta_y)$, $q = 2hab/\{d^2(a^2 + b^2)^{1/2}\}$. As $h \to \infty$, $q \to \infty$ and $\cos qz$ oscillates wildly. As this first happens when T approaches T^* from above, we see that T^* for $\theta = \pi/2$ is given by $-\alpha(T - T^*) = \gamma$. This implies that the three different ϕ values result in three distinct T^* values. In fact, it turns out that there are only these three T^* values; for $\phi \neq 0$ and $\phi \neq \pi/2$, $-\alpha(T - T^*) = 2(\zeta_x + \zeta_y)$, just as for $\phi = \tan^{-1}(b/a)$. The reason is that for ϕ unequal to either 0 or $\pi/2$, we have an equation of the form,

$$-(2m)^{-1}\mathrm{d}^2\,\psi/\mathrm{d}z^2 + \gamma_1(1 - \cos q_1 z)\psi + \gamma_2(1 - \cos q_2 z)\psi = -\alpha\psi. \tag{3.6}$$

Since as $H \to \infty$, both q_1 and $q_2 \to \infty$, and both $\cos q_1 z$ and $\cos q_2 z$ oscillate wildly. Hence, $-\alpha(T - T^*) = \gamma_1 + \gamma_2$. In short, we see that T^* is given by

$$T^*/T_c = \begin{cases} 1 - 2m\xi^2(0)/(M_y b^2) & \phi = 0 \\ 1 - 2m\xi^2(0)/(M_x a^2) & \phi = \pi/2 \\ 1 - 2m\xi^2(0)\{1/(M_x a^2) + 1/(M_y b^2)\} & \text{otherwise.} \end{cases} \tag{3.7}$$

Since T^* for $\phi = 0$ and $\pi/2$ is much larger than for the other directions, we expect that H_{c2} should show strong angular dependence near the lattice directions.

Before we include the full azimuthal dependence of the perpendicular ($\theta = \pi/2$) critical field, let us first examine Equation (3.5). This equation may be solved analytically for small and large q (i.e., small or large magnetic field). For small q, $\cos qz$ may be expanded in powers of q; $\cos qz \cong 1 - (qz)^2/2! + O(qz)^4$. The basis wavefunctions are parabolic cylinder functions, as the leading term in the potential is that of a harmonic oscillator. One may then carry out the calculation for the lowest eigenvalue order by order in perturbation theory.

For large but finite h, we may carry out the perturbation for H_{c2} above T^* by treating $1/q$ as a small parameter. To zeroth order in $1/q$, the wavefunction ψ_0 is a constant, and the lowest eigenvalue $E_0 = 0$. The leading nontrivial eigenvalue is of order $1/q^2$, yielding an H_{c2} that diverges as $(T - T^*)^{-1/2}$ from above.

Let us now examine the full ϕ dependence of H_{c2}. First, we consider the region just below T_c, where H_{c2} is small. As for the Mathieu equation, we may expand the potential in Equation (3.5) in powers of h. In leading order, we obtain the harmonic oscillator form,

$$-(2m)^{-1}\mathrm{d}^2\psi/\mathrm{d}z^2 + (1/2)(2h/d)^2(M_y^{-1}\cos^2\phi + M_x^{-1}\sin^2\phi)z^2\psi = E\psi. \tag{3.8}$$

From the lowest eigenvalue, we obtain the expected anisotropic mass result, obtained from Equation (2.21) by setting $\theta = \pi/2$,

$$H_{c2}(T, \phi) = \Phi_0\,|T - T_c|[(m/M_x)\sin^2\phi + (m/M_y)\cos^2\phi]^{-1/2}(2\pi\xi^2(0)T_c)^{-1}. \tag{3.9}$$

By keeping higher order terms in the expansion of the two cosines, using perturbation theory in powers of h, and inverting the lowest eigenvalue, we obtain an expansion in powers of $|T - T_c|/T_c$,

$$H_{c2\perp}(T, \phi) = \Phi_0\,|(T - T_c)/T_c|\{(m/M_x)\sin^2\phi + (m/M_y)\cos^2\phi\}^{-1/2}\sum_{n=0}^{\infty} a_n(\phi)|(T - T_c)/T_c|^n \tag{3.10}$$

For the case of isotropic coupling $\{\zeta_x = \zeta_y;\ \xi_\perp^2 = m\xi_\parallel^2(0)/M\}$ we find these coefficients to be

$$a_0 = 1,$$

$$a_1 = (a/4\xi_\perp{}^2)^2(\sin^4\phi + \cos^4\phi),$$

$$a_2 = (a/4\xi_\perp{}^2)^4\{(13/3)(\sin^4\phi + \cos^4\phi)^2 - (4/3)(\sin^6\phi + \cos^6\phi)\}. \tag{3.11}$$

It is easy to see that for $\phi = 0$, $\pi/2$, and $\pi/4$, these results reduce to those obtained from the Mathieu equation. It should be noted that even for isotropic coupling, there is some aximuthal anisotropy of H_{c2} (note that a_1 varies by a factor of 2). Also, since $a_1(\phi) > 0$ for all ϕ, H_{c2} exhibits positive curvature as the temperature is lowered from T, indicating the onset of decoupling of the chains. Complete decoupling occurs at T^*, of course.

We now consider the ϕ dependence of H_{c2} in the large field regime just above T^*. For $\phi = \pi/2$, Manneville found [54]

$$H_{c2}(\phi = \pi/2) \cong \Phi_0 m \xi(0) \{T_c/(T - T^*)\}^{1/2}/(\pi a^3 \sqrt{2} M), \tag{3.12}$$

where T^* was given by Equation (3.7) for $\phi = \pi/2$. Similarly, for $\phi = 0, H_{c2}(\phi = 0)$ near to T^* is given by Equation (3.12) with $a \to b, M_x \to M_y$, and T^* given by Equation (3.7) for $\phi = 0$. However, the fact that T^* is substantially lower than those values for slightly different ϕ values indicates that H_{c2} should vary substantially near to $\phi = 0, \pi/2$ in the larger field regime.

Turkevich and Klemm showed that one could solve the general ϕ problem by assuming that the arguments of the cosine potentials are commensurate [50],

$$a \sin \phi = dn_x/N; \qquad b \cos \phi = dn_y/N \tag{3.13}$$

for some integers n_x, n_y, N. By taking N extremely large, $n_{x,y}/N$ can cover all the rationals. The following procedure also works for arbitrary n_x, n_y, and N. By setting $x = hz/Nd$ and defining

$$\sigma_x = 2md^2 \zeta_x, \qquad \sigma_y = 2md^2 \zeta_y, \tag{3.14}$$

the Schroedinger equation takes the form of the two-term Hill equation [52],

$$\mathrm{d}^2 \psi/\mathrm{d}x^2 + \sigma^2 \psi + 2\lambda[\sigma_x \cos(2n_x x) + \sigma_y \cos(2n_y x)] \psi = 0, \tag{3.15}$$

where the eigenvalue is

$$\sigma^2 \equiv \lambda\epsilon = -\lambda 2md^2(\alpha + 2\zeta_x + 2\zeta_y), \tag{3.16}$$

and the small parameter $\lambda = (N/h)^2$. Since the potential is periodic, Bloch's theorem suggests we write ψ in the form [52]

$$\psi(x) = \exp(i\tau x)\{a_0 + \sum_n a_n \exp[i(p_n n_x + q_n n_y)x]\}, \tag{3.17}$$

where p_n, q_n are integers and τ is a complex number. To leading non-trivial order in λ, this equation may be solved by expanding ψ and ϵ in powers of λ,

$$\begin{aligned} \psi &= \psi_0 + \lambda\psi_1 + \lambda^2\psi_2 + \ldots \\ \epsilon &= \epsilon_0 + \lambda\epsilon_1 + \lambda^2\epsilon_2 + \ldots . \end{aligned} \tag{3.18}$$

In addition, we have the normalization condition

$$\int_0^L \mathrm{d}x\, L^{-1} |\psi|^2 = 1. \tag{3.19}$$

To zeroth order, we have $\psi_0 = 1$. To order λ,

$$\psi_1'' + \epsilon_0 + 2[\sigma_x \cos(2n_x x) + \sigma_y \cos(2n_y x)] = 0, \tag{3.20}$$

which implies

$$\epsilon_0 = 0, \qquad \psi_1 = 2\{\sigma_x(2n_x)^{-2}\cos(2n_x x) + \sigma_y(2n_y)^{-2}\cos(2n_y x)\}, \tag{3.21}$$

since ψ_1 is periodic in x. The possible additional constant of integration must equal 0 to preserve the normalization condition above. The leading non-trivial result for the eigenvalue enters in order λ^2,

$$\psi_2'' + \epsilon_1 + 2[\sigma_x \cos(2n_x x) + \sigma_y \cos(2n_y x)]\,\psi_1 = 0, \tag{3.22}$$

which implies

$$\epsilon_1 = -(1/2)\{(\sigma_x/n_x)^2 + (\sigma_y/n_y)^2\} \tag{3.23}$$

in order that ψ_2 be periodic. We thus have

$$\lambda\epsilon - \lambda^2\epsilon_1 + \ldots = 0, \tag{3.24}$$

or

$$\lambda = \epsilon/\epsilon_1 \tag{3.25}$$

to this order, where ϵ is given by Equation (3.16). Since both sides of the equation are proportional to N^2, we may solve for h and hence H_{c2},

$$H_{c2}(T, \phi) \cong \Phi_0\xi(0)m(\sqrt{2}\pi)^{-1}\{T_c/(T - T^*)\}^{1/2}\{(M_x a^3 \sin\phi)^{-2} +$$

$$+ (M_y b^3 \cos\phi)^{-2}\}^{1/2} \tag{3.26}$$

near to T^*, where T^* is given by Equation (3.7) for $\phi \neq 0, \pi/2$. We note that this expansion diverges as $\phi \to 0, \pi/2$, as expected. It should be noted that at $\phi = \tan^{-1}(b/a)$, it underestimates the exact expression obtained from the Mathieu equation by the factor $\{1 + 2\eta/(1 + \eta^2)\}^{1/2}$, where $\eta = M_x a^2/M_y b^2$ (for isotropic coupling, this factor is $\sqrt{2}$). This factor arises from the commensurability of n_x and n_y; it can be understood by setting $n_x = n_y$ before performing the calculation for ϵ. It is easily seen that instead of Equation (3.25) [with ϵ_1 given by Equation (3.23)] for the eigenvalue λ, we obtain a similar equation with ϵ_1 given by

$$\epsilon_1 = -(\sigma_x + \sigma_y)^2/(2n^2). \tag{3.27}$$

Turkevich and Klemm showed that by mapping the eigenvalue equation onto a two-dimensional lattice problem and by summing over all equivalent starting points on the lattice, an expansion could be found which is exact near the desired (commensurate) value of ϕ [50]. Nevertheless, the procedure described here gives the correct T^* and the correct power of the divergence of H_{c2} above T^*. Similarly, in the vicinity of $\phi = 0$ or $\pi/2$, an infinite class of divergent terms must be summed in order to obtain a finite answer. The results for the temperature dependence of H_{c2} for several azimuthal angles is shown in Figure (17) for the special case of isotropic coupling. The very strong azimuthal dependence near to $\phi = 0$ should be noted.

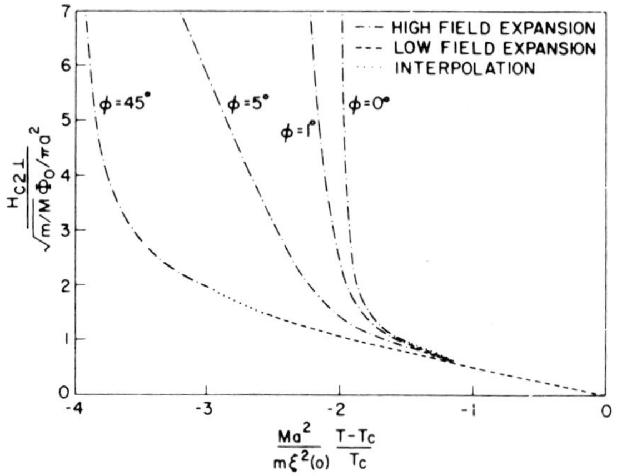

Fig. 17. Perpendicular upper critical field $H_{c2\perp}(T)$ for various azimuthal angles, for a square chain lattice. The azimuthal dependence is symmetric about $\phi = 45°$. There is pronounced azimuthal dependence near the lattice direction $\phi = 0°$. From [50].

3.3. THE FIELD PARALLEL TO THE CHAINS

We now consider the case of the magnetic field parallel to the superconducting chains. Using the vector potential as given by Equation (3.2) with $\theta = 0$, we minimize the free energy with respect to variations in the order parameter ψ, just as in leading up to Equation (3.3) for the perpendicular field direction. As the resulting potential is independent of l and z, we form the respective Fourier series and Fourier transform in those variables, setting the wavevectors Q_x and $Q_y = 0$, as we did in obtaining Equation (3.4) for the perpendicular direction. We obtain,

$$(\epsilon - 2\zeta_y \cos k\chi)\psi_k - \zeta_x(\psi_{k-1} + \psi_{k+1}) = 0, \tag{3.28}$$

where $\epsilon = \alpha + 2\zeta_x + 2\zeta_y$ is the eigenvalue of the difference equation, and the dimensionless variable $\chi = 2hab/d^2$. This difference equation may be solved in a manner analogous to the Mathieu equation. We note that ϵ and ψ are periodic in χ. Hence, ψ may be written in the form

$$\psi_k = \exp(ik\tau) \sum_{n=0}^{\infty} a_n \cos(nk\chi). \tag{3.29}$$

Substituting Equation (3.29) into Equation (3.28), equating the resulting coefficients of $\cos(mk\chi)$, and setting $\tau = 0$ (to give the lowest eigenvalue), one obtains the set of equations,

$$\epsilon a_0 = \zeta_y a_1 + 2\zeta_x a_0 \tag{3.30a}$$

$$\epsilon a_1 = 2\zeta_y a_0 + \zeta_y a_2 + 2\zeta_x a_1 \cos\chi \tag{3.30b}$$

$$\epsilon a_n = \zeta_y (a_{n-1} + a_{n+1}) + 2a_n \zeta_x \cos(n\chi), \quad n > 1. \tag{3.30c}$$

To solve this system of equations, we start with the top equation, writing a_0 in terms of a_1, which we write in terms of a_0 and a_2, etc. More explicitly, from Equation (3.30a) and (3.30b), we have,

$$(\epsilon - 2\zeta_x)a_0 = \zeta_y a_1 = \zeta_y^2(2a_0 + a_2)/(\epsilon - 2\zeta_x \cos\chi), \tag{3.31}$$

which implies

$$\{\epsilon - 2\zeta_x - 2\zeta_y^2/(\epsilon - 2\zeta_x \cos\chi)\}a_0 = a_2\zeta_y^2/(\epsilon - 2\zeta_x \cos\chi) = 0. \tag{3.32}$$

We then solve for a_2 in terms of a_0 and a_3 (by using Equation (3.30a) to write a_2 in terms of a_3 and a_1, and then using Equation (3.30b) to write a_1 in terms of a_0 and a_2. One may then solve for a_2 in terms of a_0 and a_3),

$$a_2 = \frac{\zeta_y\{a_3 + 2\zeta_y a_0/(\epsilon - 2\zeta_x \cos\chi)\}}{\{\epsilon - 2\zeta_x \cos 2\chi - \zeta_y^2/(\epsilon - 2\zeta_x \cos\chi)\}}\,. \tag{3.33}$$

This expression may now be used in combination with Equation (3.32) to give a_0 in terms of a_3. This procedure may be carried on ad infinitum, and yields the continued fraction equation for the eigenvalue ϵ,

$$\epsilon - 2\zeta_x - \cfrac{2\zeta_y^2}{\epsilon - 2\zeta_x\cos\chi - \cfrac{\zeta_y^2}{\epsilon - 2\zeta_x\cos 2\chi - \cfrac{\zeta_y^2}{\epsilon - 2\zeta_x \cos 3\chi - \ldots}}} \tag{3.34}$$

This continued fraction must be solved for the eigenvalue $\epsilon(\chi)$ by truncating the denominator at some point, and then inverting the resulting expression to give χ (and hence H_{c2}) in terms of ϵ (and hence T). Turkevich and Klemm [50] performed this calculation numerically, obtaining the temperature dependence of H_{c2} shown in Figure 18. Note that the periodicity of $\epsilon(\chi)$ results in the figure being periodic in H_{c2} for a fixed temperature. Each repetition of the $T(H_{c2})$ curve arises from an additional flux quantum being placed in each unit cell of the chain lattice. The additional structure (or kinks) in the H_{c2} curve arises from a non-integral, commensurate fitting of the flux lattice into the chain lattice. The most striking anomaly is for one flux quantum in two unit cells. The next most important ones are for one or two flux quanta in three unit cells, and so on. However,

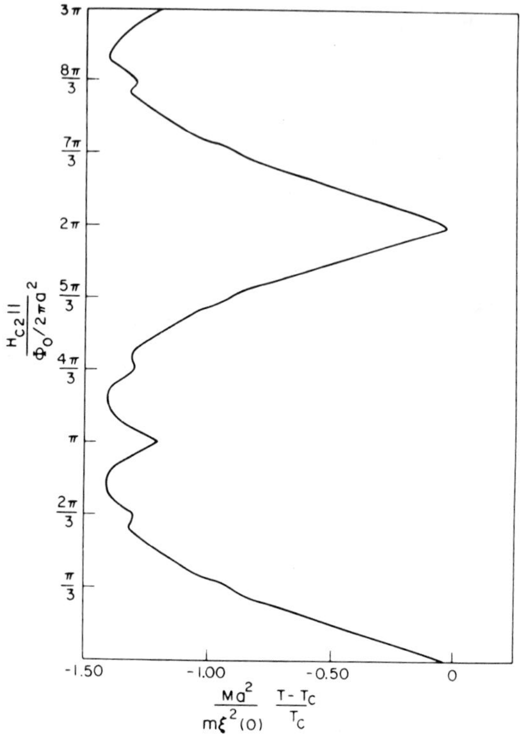

Fig. 18. Parallel upper critical field $H_{c2\parallel}(T)$. This is periodic in H every time one flux quantum is inserted through the chain lattice. The additional kinks arise when a flux quantum is inserted into a section of the chain lattice larger than a single unit cell. From [50].

it should be noted that the repeated structure will disappear when Pauli pairbreaking is included in the theory [41], just as the apparent divergence of H_{c2} at T^* disappears in the layered compounds.

In a layered superconductor, the upper critical field is found from the usual pairbreaking equation [25],

$$\ln(T/T_{c0}) + \psi(1/2 + \alpha_{pb}/2\pi T) - \psi(1/2) = 0, \tag{3.35}$$

where $\psi(x)$ is the digamma function, and the pairbreaking parameter α_{pb} is given by

$$\alpha_{pb} = \epsilon + \tau_{so}(\mu_B H)^2, \tag{3.36}$$

with τ_{so} the spin-orbit scattering time (for two-dimensional scattering), and ϵ the eigenvalue of the Schroedinger equation with both harmonic and periodic potentials,

$$\{D[-\mathrm{d}^2/\mathrm{d}x^2 + (2eHx\cos\theta)^2] + J^2\tau[1 - \cos(2eHsx\sin\theta)]\}\phi = 2\epsilon\phi, \tag{3.37}$$

with $D = v^2\tau/2$ the two-dimensional diffusion constant and J the interlayer electron tunneling energy [55]. As an indication of the role of Pauli pairbreaking and spin-orbit scattering, Figure 19 exhibits the expected behavior in a layered superconductor in a parallel field.

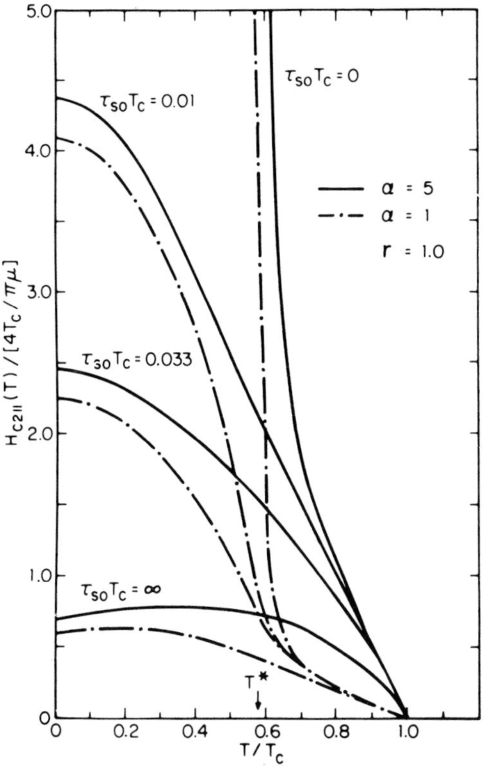

Fig. 19. Critical fields of layered compounds for modest ($\alpha = 1$) and strong ($\alpha = 5$) Pauli paramagnetic limiting and for various spin-orbit scattering rates. From [25].

4. Superconducting Networks

4.1. Some Puzzling Experimental Results

Although there are examples of materials that can be made in a crystalline form, some materials that exhibit quasi-one-dimensional superconductivity appear to be composed of bundles of fibers. Examination of $(SN)_x$ under a scanning electron microscope has revealed the twisted, strandlike nature of the material [1]. This non-crystalline feature of the material results in critical field curves that are anomalous in their temperature dependence. The anomalous temperature dependence of H_{c2} in $(SN)_x$ is exhibited in Figure 9. For comparison, the same quantity is also displayed for the brominated material $(SNBr_{0.4})_x$, which behaves as an ordinary anisotropic bulk material.

Another puzzling experimental result is the low temperature superconductivity observed in $NbSe_3$. This material is known to have two incommensurate charge-density waves, setting in at the temperatures 145 and 59 K, respectively [37]. However, low temperatures resistivity studies have shown a dramatic decrease in the resistivity [3], as pictured in Figure 20. More recent investigations have indicated that not all $NbSe_3$

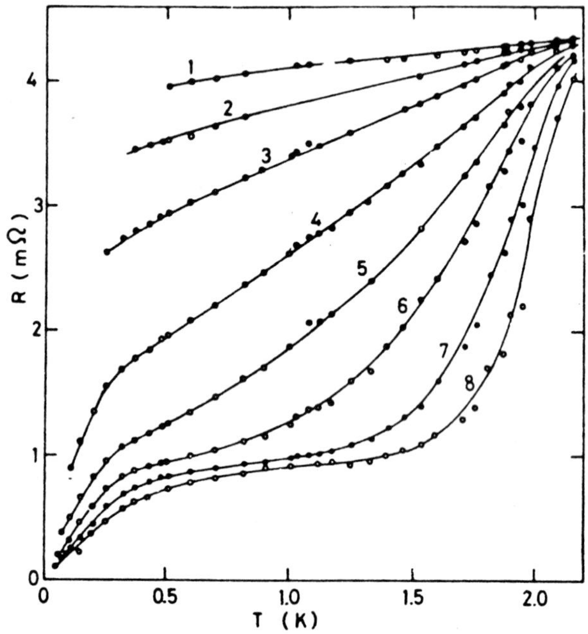

Fig. 20. Resistance of $NbSe_3$ for sample A (with $R_{300}/R_{40} = 80$) *vs* temperature for different current densities. In A/mm^2, the current densities from Nos. 1–8 are: 0.52, 0.26, 0.13, 0.052, 0.026, 0.013, 0.0052, and 0.0013. From [12].

samples exhibit this superconducting-like behavior. Apparently, single crystal samples do not show the behavior, whereas samples composed of many strands do [22]. This dramatic sample variation of the low temperature resistivity is pictured in Figure 21.

Very recently, yet another interesting feature has been observed in the highly anisotropic superconductor $TaSe_3$. Although this material has been known for some time to be a superconductor, and exhibits the angular and azimuthal dependence of the upper critical field characteristic of an anisotropic bulk material (e.g. Figure 5 and 6), it also exhibits extraordinary, novel resistivity behavior for low currents near to T_c, as pictured in Figure 22 [21]. These experiments are not yet understood, although the magnetic field dependence of the resistivity peak and the high anisotropy of the material are suggestive of a spin-density wave (e.g. see Figure 1), a fibrous structure such as has been observed in NbSe, has not yet been ruled out. It will be interesting to see how this material 'unravels'.

In this section, we will summarize the theoretical attempts to treat the strand-like nature of some of the materials. The most comprehensive treatment to date is by S. Alexander; hence most of what follows will be based upon his work [56].

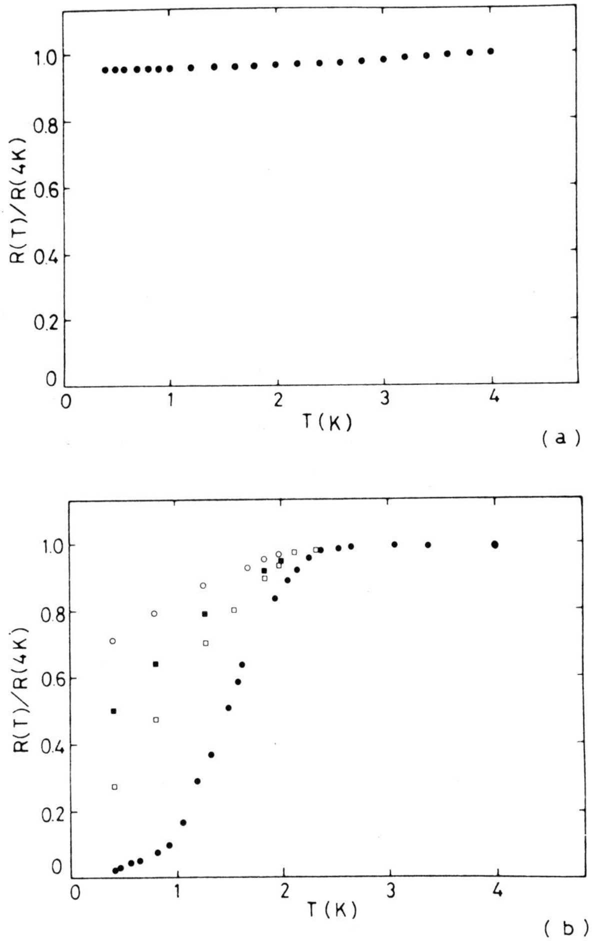

Fig. 21. Low temperature resistivity of different samples of $NbSe_3$. (a) Single crystal (non-superconducting). (b) material consisting of many fibers (exhibits apparent superconductivity). The different curves are from top to bottom for decreasing current densities, analogous to Figure 20. After [22].

4.2. THE SINGLE STRAND

The simplest model of a filamentary superconductor consisting of a number of strands is that of a single strand. The upper critical field of a single strand has been found in the Ginzburg–Landau regime (i.e., near to T_c) by several authors [29, 31, 50]. One might at first think that the upper critical field would be exactly that for an anisotropic bulk superconductor, as at H_{c2}, H and B are equal and there are no demagnetization effects. However, for filaments with a maximum cross-sectional diameter much less than the coherence length, the restricted geometry can influence the temperature and angular dependence of the upper critical field in a manner analogous to that of a thin film. For simplicity, the effective mass perpendicular to the highly conducting axis is assumed

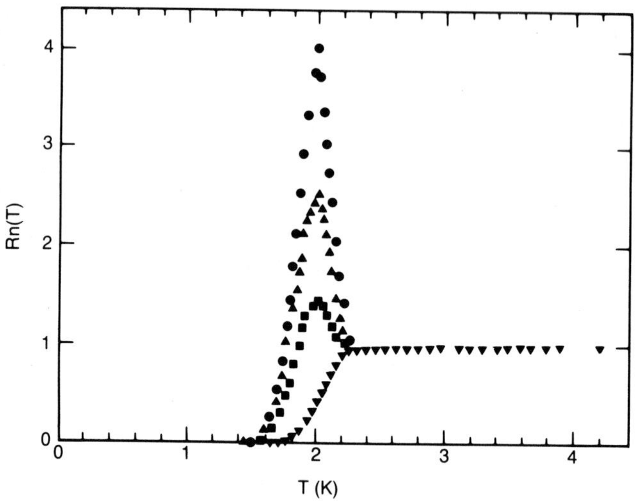

Fig. 22. Normalized resistance $R(T)/R(4.2\text{ K})$ for a sample of $TaSe_3$ with room temperature resistance ratio = 167, as a function of current density. From top to bottom, the current density (A/cm^2) is 0.4, 0.7, 1.5, and 2.2, respectively. After [21].

independent of azimuthal angle, $m_x = m_y = M$, $m_z = m$. The Ginzburg–Landau free energy appropriate for an uniaxially anisotropic superconductor of cylindrical shape is given by

$$F = \int_0^L \mathrm{d}z \int_0^R \rho\, \mathrm{d}\rho \int_0^{2\pi} \mathrm{d}\theta\ \{\alpha|\psi|^2 + (\beta/2)|\psi|^4 + h^2/(8\pi) + $$

$$+ (2m)^{-1}\left|\left(-i\frac{\partial}{\partial z} - 2eA_z\right)\psi\right|^2 + (2M)^{-1}|(-i\nabla_\perp - 2eA_\perp)\psi|^2\}. \tag{4.1}$$

A simplifying choice of the vector potential in cylindrical coordinates is

$$\mathbf{A} = (A_\rho, A_\theta, A_z) = H\rho\{0, (1/2)\cos\theta_0, -\sin\theta_0 \sin(\theta - \phi_0)\}, \tag{4.2}$$

where the magnetic field has been taken to be in the direction

$$\mathbf{B} = H(\sin\theta_0 \cos\phi_0, \sin\theta_0 \sin\phi_0, \cos\theta_0). \tag{4.3}$$

The relevant boundary condition is that there be no supercurrent across the boundary

$$\frac{\partial\psi}{\partial\rho}\Big|_R = \frac{\partial\psi}{\partial z}\Big|_{0,L} = 0. \tag{4.4}$$

Variation of the free energy [Equation (4.1)] with respect to ψ^* yields,

$$-(2M)^{-1}\left\{\frac{\partial^2 \psi}{\partial \rho^2} + \rho^{-1}\frac{\partial \psi}{\partial \rho} + \rho^{-2}\frac{\partial^2 \psi}{\partial \theta^2} + (M/m)\frac{\partial^2 \psi}{\partial z^2}\right\} +$$

$$+ (ieH/M)\left[\cos\theta_0 \frac{\partial \psi}{\partial \theta} + (M/m)2\rho \sin\theta_0 \sin(\theta - \phi_0)\frac{\partial \psi}{\partial z}\right] +$$

$$+ (2e^2H^2\rho^2/M)[(1/4)\cos^2\theta_0 + \sin^2\theta_0 \sin^2(\theta - \phi_0)]\,\psi = -\alpha\psi = E\psi. \quad (4.5)$$

Equation (4.5) is a Schroedinger equation, and may be solved for small magnetic field by perturbation theory. The unperturbed problem is

$$-(2M)^{-1}\left[\rho^{-1}\frac{\partial}{\partial \rho}\left(\rho\frac{\partial \psi}{\partial \rho}\right) + \rho^{-2}\frac{\partial^2 \psi}{\partial \theta^2} + (M/m)\frac{\partial^2 \psi}{\partial z^2}\right] = E_0\psi, \quad (4.6)$$

which is just the Laplace equation in cylindrical coordinates (provided the z axis is appropriately scaled to remove the anisotropy). The unperturbed eigenfunctions satisfying the boundary conditions [Equation (4.4)] are of the form

$$\psi_{kln} = \exp(ikz)\exp(ilq)J_l(x_{ln}/R), \quad (4.7)$$

where x_{ln} is the nth zero of the first derivative of the lth Bessel function, corresponding to the energies

$$E_0 = k^2/2m + (x_{ln}/R)^2/2M. \quad (4.8)$$

The lowest eigenvalue is for $k = 0$ and $x_{0,0} = 0$, corresponding to a constant order parameter across the strand. The remainder of the left-hand side of Equation (4.5) can be treated as a perturbation. To first order in H, there is no correction to the eigenvalue E. To second order in H, however, there is a correction arising from the part of the perturbation independent of spatial derivatives of ψ, yielding

$$E_2 = (e^2H^2R^2/2M)[(1/2)\cos^2\theta_0 + (M/m)\sin^2\theta_0]. \quad (4.9)$$

This second-order energy may be inverted to obtain the upper critical field

$$H_{c2} = M\Phi_0(m\pi R\xi(0))^{-1}\,|2(T - T_c)/T_c|^{1/2}(\cos^2\theta_0 + (M/m)\sin^2\theta_0)^{-1/2}. \quad (4.10)$$

We note that the angular dependence of H_{c2} in an isolated strand is very similar to that expected for an anisotropic bulk material (i.e. Equation (2.21)), and could be interpreted as identical to it, with a change in the definition of the effective mass anisotropy by a factory of 2. This angular dependence is also much more gradual than the cusp at $\theta = 0$ given by the corresponding formula due to Tinkham for thin films [57]. In the thin-film case, the normal vortex cores can penetrate the film for a perpendicularly applied field, but get progressively 'squeezed out' as the external field is made more parallel. This results in a divergence of H_{c2} for an infinitely thin film. In the above thin-filament case, the normal vortex cores have been 'squeezed out' for all orientations of the external field. Thus in the limit $R \to 0$, H_{c2} diverges for all orientations. Note also that the temperature dependence near to T_c is proportional to $|T - T_c|^{1/2}$ for all angles, and hence exhibits downward curvature as the temperature is lowered, a result identical to the temperature dependence of H_{c2} for thin films [57].

4.3. General Network Equations

We now consider systems of superconducting filaments. We imagine that a particular material consists of a disordered lattice of chains. This disorder can take on many forms. Some chains may lie in an ordered fashion relative to the nearby chains. Others may be physically separated from the nearest chains, due perhaps to vacancies of chains. Still others may be twisted, ending with dead ends away from other chains. It may happen that the scale of chain vacancies is well-defined, or there may be no particular length scale, i.e., essentially all length scales less than the sample size may be present. Alexander [56] has studied these types of disorder by considering various networks of thin superconducting wires (or chains), as was first proposed by de Gennes and Deutscher *et al.* [58, 59]. They treated the linearized Ginzburg–Landau equations on networks of thin filaments by deriving linear-difference equations for the value of the order parameter at the junctions of the net, using the solutions along the filaments. He then solved for the upper critical field for several finite networks and for the Sierpinski gasket, shown in Figure 23. The Sierpinski gasket represents a simplified model for a lattice of filaments

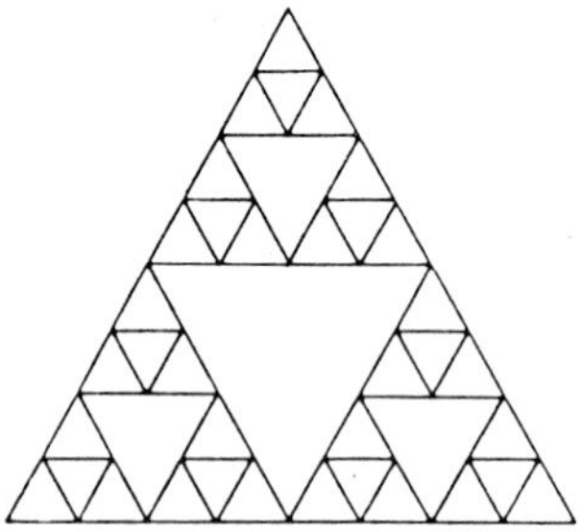

Fig. 23. A finite section of the triangular Sierpinski gasket. After [56].

with filament vacancies of all length scales, and leads to a new length scale proportional to a power of the bare (single strand) coherence length. The dead ends lead to a 'mass' or T^* renormalization.

The Ginzburg–Landau free energy for an infinitely thin strand at the upper critical field may be written [62]

$$F = \int ds\ \{A|\psi|^2 + (1/2)B|\psi|^4 + C|\hat{s}\cdot(i\nabla - 2e\mathbf{A})\psi|^2\}, \tag{4.11}$$

where $\hat{s}$ is a unit vector tangential to the stand, and the order parameter $\psi(s)$ is a function of the position along the strand. Equation (4.11) is obtained from Equation (4.1) by letting $R \to 0$, and the parameters A, B, C are proportional to α, β, and $(2m)^{-1}$, respectively. Minimizing Equation (4.11) with respect to variations in ψ leads to

$$\left\{-1/\xi_s^2 + \left(i\frac{\partial}{\partial s} - a\right)^2 + b|\psi|^2\right\}\psi = 0, \tag{4.12}$$

where $a = 2e\hat{s}\cdot\mathbf{A}$ is the reduced vector potential, $b = B/C$, and

$$1/\xi_s^2 = -A/C \propto (T_c - T)/T D_0, \tag{4.13}$$

where C has been taken to be proportional to the diffusion constant D_0 on the strand (i.e., the dirty limit has been assumed) [23]. To describe a network, each strand obeys the above GL equation, plus there are the appropriate boundary conditions at the junctions. We first generalize the order parameter ψ to a network, denoting $\psi_{ij}(s_{ij})$ as the order parameter at position s_{ij} on the strand with length l_{ij} between the connecting points (or junctions) s_i and s_j. Continuity of the order parameter at the junctions requires

$$\psi_{ij}(s_i) = \Delta_i, \qquad \psi_{ij}(s_j) = \Delta_j. \tag{4.14}$$

The surface term obtained from the free energy variation is of the form

$$\sum_j \left[\left(i\frac{\partial}{\partial s_{ij}} - a_{ij}\right)\psi_{ij}(s_{ij})\right]_{s_i} = 0, \tag{4.15}$$

where a_{ij} is the vector potential on the strand between junctions i and j at the point s_{ij}, and all derivatives are in the outgoing directions on the respective branches. The summation is over all branches ending at the junction at s_j. The problem as stated by Equation (4.12–4.15) is well-posed, but the non-linearity in Equation (4.12) is important only for systems that are so inhomogeneous that, for a given external field H near to the macroscopic H_{c2}, some parts of the sample are far from their local H_{c2}. Hence, for most systems, it suffices to neglect the term in Equation (4.13) cubic in ψ. The linearized Equation (4.13) combined with the boundary conditions [Equation (4.14)] results in the solution

$$\psi_{ij}(s_{ij}) = \exp(-ia_{ij}s_{ij})\,[\Delta_i \sin(\theta_{ij} - \phi_{ij}) + \Delta_j \exp(i\gamma_{ij})\sin\phi_{ij}]/\sin\theta_{ij}, \tag{4.16}$$

where

$$\theta_{ij} = l_{ij}/\xi_s, \qquad \phi_{ij} = s_{ij}/\xi_s, \tag{4.17}$$

and

$$\gamma_{ij} = \int a_{ij}\,\mathrm{d}s_{ij} \cong a_{ij}l_{ij}. \tag{4.18}$$

This equation combined with the 'Kirchoff' condition (Equation (4.14)) gives the set of coupled linear equations [58]

$$-\Delta_i \sum_j \cot\theta_{ij} + \sum_j \left[\Delta_j \exp(i\gamma_{ij})/\sin\theta_{ij}\right] = 0. \tag{4.19}$$

This may be rewritten as

$$m_i \Delta_i / \xi_s^2 + \sum_j d_{ij} (\eta_{ij} \Delta_j - \Delta_i) = 0, \tag{4.20}$$

where $\eta_{ij} = \exp(i\gamma_{ij})$ and the 'bond diffusivity' d_{ij} and mass m_i associated with the ith junction are given by

$$d_{ij} = (\xi_s \sin \theta_{ij})^{-1} \tag{4.21}$$

and

$$m_i = \xi_s \sum_j \tan(\theta_{ij}/2). \tag{4.22}$$

The above equations have been studied by Alexander for a variety of network geometries [56].

4.4. FINITE NETWORKS

The first and simplest example considered by Alexander is that of a single superconducting ring of circumference L. A vortex can be formally introduced on the ring, and Equation (4.19) becomes [56]

$$(-2 \cot \theta + 2 \cos \gamma / \sin \theta) \Delta = 0, \tag{4.23}$$

which implies

$$\cos \theta = \cos \gamma, \tag{4.24}$$

where $\theta = L/\xi_s$ and γ is the line integral of a around the ring, proportional to the flux through the ring. Equation (4.23) implies a solution (and hence an upper critical field) when

$$L/\xi_s < \pi. \tag{4.25}$$

For a larger ring, the vortex fits inside the ring, and hence does not destroy the superconductivity. If Equation (4.25) can be satisfied for some temperature > 0, then there will be a temperature T^* at which H_{c2} would diverge in the absence of other pairbreaking effects, given by

$$\xi_s(T^*) = L/\pi. \tag{4.26}$$

Alexander next considered the case of a single ring of circumference L with an open side branch of length l attached to it. The two junctions are labelled 1 and 2 at the point where the branch is connected and at the end of the branch, respectively. The two equations obtained from Equation (4.19) are [56]

$$-\Delta_2 \cot\theta' + \Delta_1 \exp(i\gamma_{21})/\sin\theta' = 0 \tag{4.27}$$

and

$$-\Delta_1(2\cot\theta + \cot\theta') + 2\Delta_1 \cos\gamma/\sin\theta + \Delta_2 \exp(i\gamma_{12})/\sin\theta' = 0, \tag{4.28}$$

where $\theta' = l/\xi_s$. Combination of the above two equations and using the fact that $\gamma_{12} = -\gamma_{21}$, one finds

$$\cos\theta - (1/2)\sin\theta \tan\theta' = \cos\gamma. \tag{4.29}$$

Near to T_c, one may expand for small θ, θ', and γ, and one finds

$$\gamma^2 = L(L + l)/\xi_s^2, \tag{4.30}$$

which implies that the critical field is increased by the addition of the side branch. The equation from which T^* is determined is Equation (4.29) with the right hand side replaced by -1. As the side branch gets longer, T^* is increased from its single-ring value [Equation (4.26)], since the coherence length, proportional to the vortex core radius, is decreased.

It is easy to generalize the ring with one side branch to a ring with many side branches. Consider a ring of circumference L with N equally spaced side branches of length l, as pictured in Figure 24. We label the junctions on the ring $i = 1, \ldots, N$, and the dead ends

Fig. 24. Ring with N equally spaced dead-end side branches. After [56].

the corresponding $i' = 1', \ldots, N'$. The equations obtained from Equation (4.19) for the ith and i'th junctions are [56]

$$-2\cot\theta\Delta_i + \eta^*\Delta_{i-1}/\sin\theta + \eta\Delta_{i+1}/\sin\theta - \cot\theta'\Delta_i + \eta'\Delta_{i'}/\sin\theta' = 0 \tag{4.31}$$

and

$$\Delta_{i'} \cot\theta' - \eta'^*\Delta_i/\sin\theta' = 0, \tag{4.32}$$

where $\theta = L/\xi_s$, $\theta' = l/\xi_s$, $\eta = \exp(i\gamma_{ii+1}) = \exp(i\gamma/N)$ where γ is the flux in the ring, and $\eta' = \exp(i\gamma_{ii'})$. Using Equation (4.32) to solve for $\Delta_{i'}$ in terms of Δ_i, and noting that the periodicity of Δ_i requires Δ_i to be of the form,

$$\Delta_j = \Delta \exp(2\pi imj/N), \tag{4.33}$$

one finds immediately

$$\cos[(2\pi m + \gamma)/N] = \cos(\theta/N) - (1/2)\sin(\theta/N)\tan\theta'. \tag{4.34}$$

We note that Equation (4.34) reduces to Equation (4.29) for $N = 1$. For large N, one may expand Equation (4.34) and obtains [59]

$$(2\pi m + \gamma)^2 = L(L + Nl)/\xi_s^2, \tag{4.35}$$

which is valid for small θ, θ', and for γ and winding number $m \ll N$. By comparison with Equation (4.30), we see that the effective coherence length (proportional to the radius of the vortex cores) is progressively reduced as more side chains are added.

Although in the preceding example, the coherence length was found to be reduced by the addition of the dead-end side chains, the order parameter Δ_i was found to be uniform for winding number 0. For $m \neq 0$, the phase of the order parameter was found to vary from junction to junction. This winding number m is related to flux quantization, which is also present in the single ring previously considered. However, in that system there was only one junction on the ring, so that this destruction of the phase coherence was not present. These two effects of the dead ends are quite general: the coherence length is reduced and the phase coherence is destroyed. Alexander has demonstrated the generality of this statement by considering a 'transmission line' strand with $N - 1$ sidebranches of variable length located at arbitrary positions along the main strand. One would obtain this configuration by cutting the ring pictured in Figure 24 between two side chains, and making the lengths and positions of the side chains variable. The essential result of his calculation is that the diffusivity and the mass are both renormalized [56],

$$m/d = (L/\xi_L)^2, \tag{4.36}$$

where ξ_L is the effective coherence length for the line on the scale L.

A distinctly different type of finite lattice is the double loop pictured in Figure 25.

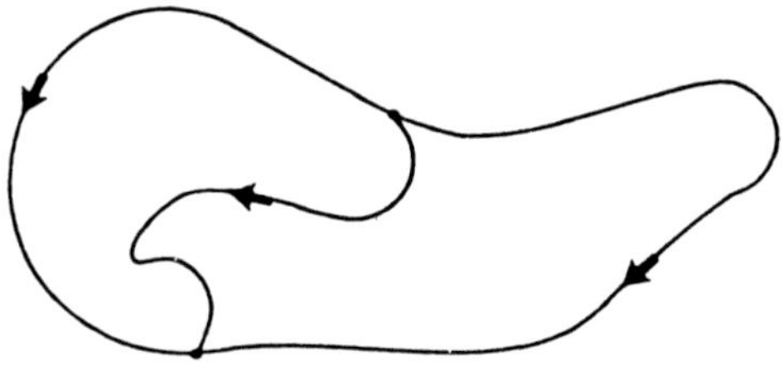

Fig. 25. Double loop. After [56].

Since the general situation is rather complicated, Alexander considered only special cases. In the first case, the flux through each half loop was taken equal to γ, and the length of the left and right paths from the upper to the lower vertex equal to L. The length of the middle path is taken to be L', and setting $\theta = L/\xi_s$ and $\theta' = L'/\xi_s$, one finds from Equation (4.19) [56],

$$2 \cot \theta + \cot \theta' = \pm[2 \cos \gamma / \sin \theta + 1/ \sin \theta'] . \tag{4.37}$$

There are two situations. In the first situation, which is appropriate for θ' near to 0, one has

$$\cos \theta - (1/2) \sin \theta \tan(\theta'/2) = \cos \gamma, \tag{4.38}$$

which corresponds to two independent half loops, each with a sidebranch of length $L'/2$. We note that when $\theta = \theta' \ll 1$, we have two (half) loops with flux γ and a (full) loop with no flux. In the second case, appropriate for θ' near to π, we have

$$\cos \theta + (1/2) \sin \theta \cot(\theta'/2) = -\cos \gamma, \tag{4.39}$$

which reduces to Equation (4.38) by $\theta \to \pi - \theta, \theta' \to \pi - \theta'$.

The second special case of the double loop considered by Alexander is that of all three paths from one junction to the other having the same length L, and the fulx in the left and right half loops being γ and 2γ, respectively, so that the total flux is 3γ. One finds [56]

$$9 \cos^2 \theta = 3 + 2(\cos \gamma + \cos 2\gamma + \cos 3\gamma). \tag{4.40}$$

Since this equation is cubic in $\cos \gamma$, there are several possibilities. First, for $\theta < 0.94$, there is only one critical field, as usual. However, for

$$0.94 < \theta < 1.23 \tag{4.41}$$

there are three critical fields. Superconductivity disappears at H_1, reappears at H_2, and disappears again at H_3. In the range

$$1.23 < \theta < 1.37 \tag{4.42}$$

there are two critical fields. Finally, for

$$1.37 < \theta \leq \pi/2 \tag{4.43}$$

there is no critical field. At the isolated points $\theta = 0.94$ and 1.37, there are two and one solutions, respectively. This pattern is symmetric about $\theta = \pi/2$. Thus, deformations in the lattice containing loops with different amounts of magnetic flux can give rise to 'reentrant' superconductivity, at least regarding the upper critical field.

4.5. THE SIERPINSKI GASKET

The above networks considered were all finite in size, and are not easily generalized to infinite networks of a type that might simulate the effects of percolating clusters. However, a model for a network which can be solved and which contains some of the features of a percolating cluster is that of the Sierpinski gasket (Figure 23) [60, 61]. This model is 'self-similar', that is, it looks the same on many different length scales. The smallest length scale is a, the side of one of the smallest triangles in the gasket pictured. The smallest network on this length scale consists of an equilateral triangle composed of three triangles of size a placed corner to corner in such a fashion that a triangular hole of equal size is formed in the center. This finite network can then be used to construct a network of the next hierarchical size: three such networks are placed together as before, yielding a triangular hole in the center the size of the previous network scale.

Let us first consider the smallest network, consisting of three triangles of size a (on a side) connected together at their vertices, yielding an equilateral triangle of size $2a$ with junctions in the middle of each side. Let us label the junctions 1, 2, and 3, respectively. Use of Equation (4.19) results in the matrix equation [59],

$$p_0 T_{123} \Delta = 0, \tag{4.44}$$

where

$$p_0 = (\sin \theta_0)^{-1}, \tag{4.45}$$

the vector Δ has components Δ_1, Δ_2, and Δ_3, respectively, and the T matrix is given by

$$T_{123} = \begin{bmatrix} -2t_0 & \eta_{12} & \eta_{13} \\ \eta_{21} & -2t_0 & \eta_{23} \\ \eta_{31} & \eta_{32} & -2t_0 \end{bmatrix}, \tag{4.46}$$

where $t_0 = \cos \theta_0$ and $\eta_{ij} = \exp(i\gamma_{ij})$ are defined as before, with $\theta_0 = a/\xi_s$.

The simple network may be extended to include coupling to the adjacent networks by replacing the right hand side of Equation (4.44) by a vector F. Alexander has considered this problem in detail, by iterating to successively larger networks. He finds that after n iterations, the recursion relations for t and p become [56]

$$t_{n+1} = t_n(4t_n - 3) \tag{4.47}$$

and

$$p_{n+1}/p_n = (2t_n + 1)/[(4t_n + 1)(2t_n - 1)]. \tag{4.48}$$

As can be immediately seen from Equation (4.47), there are two fixed points $t = 0$ and $t = 1$. Although both of these fixed points are unstable, the $n = 1$ fixed point is relevant to the resistor network, and is appropriate for small θ_0. The flux through the gasket is essentially just the added flux through all of the single triangles,

$$\gamma_n \cong 4^n \gamma_0, \tag{4.49}$$

where the single triangle flux $\gamma_0 = \eta_{12}\eta_{23}\eta_{31}$.
The 'angle' θ is renormalized, however,

$$\theta_n \cong 5^n \theta_0. \tag{4.50}$$

Alexander showed that the above recursion relations could be used to solve for a finite gasket of size [56]

$$L = 2^n a, \tag{4.51}$$

by renormalizing the gasket to a single triangle with parameters t_n and γ_n. Since there will be no external interactions with other vertices, the matrix equation is as given by Equation (4.44). Thus, the condition for a solution is

$$\det T_n = 0 = -8t_n^3 + 6t_n + 2\cos\theta_n. \tag{4.52}$$

Since Equation (4.52) has solutions for t_n in the range -1 to 1, we may write $t_n = \cos\theta_n$, and we find

$$\cos(3\theta_n) = \cos(\gamma_n). \tag{4.53}$$

Assuming we are near to the $n = 1$ fixed point, Equation (4.53) may be expanded for small θ_n and γ_n,

$$9\theta_n^2 \cong \gamma_n^2. \tag{4.54}$$

First, by defining θ_n to be

$$\theta_n = L/\overline{\xi}_s, \tag{4.55}$$

by using Equations (4.50) and (4.51) and the definition of θ_0, we have

$$\overline{\xi}_s^2 \cong (4/5)^n \xi_s^2 \propto L^{-\theta} \xi_s^2, \tag{4.56}$$

where

$$\theta = -2 + \ln 5/\ln 2 \cong 0.32. \tag{4.57}$$

Then, using Equation (4.54) and using the fact that $\gamma_0 = \sqrt{3}\,\pi H a^2/(2\Phi_0)$, where Φ_0 is the flux quantum, we have

$$H_{c^2} \cong \Phi_0/(L\overline{\xi}_s) \cong \Phi_0/(\xi_s L^{1-\theta/2}). \tag{4.58}$$

In the above equations, it has been assumed that the gasket is self-similar over its entire extent, i.e., on the scale L. For less perfect gaskets, which are self-similar on a scale $\lambda_s < L$, the linearized Ginzburg–Landau equations will break down, as the inhomogeneities will allow for variations in the order parameter on the scale λ_s. We then have Equation (4.56) with λ_s replacing L, and setting $\theta_s = \lambda_s/\bar{\xi}_s$, the requirement that θ_s be of order of magnitude unity gives

$$\lambda_s \propto \xi_s{}^{2/(2+\theta)}. \tag{4.59}$$

This length sets an upper bound on the upper critical field, obtained by setting $L \cong \lambda_s$ in Equation (4.56), and using Equation (4.59) for ξ_s in terms of λ_s,

$$H_s \cong \Phi_0/\lambda_s^2. \tag{4.60}$$

A lower limit for H_{c2} is simply obtained by assuming one flux quantum penetrates the entire sample,

$$H_L \cong \Phi_0/L^2. \tag{4.61}$$

By comparing Equations (4.60) and (4.61), we see that the upper critical field as a function of L can generally be written

$$H_{c2}(L) \propto \Phi_0(L/\lambda_s)^y/L^2 \tag{4.62}$$

for $L > \lambda$, where $y < 2$ in order to satisfy the upper and lower limits. This implies that H_{c2} is a decreasing function of L in this regime.

4.6. Percolation Clusters

Finally, Alexander attempted to relate the above results to the case of percolation clusters of superconducting chains. He compared his results to the Skal–Shklovski–deGennes (SSG) model of percolation clusters [62] containing dead ends but no loops on a scale smaller than ξ_p. First, he neglected the effect of the dead ends. In the SSG model, the percolation length ξ_p is divergent at the percolation concentration p_c,

$$\xi_p \propto (p - p_c)^{-\nu}, \tag{4.63}$$

and the string length (or length of the backbone near to percolation) is

$$l_p \propto (p - p_c)^{-\zeta}, \tag{4.64}$$

where the exponent ζ is related to the conductivity exponent t [61],

$$\sigma \propto (p - p_c)^{-t}, \qquad \zeta = t - (d-2)\nu, \tag{4.65}$$

where d is the dimensionality. For a percolation cluster, the relevant lattice scale is ξ_p instead of a, so the appropriate γ is

$$\gamma \cong 2\pi H \xi_p^2/\Phi_0. \tag{4.66}$$

The strands connecting the junctions are not necessarily a distance ξ_p apart, but are a greater distance l_p apart, on the average, due to their wildly meandering paths. The relevant 'angle' to use in Equation (4.19) is

$$\theta = l_p/\xi_s \tag{4.67}$$

Assuming a square lattice such as that considered in Section 3, we find that for small $\theta, \gamma \cong \theta^2$, which implies

$$H_{c2} \propto (l_p/\xi_p)^2 \Phi_0/\xi_s^2. \tag{4.68}$$

This implies that there is an effective coherence length $\bar{\xi}_s$ such that $H_{c2} \propto \xi_s^{-2}$,

$$\bar{\xi}_s = (\xi_p/l_p)\xi_s. \tag{4.69}$$

When dead ends are considered, the effective coherence length is further modified. From Equation (4.36), setting $d = l_p^{-1}$ and $L = \xi_p$

$$\bar{\xi}_s \cong \xi_p (l_p m_p)^{-1/2}, \tag{4.70}$$

where the 'mass' on the percolation lattice of scale ξ_p is given by

$$m_p \propto (p - p_c)^{\beta} \xi_p^d, \tag{4.71}$$

where the exponent β is analogous with other critical phenomena in which the order parameter (e.g., spin density) exponent is β [61]. This implies

$$\bar{\xi}_s \cong \xi_s (p - p_c)^{(t-\beta)/2}, \tag{4.72}$$

provided that $\bar{\xi}_s$ and λ_s are both greater than ξ_p, where

$$\theta = (t - \beta)/\nu \tag{4.73}$$

in the expression (Equation (4.59)) for λ_s. With this effective coherence length, the upper critical field is

$$H_{c2} \propto (\Phi_0/\pi\xi_s^2)(p - p_c)^{\beta - t}. \tag{4.74}$$

This expression is in reasonably good agreement with the experimental measurements of Deutscher *et al.* [59] on small particles in the small ξ_p regime, where $\xi_p < \lambda_s$. When $\xi_p \cong \lambda_s$, however, Alexander suggests that there should be a crossover regime, and that for larger ξ_p, H_{c2} should behave as

$$H_{c2} \sim \Phi_0/\xi_p^2, \tag{4.75}$$

although this is really only a lower limit result. In any event, the experiments show that there is a crossover, but that H_{c2} continues to increase as ξ_p increases, contrary to Alexander's arguments. He considers patching up the theory by including loops in a more reasonable fashion, but concludes that even such patch-ups will not give the experimentally observed behavior close to percolation [56].

5. Summary

We have seen that the expected behavior of quasi-one-dimensional superconductors is quite different from what one would expect in ordinary bulk materials. The temperature dependence of the upper critical field should show upward curvature indicative of dimensional crossover, regardless of the angle of the applied field, provided that the anisotropy is sufficiently great. For the field parallel to the chain lattice, the temperature dependence of H_{c2} should also exhibit kinks associated with a commensurate fitting to the vortex lattice into the chain lattice. These kinks are most likely to be pronounced for a fitting of a small number (e.g., 1 or 2) of flux quanta into a slightly larger small number of unit cells. If there is sufficient disorder in the material, similar anomalies in the upper critical field curves could arise from the fitting of flux quanta into rings or loops of the strands. However, if the strands form loops and rings of many different length scales, then there would be so many possibilities for the fitting of vortices into the rings or loops that the effective temperature dependence of H_{c2} would be altered to a non-trivial power of $T_c - T$. Near to the percolation threshold, the extreme inhomogeneities of the sample allow for an order parameter that is spatially non-uniform; some regions will have a very small order parameter near to H_{c2}, while others have a local order parameter that is non-vanishing at H_{c2}. This picture seems to be appropriate for materials such as $NbSe_3$, which is now fairly well-established as consisting of bundles of fibers, at least in the materials that exhibit superconductivity [22].

To date, no materials have been made with sufficient intrinsic anisotropy in addition to sufficient crystallinity to observe the kinks in the parallel upper critical field that have been predicted. It remains to be seen if existing materials can be structurally perfected, or if more anisotropic materials can be produced.

As regards the lower critical field, there does not appear to be much difference between a highly anisotropic bulk superconductor with filamentary symmetry and a quasi-one-dimensional superconductor. For a material with sufficient anisotropy such that the kinks in the temperature dependence of the parallel upper critical field are observable, the angular dependence of the lower critical field should exhibit a kink, corresponding to the vortex cores preferring to lie parallel to the easy axis (i.e., parallel to the chains). Such behavior is not expected in a material with an effective mass anisotropy of roughly 10 or less, however.

It is worth noting that measurements of the lower critical field have been plagued with the difficulty in determining the field at which the departure from the linear magnetization versus field curve occurs. Unlike the case of $NbSe_2$, for which this field is rather well-defined [30], $(SN)_x$ did not appear to exhibit a well-defined H_{c1} [8]. The strand-like nature of that and other materials suggests that there are many possible easy ways in which to insert a single vortex into the material. For example, if the strands make a loop, it is easier for the flux to penetrate that loop than the bulk of the sample. Unfortunately, calculations of the lower critical field for such geometries are generally non-trivial: the order parameter is large and non-uniform, and hence the GL equations are non-linear.

The organic material $(TMTSF)_2ClO_4$, however, has been made sufficiently crystalline that H_{c1} could reliably be determined [46]. Nevertheless, even in that material, the difficulty of measuring the lower critical field precisely proved to be sufficient as to preclude the determination of the existence of a kink in its angular dependence.

Finally, it is hoped that new materials will continue to be made, and existing materials will continue to be made better. It is quite likely, as we have seen, that novel behavior should be observed.

Acknowledgement

The author would like to thank Dr Wendy W. Fuller for her kind help in compiling the list of references.

References

1. R. L. Greene, G. B. Street, and L. J. Suter, *Phys. Rev. Lett.* **34**, 577 (1975).
2. L. J. Azevedo, W. G. Clark, G. Deutscher, R. L. Greene, G. B. Street and L. J. Suter, *Solid State Commun.* **19**, 197 (1976).
3. P. Monceau, J. Peyrard, J. Richard, and P. Molinié, *Phys. Rev. Lett.* **39**, 161 (1977).
4. T. Sambongi, M. Yamamoto, K. Tsutsumi, Y. Shiozaki, K. Yamaya, and Y. Abe, *J. Phys. Soc. Japan* **42**, 1421 (1977).
5. M. Yamamoto, *J. Phys. Soc. Japan* **45**, 431 (1978).
6. R. H. Dee, D. H. Dollard, B. G. Turrell, and J. F. Carolan, *Solid State Commun.* **24**, 469 (1977).
7. R. H. Dee, A. J. Berlinsky, J. F. Carolan, E. Klein, N. J. Stone, and B. G. Turrell, *Solid State Commun.* **22**, 303 (1977).
8. R. H. Dee, J. F. Carolan, B. G. Turrell, and R. L. Greene, *Phys. Rev. B***22**, 174 (1980).
9. R. A. Buhrman, C. M. Bastuscheck, J. C. Scott, and J. D. Kulick, in *Inhomogeneous Superconductors* – 1979, edited by D. V. Gubser, T. L. Francavilla, S. A. Wolf, and J. R. Leibowitz, (New York, American Institute of Physics, 1980), p. 207.
10. W. W. Fuller, P. M. Chaikin, and N. P. Ong, *Solid State Commun.* **30**, 689 (1979).
11. P. Haen, G. Waysand, G. Boch, A. Waintal, P. Monceau, N. P. Ong, and A. M. Portis, *J. de Phys.* (Paris) **37**, 179 (1976).
12. P. Haen, J. M. Mignot, P. Monceau, and M. Núñez Regueiro, *J. de Phys.* (Paris) **39**, C6–703 (1978).
13. P. Haen, F. Lapierre, P. Monceau, M. Núñez Regueiro, and M. Richard, *Solid State Commun.* **26**, 725 (1978).
14. J. F. Kwak, R. L. Greene, and W. W. Fuller, *Phys. Rev. B***20**, 2658 (1979).
15. R. L. Civiak, C. Elbaum, L. F. Nichols, H. I. Kao, and M. M. Labes, *Phys. Rev.* **B14**, 5413 (1976).
16. R. L. Greene and G. B. Street, in *Chemistry and Physics of One-Dimensional Metals*, edited by H. J. Keller (Plenum, New York) (1977).
17. G. B. Street and R. L. Greene, *IBM J. Res. Dev.* **21**, 99 (1977).
18. R. H. Dee, J. F. Carolan, B. G. Turrell, R. L. Greene, and G. B. Street, *J. de Phys.* (Paris) **39**, C6–444 (1978).
19. M. Potel, R. Chevrel, M. Sergent, J. C. Armici, M. Decroux, and Ø. Fischer, *J. Solid State Chem.* **35**, 286 (1980).
20. J. C. Armici, M. Decroux, Ø. Fischer, M. Potel, R. Chevrel, and M. Sergent, *Solid State Commun.* **33**, 607 (1980).

21. Y. Tajima and K. Yamaya, *J. Phys. Soc. Jpn.* **53**, 495 (1984).
22. K. Kawabata and M. Ido, *Solid State Commun.* **44**, 1539 (1982).
23. A. A. Abrikosov, *Sov. Phys. JETP* **5**, 1174 (1957); A. L. Fetter and P. C. Hohenberg, in *Superconductivity*, edited by R. D. Parks (Marcel Dekker, New York, 1969), p. 817.
24. D. Saint-James, G. Sarma, and E. J. Thomas, *Type-II Superconductivity* (Pergamon Press, New York, 1969); P. G. de Gennes, *Superconductivity of Metals and Alloys* (Benjamin, New York, 1966); A. A. Abrikosov, L. P. Gor'kov, and I. E. Dzyaloshinski, *Methods of Quantum Field Theory in Statistical Physics* (Prentice-Hall, Englewood Cliffs, New Jersey, 1963).
25. R. A. Klemm, A. Luther, and M. R. Beasley, *Phys. Rev. B***12**, 877 (1975); L. N. Bulaevskii, *Zh. Eksp. Teor. Fiz.* **65**, 1278 (1975).
26. D. E. Prober, R. E. Schwall, and M. R. Beasley, *Phys. Rev. B***21**, 2717 (1980); R. C. Morris and R. V. Coleman, *Phys. Rev. B*7, 991 (1973).
27. S. T. Ruggiero, T. W. Barbee, Jr., and M. R. Beasley, *Phys. Rev. Lett.* **45**, 1299 (1980).
28. This has been discussed by many authors. For a review, see V. J. Emery, in *Highly Conducting One-Dimensional Solids*, edited by J. Devreese, R. Evard, and V. van Doren (Plenum, New York, 1979).
29. J. Bardeen, *Rev. Mod. Phys.* **34**, 667 (1962).
30. M. W. Denhoff and S. Gygax, *Phys. Rev. B***25**, 4479 (1982).
31. E. Z. da Silva and B. L. Gyorffy, *Phys. Rev. B***20**, 147 (1979).
32. R. A. Klemm and H. Gutfreund, *Phys. Rev. B***14**, 1086 (1976).
33. H. Horowitz, H. Gutfreund, and M. Weger, *Phys. Rev. B***9**, 1246 (1974); *ibid*, **12**, 3174 (1975).
34. R. Sooryakumar and M. V. Klein, *Phys. Rev. Lett.* **45**, 45 (1980).
35. P. B. Littlewood and C. M. Varma, *Phys. Rev. Lett.* **47**, 811 (1981); *Phys. Rev. B***26**, 4883 (1982).
36. D. E. Moncton, J. D. Axe, and F. J. DiSalvo, *Phys. Rev. B***16**, 801 (1977); M. Barmatz, L. R. Testardi, and F. J. DiSalvo, *Phys. Rev. B***12**, 4367 (1975).
37. P. Monceau, N. P. Ong, A. M. Portis, A. Meerschaut, and J. Rouxel, *Phys. Rev. Lett.* **37**, 602 (1976).
38. J. Bardeen, L. N. Cooper, and J. R. Schrieffer, *Phys. Rev.* **108**, 1175 (1957).
39. R. A. Klemm and J. R. Clem, *Phys. Rev. B***21**, 1868 (1980).
40. D. R. Tilley, *Proc. Phys. Soc. London* **85**, 1977 (1965).
41. B. S. Chandrasekhar, *Appl. Phys. Lett.* **1**, 7 (1962); A. M. Clogston, *Phys. Rev. Lett.* **9**, 266 (1962); T. P. Orlando and M. R. Beasley, *Phys. Rev. Lett.* **46**, 1598 (1981).
42. A. A. Abrikosov and L. P. Gor'kov, *Zh. Eks. i Teor. Fiz.* **39**, 1781 (1960). (*Soviet Phys. JETP* **12**, 1243 (1961)).
43. C.-R. Hu, *Phys. Rev. B*6, 1756 (1972).
44. Note that there is an error in sign in Equation (58) of ref. 39. This implies that the dependence of H_{c1} upon the angle H makes with the normal to the $\hat{z}$ axis is as shown in Figure 11, not as in Figure 5 of ref. 39. This error also has implications for the dependence of the lower critical field upon the external field angle for a material of filamentary shape. The correct figure is shown in Figure 12. The author is grateful to H. Schwenk for pointing out this error.
45. J. A. Osborn, *Phys, Rev.* **67**, 351 (1945); R. A. Klemm, *J. Low Temp. Phys.* **39**, 589 (1980); see ref. 44.
46. H. Schwenk, K. Andres, and F. Wudl, (to be published).
47. V. G. Kogan, *Phys. Rev. B***24**, 1572 (1981).
48. V. G. Kogan and J. R. Clem, *Phys. Rev. B***24**, 2497 (1981).
49. W. Lawrence and S. Doniach, in *Proceedings of the 12th International Conference on Low Temperature Physics*, edited by Eizo Kanda (Academic, Kyoto, 1971) p. 361.
50. L. I. Turkevich and R. A. Klemm, *Phys. Rev. B***19**, 2520 (1979).
51. Note that we have defined the azimuthal angle in the conventional fashion, consistent with Equation (2.2), but differing from ref. 50 by $\phi \rightarrow \pi/2 - \phi$.
52. E. T. Whittaker and G. H. Watson, *A Course of Modern Analysis* (Cambridge University, Cambridge, 1946) Sec. 19.12; W. Magnus and W. Winkler, *Hill's Equation* (Interscience, New York, 1966).

53. N. W. McLachlan, *Theory and Application of Mathieu Functions* (Dover, New York, 1964).
54. P. Manneville, *J. de Phys.* (Paris) **36**, 701 (1975).
55. Note that the spin-orbit scattering rate differs by a factor of 2/3 from that in Klemm, Luther, and Beasley [25] This result is correct for two-dimensional scattering.
56. S. Alexander, *Phys. Rev. B*27, 1541 (1983).
57. M. Tinkham, *Phys. Rev.* **129**, 2413 (1963).
58. P. G. de Gennes, *C. R. Acad. Sci. B*292, 279 (1981); *ibid. B*292, 9 (1981).
59. G. Deutscher, I. Grave, and S. Alexander, *Phys. Rev. Lett.* **48**, 1497 (1982); G. Deutscher, O. Entin-Wohlman, S. Fishman, and Y. Shapira, *Phys. Rev. B*21, 5041 (1980); O. Entin-Wohlman, A. Kapitulnik, and Y. Shapira, *Phys. Rev. B*24, 6464 (1981).
60. B. B. Mandelbrot, *The Fractal Geometry of Nature* (Freeman, San Francisco, 1983), and *Fractals: Form, Chance and Dimension* (Freeman, San Francisco, 1977).
61. Y. Gefen, B. B. Mandelbrot, and A. Aharony, *Phys. Rev. Lett.* **45**, 855 (1980); Y. Gefen, A. Aharony, B. B. Mandelbrot, and S. Kirkpatrick, *Phys. Rev. Lett.* **47**, 1771 (1981).
62. A. S. Skal and B. I. Shklovskii, *Fiz. Tekh. Poluprovodn.* **8**, 1586 (1974) (*Sov. Phys.* – Semicond **8**, 1029 (1975)); P. G. de Gennes, *J. de Phys.* (Paris) *Lett.* **37**, L1 (1976).

INDEX OF NAMES

INDEX OF SUBJECTS